Adaptive Control of Systems with Actuator Failures

Springer

*London
Berlin
Heidelberg
New York
Hong Kong
Milan
Paris
Tokyo*

Gang Tao, Shuhao Chen, Xidong Tang and Suresh M. Joshi

Adaptive Control of Systems with Actuator Failures

With 50 Figures

Springer

Gang Tao, PhD
Department of Electrical and Computer
  Engineering
University of Virginia
Charlottesville, USA

Xidong Tang
Department of Electrical and Computer
  Engineering
University of Virginia
Charlottesville, USA

Shuhao Chen
Department of Electrical and Computer
  Engineering
University of Virginia
Charlottesville, USA

Suresh M. Joshi, PhD
Dynamics and Control Branch, NASA
Langley Research Center
Hampton, VA, USA

British Library Cataloguing in Publication Data
Adaptive control of systems with actuator failures
    1. Adaptive control systems.   2. Actuators   3. Fault-tolerant
    computing
    I. Tao, Gang
    629.8′36
    ISBN 1852337885

A catalog record for this book is available from the Library of Congress.

ISBN 1-85233-788-5   Springer-Verlag London Berlin Heidelberg
Springer-Verlag is part of Springer Science+Business Media
springeronline.com

Typesetting: Electronic text files prepared by authors
69/3830-543210 Printed on acid-free paper SPIN 10945681

Preface

Actuator failures in control systems may cause severe system performance deterioration and even lead to catastrophic closed-loop system instability. For example, many aircraft accidents were caused by operational failures in the control surfaces, such as rudder and elevator. For system safety and reliability, such actuator failures must be appropriately accommodated. Actuator failure compensation is an important and challenging problem for control systems research with both theoretical and practical significance.

Despite substantial progress in the area of actuator failure compensation, there are still many important open problems, in particular those involving system uncertainties. The main difficulty is that the actuator failures are uncertain in nature. Very often it is impossible to predict in advance *which* actuators may fail during system operation, *when* the actuator failures occur, *what* type and *what* values of the actuator failures are. It may also be impractical to determine such actuator failure parameters after a failure occurs. It is appealing to develop control schemes that can accommodate actuator failures without explicit knowledge of the occurrences of actuator failures and the actuator failure values. Adaptive control, which is capable of accommodating system parametric, structural, and environmental uncertainties, is a suitable choice for such actuator failure compensation schemes.

This book presents our recent research results in designing and analyzing adaptive control schemes for systems with unknown actuator failures and unknown parameters. The main feature of the adaptive actuator failure compensation approach developed in this book is that no explicit fault detection and diagnosis procedure is used for failure compensation. An adaptive law automatically adjusts the controller parameters based on system response errors, so that the remaining functional actuators can be used to accommodate the actuator failures and systems parameter uncertainties.

The book is in a comprehensive and self-contained presentation, while the developed theory is in a general framework readily applicable to specific practical adaptive actuator failure compensation problems. The book can be used as a technical reference for graduate students, researchers, and engineers from fields of engineering, computer science, applied mathematics, and others who have a background in linear systems and feedback control at the undergraduate level. It can also be studied by interested undergraduate students for their thesis projects.

This book is focused on adaptive compensation of actuator failures characterized by the failure model that some unknown control inputs may get stuck at some unknown fixed (or varying) values at unknown time instants and cannot be influenced by the control signals. The type of fixed-value actuator failures, referred to as "lock-in-place" actuator failures, is an important type of actuator failures and is often encountered in many critical control systems. For example, in aircraft flight control systems, the control surfaces may be locked in some fixed places and hence lead to catastrophic accidents. Varying value failures can occur, for example, due to hydraulics failures that can produce unintended movements in the control surfaces of an aircraft.

For actuator failure compensation, a certain redundancy of actuators is needed. For a system with multiple actuators, one case is that all actuators have the same physical characteristics; for example, they are segments of a multiple-segment rudder or elevator for an aircraft. For this case, a reasonable (natural) design for the applied control inputs is one with equal or proportional actuation for each actuator, that is, all control inputs are designed to be equal or proportional to each other. This actuation scheme is employed throughout the book, except for Chapter 5, where a multivariable design is used for the case when the actuators are divided into several groups and each group has actuators of the same physical characteristics (for example, an aircraft has a group of four engines and a group of three rudder segments), and within each group, an equal or proportional actuation is used.

With 12 chapters, the book systematically develops adaptive state tracking and output tracking control schemes for systems with parameter and actuator failure uncertainties. Designs and analysis for both linear systems and nonlinear systems with unknown actuator failures are covered. Key issues for adaptive actuator failure compensation, namely, design condition, controller structure, error equations, adaptive laws for updating the controller parameters, analysis of stability and tracking properties, are given in detail. Extensive simulation results are presented to verify the desired closed-loop system performance. This work is aimed at developing a theoretical framework for adaptive control of systems with actuator failures, to provide guidelines for designing control systems with guaranteed stability and tracking performance in the presence of system parameter uncertainties and failure uncertainties.

Chapter 1 presents some background material. Basic concepts and fundamental principles of adaptive control systems are introduced. The actuator failure compensation problems for linear systems and nonlinear systems are formulated. An overview of several existing actuator failure compensation design methods, including multiple models, switching and tuning designs, fault diagnosis designs, adaptive designs, and robust designs, is also given.

Chapters 2–8 address the adaptive actuator failure compensation problems for linear time-invariant systems with unknown actuator failures. Chapter 2 presents several model reference state feedback state tracking designs. For a linear time-invariant system with m actuators, the adaptive actuator failure compensation problem for up to $m - 1$ unknown actuator failures is investigated. Designs for three types of actuator failures: "lock-in-place," parametrizable time-varying, and unparametrizable time-varying, are developed. Conditions and controller structures for achieving plant-model state matching, adaptive laws for updating the controller parameters, and analysis of closed-loop stability and asymptotic state tracking properties are addressed in ·a unified and comprehensive framework. State feedback actuator failure compensation designs for a class of multi-input systems are also derived. A more general case of up to $m - q$ $(q \geq 1)$ unknown actuator failures is then addressed. Necessary and sufficient conditions for actuator failure compensation are derived. It is shown that the number of fully functional actuators is crucial in determining the actuation range that specifies the compensation design conditions in terms of system actuation structures. Such conditions are required for both a nominal design using system and failure knowledge and an adaptive design without such knowledge. An adaptive actuator failure compensation control scheme based on such system actuation conditions is developed for systems with unknown dynamics parameters and unknown "lock-in-place" actuator failures. Simulation results are presented to verify the desired system performance with failure compensation.

Chapter 3 investigates the state feedback output tracking problem for single-output linear time-invariant systems with any up to $m - 1$ uncertain failures of the total m actuators. In particular, adaptive rejection of the effect of certain unmatched input disturbances on the output of a linear time-invariant system is addressed in detail. A lemma that presents a novel basic property of linear time-invariant systems is derived to characterize system conditions for plant-model output matching. An adaptive disturbance rejection control scheme is developed for such systems with uncertain dynamics parameters and disturbances. This adaptive control technique is applicable to control of systems with actuator failures whose failure values, failure time instants, and failure patterns are unknown. A solution capable of accommo-dating the "lock-in-place" and time-varying actuator failures in the presence of any up to $m-1$ uncertain failures of the total m actuators is presented to this adaptive actuator failure compensation problem. The developed adaptive actuator failure compensation schemes ensure closed-loop stability and asymptotic output tracking despite the uncertainties in actuator failures and system parameters. Simulation results verify the desired system performance in the presence of unknown actuator failures.

Chapter 4 develops a model reference adaptive control scheme using output feedback for output tracking for linear time-invariant systems with unknown actuator failures. An effective output feedback controller struc-ture is proposed for actuator failure compensation. When implemented with true matching parameters, the controller achieves desired plant-model out-put matching, and when implemented with adaptive parameter estimates, the controller achieves closed-loop stability and asymptotic output tracking, which is also verified by simulation results. Compensation of varying failures is achieved based on an output matching condition for a system with multiple inputs whose actuation vectors may be linearly independent.

Chapter 5 deals with the output tracking problem for multi-output linear time-invariant systems using output feedback. Two adaptive control schemes based on model reference adaptive control are developed for a class of multi-input multi-output systems with unknown actuator failures. An effective con-troller structure is proposed to achieve the desired plant-model output match-ing when implemented with matching parameters. Based on design conditions on the controlled plant, which are also needed for nominal plant-model out-put matching for a chosen controller structure, two adaptive controllers are proposed and stable adaptive laws are derived for updating the controller parameters when system and failure parameters are unknown. All closed-loop signals are bounded and the system outputs track some given reference outputs asymptotically, despite the uncertainties in failures and system pa-rameters. Simulation results are presented to demonstrate the performance of the adaptive control system in the presence of unknown rudder and aileron failures in an aircraft lateral dynamic model.

Chapter 6 studies adaptive pole placement control for linear time-invariant systems with unknown actuator failures, applicable to both minimum and nonminimum phase systems. A detailed analysis shows the existence of a nominal controller (when both system and actuator failure parameters are known) that achieves the desired pole placement, output tracking, and closed-loop signal boundedness. For that case when both system and failure parameters are unknown, an adaptive control scheme is developed. A simulation study with a linearized lateral dynamic model of the DC-8 aircraft is presented to verify the desired actuator failure compensation performance.

Chapter 7 applies several adaptive control schemes developed in the previous chapters to a linearized longitudinal dynamic model of a transport aircraft model. The tested adaptive schemes include state feedback design for state tracking, state feedback design for output tracking, and output feedback design for output tracking. Various actuator failures are considered. Extensive simulation results for different cases are presented to demonstrate the effectiveness of the adaptive actuator failure compensation designs.

Chapter 8 presents a robust adaptive control approach using output feedback for output tracking for discrete-time linear time-invariant systems with uncertain failures of redundant actuators in the presence of the unmodeled dynamics and bounded output disturbance. Technical issues such as plant-model output matching, adaptive controller structure, adaptive parameter update laws, stability and tracking analysis, and robustness of system performance are solved for the discrete-time adaptive actuator failure compensation problem. A case study is conducted for adaptive compensation of rudder servomechanism failures of a discrete-time Boeing 747 dynamic model, verifying the desired adaptive system performance.

Chapters 9–11 deal with actuator failure compensation problems for nonlinear systems. Chapter 9 formulates such problems and develops adaptive control schemes for feedback linearizable systems. Different structure conditions that characterize different classes of systems amenable to actuator failure compensation are specified, with which adaptive state feedback control schemes are developed for systems with uncertain actuator failures.

Chapter 10 addresses actuator failure compensation problems for nonlinear systems that can be transformed into parametric-strict-feedback form with zero dynamics. Two main cases are studied for adaptive actuator failure compensation: systems with stable zero dynamics, and systems with extra controls for stabilization. Design conditions on systems admissible for actuator failure compensation are clarified. Adaptive state feedback control schemes are developed, which ensure asymptotic output tracking and closed-loop signal boundedness despite the uncertainties in actuator failures as well as in system parameters. An adaptive control scheme is applied to a twin otter aircraft longitudinal nonlinear dynamics model in the presence of unknown failures in a two-segment elevator servomechanism. Simulation results verify the desired adaptive actuator failure compensation performance.

Chapter 11 presents an adaptive control scheme that achieves stability and output tracking for output-feedback nonlinear systems with unknown actuator failures. A state observer is designed for estimating the unavailable system states, based on a chosen control strategy, in the presence of actuator failures with unknown failure values, time instants, and pattern. An adaptive controller is developed by employing a backstepping technique, for which parameter update laws are derived to ensure asymptotic output tracking and closed-loop signal boundedness, as shown by detailed stability analysis. An extension of the developed adaptive actuator failure compensation scheme to nonlinear systems whose dynamics are state-dependent is also given to accommodate a larger class of nonlinear systems. An application to controlling the angle of attack of a nonlinear aircraft model in the presence of elevator segment failures is studied, with simulation results presented to illustrate the effectiveness of the failure compensation design.

Chapter 12 presents concluding remarks and suggests a list of theoretical and practical topics for further research in this area of adaptive control.

To help the readers understand the basic designs of adaptive control in the absence of actuator failures, the book includes an appendix that presents the schemes of model reference adaptive control using state feedback for state tracking, state feedback for output tracking, output feedback for output tracking, and multivariable design, as well as adaptive pole placement control. Key issues such as *a priori* system knowledge, controller structure, plant-model matching, adaptive laws, and stability are addressed.

This book describes adaptive actuator failure compensation approaches for effectively controlling uncertain dynamic systems with uncertain actuator failures. It addresses the theoretical issues of actuator failure models, controller structures, design conditions, adaptive laws, and stability analysis, with extensive simulation results on various aircraft system models. Design guidelines provided here may be used to develop advanced adaptive control techniques for control systems with controller adaptation and failure compensation capacities to improve reliability, maintainability, and survivability. The research leading to this book was supported by the National Aeronautics and Space Administration (NASA). However, the views and contents of this book are solely those of the authors and not of NASA.

Acknowledgements

We would like to express our thanks to Professors Karl Åström, Petros Ioannou, Petar Kokotović, Frank Lewis, and Kumpati Narendra for their knowledge and encouragement, to Dr. Jovan Boskovic for his inspiring work, to Professor Marios Polycapou for his help, to Professor Jack Stankovic for his interest and support, to Professors Michael Demetriou and Hong Wang for their comments, to Dr. Xiao-Li Ma for her contribution to Chapter 2, to Mr. Juntao Fei for his contribution to Chapter 8, to Mr. Richard Hueschen for his useful discussion about transport aircraft dynamics and actuator configurations, to Drs. Emin Faruk Kececi and Avinash Taware for their discussion, to Professors Zong-Li Lin and Steve Wilson for their support, and to the anonymous reviewers for their comments, which all have been continually motivating and highly beneficial to our related research, whose results have been reported in this book.

The first three authors wish to gratefully acknowledge the support by the NASA Langley Research Center to this work.

We are especially grateful to our families for their love and their support to our research work, which made this project possible.

Gang Tao, Shuhao Chen, and Xidong Tang
Charlottesville, Virginia, USA

Suresh M. Joshi
Hampton, Virginia, USA

Table of Contents

Chapter 1

Introduction

Adaptive control systems adjust controller parameters using system response errors to obtain desired performance. Adaptive control systems are capable of accommodating system parametric, structural, and environmental uncertainties caused by payload variation or system aging, component failures, and external disturbances. There have been significant advances in both theory and applications of adaptive control (see, for example, [3], [6], [7], [22], [23], [27], [28], [39], [40], [45], [53], [55], [62], [63], [66], [67], [68], [71], [72], [75], [78], [79], [88], [90], [94], [95], [100], [101], [105], [110], [123], [125], [131], [132], [149]). The field of adaptive control continues to develop rapidly with the emergence of new challenging problems and their innovative solutions. One such challenging problem is adaptive control of systems with actuator failures, which has many applications such as aircraft and spacecraft flight control systems, process control systems, and power systems as well.

This book is devoted to the development of effective adaptive control schemes that can accommodate actuator failures with both system parameter and actuator failure uncertainties. In this chapter, we introduce some background material, including basic concepts of adaptive control systems, actuator failure compensation problems under consideration, and a literature review of existing actuator failure compensation designs.

1.1 Actuator Failure Compensation

Actuator failures can be uncertain, that is, it is not known when, in what manner, and how many actuators fail. For example, some unknown inputs may be stuck at some unknown fixed values at unknown time instants. A number of aircraft accidents were caused by actuator failures, such as the horizontal stabilizer or the rudder being stuck in an unknown position, leading to catastrophic failures. Actuator failure compensation is an important and challenging problem for control systems research with both theoretical and practical significance.

1.1.1 Literature Overview

In recent years, the actuator failure compensation problem has been studied via several different approaches. There have been a number of results in the literature on control of systems with failures. Typical design methods include multiple-model, switching, and tuning designs, adaptive designs, fault detection and diagnosis designs, and robust control designs.

Multiple-Model, Switching, and Tuning. For control of systems with component failures, one class of designs is based on multiple-model, switching, and tuning and has been applied to reconfigurable flight control [16], [17], [20], [46], [152]. The basic idea of multiple-model, switching, and tuning designs is assuming that the controlled system (plant) belongs to a set of plant models. For each model, a controller is designed to achieve the control objective. During system operation, these models run in parallel with the plant, and if one actuator fails, the switching mechanism will find the best-matched model and switch to the appropriate controller. The multiple-model, switching, and tuning design has several forms [96]: one based on all-fixed plant models, one based on all-adaptive plant models, one based on fixed models and one adaptive model, and one based on fixed models with one free-tuning and one reinitialized adaptive model. An all-fixed design needs sufficient density of models in the set of plant models, as explained in [96].

Adaptive Designs. Another type of control designs are indirect or direct adaptive control based schemes [1], [13], [17], [18], [19], [91]. In [1], an indirect adaptive LQ controller is used to accommodate failures in the pitch control channel or the horizontal stabilizer, leading to performance improvement. In [13], several indirect and direct adaptive control algorithms are presented for control of aircraft with a failure characterized by a locked left horizontal tail surface. An adaptive controller is used to accommodate the system dynamics change caused by such a failure. In [17] and [18], indirect adaptive control schemes are used for compensation of loss of effectiveness of control surfaces. In [19], an adaptive algorithm is used for control of a dynamic system with known dynamics but unknown actuator failures. The control law for the known dynamics is based on a model matching design, while the compensation for actuator failure is based on an adaptive tuning of actuation parameter matrices. A model following adaptive design for failure compensation is presented in [91], which achieves output tracking for some multi-output systems.

Designs based on *indirect* adaptive control first estimate the system and failure parameters and then implement control law reconfiguration employing the functioning actuators. *Direct* adaptive control based designs do not explicitly involve system and failure parameter estimation, and instead they adaptively update control reconfiguration parameters online.

In this book, we develop a general framework for direct adaptive control of systems with both uncertain parameters and uncertain actuator failures, with comparison to indirect designs, and present various adaptive schemes for different control designs and performance requirements.

Adaptive reconfigurable flight control designs using neural networks have been developed for aircraft systems with failures [21], [49], [50], [69], [74], [108], [142]. Unlike the neural networks based adaptive designs, our adaptive failure compensation control designs is model-based, that is, the nominal system structural information is incorporated into adaptive failure compensation designs, to analytically ensure system *stability and tracking* properties. For applications, these two adaptive approaches can be further combined to achieve desired system performance.

Fault Detection and Diagnosis. The fault detection and diagnosis approach [24], [35], [41], [59], [64], [83], [106], [135], [139], [140] has also been used for control of systems with component failures. Related results also include those in [9], [144], [150], [151], using fault-tolerant control designs; in [29] and [58], using identification of multiplicative faults based on parameter estimation techniques; in [15], using function approximations for control and adaptive law design; in [2], [47], [61], [80], [85], [103], [140], using residual generation techniques for fault detection and diagnosis; and in [82], [97], [143], [145], using other design and analysis techniques.

Robust Control Designs. Robust control designs, which can deal with parameter variations and model uncertainties, have also been used to accommodate certain presumed component failures by treating them as uncertainties. As a result, system stability can be guaranteed and an acceptable closed-loop performance can be maintained in the presence of actuator failures. Typical robust control techniques used in the design of reliable control systems are H_∞ controller [137], [147], linear quadratic regulator (LQR) [76], [136], [148], linear matrix inequality [38], [77], and eigenstructure assignment [154]. The robust control based fault-tolerant designs use fixed parameter controllers, which are for the worst-case failures and do not adapt to changes of system failure pattern and failure values.

This subsection gives only a subset of the results on system (actuator) failure compensation. Interested readers are referred to the survey papers [10], [11], [102], [107], [112], [153] and the references therein for a more complete picture for recent development of fault-tolerant control systems.

1.1.2 Research Motivation

Although there have been many advances in control of systems with unknown actuator failures, many open and challenging problems still exist. An effective actuator failure compensation approach is needed to handle both system parameter uncertainties and actuator failure uncertainties.

Adaptive control designs are able to handle uncertainties in both system dynamics and actuator failures that can occur during system operation. Such failures are often uncertain in time, value, and pattern, that is, when, how much, and which actuators fail. Compared with multiple-model, switching, and tuning designs and fault diagnosis designs, adaptive failure compensation control designs have simpler controller structures. Only one adaptive controller is used to accommodate the system dynamics change caused by actuator failures. Adaptive actuator failure compensation designs adaptively adjust controller parameters using system response errors to achieve desired performance. Adaptive actuator failure compensation schemes do not rely on the knowledge of actuator failures, while they also do not rely on the knowledge of the controlled system, as compared with robust control designs.

An important feature of adaptive failure compensation is that such a design is able to adapt to the changes of system failure pattern and failure values, so that in addition to stability, asymptotic tracking of a reference signal is ensured, despite the system and failure uncertainties.

Our research has been focused on the development of a new general theoretical framework for adaptive actuator failure compensation to provide guidelines for designing control systems with guaranteed stability and tracking performance in the presence of system parameter and failure uncertainties. Solutions have been derived for some key issues such as controller structures, matching conditions, error models, adaptive laws, and stability and performance analysis, for uncertain systems (linear, multivariable, nonminimum phase, or nonlinear) with uncertain actuator failures. The desired performance of the developed adaptive actuator failure compensation schemes has been verified on various linear and nonlinear aircraft dynamic models with uncertain elevator, stabilizer or rudder segment failures.

1.2 Adaptive Control System Concepts

In this section, we give a brief overview of adaptive control systems. The detailed description of adaptive control systems can be found in many textbooks, for example, [6], [45], [55], [71], [73], [95], [110], [118], and others.

Adaptive control systems are capable of accommodating systems with parametric, structural, and environmental uncertainties. An adaptive controller, which can effectively deal with systems with unknown parameters, is formed by combining an online parameter estimator, which provides estimates of unknown system parameters at each time instant, with a control law that has two basic parts: structure and parameters. The controller structure is designed based on the known parameter case, while the controller parameters are updated from an adaptive law called a parameter estimator. Such a controller parameter adaptation can be realized in two ways: indirectly and directly. For indirect adaptive control, the unknown parameters of a controlled system (plant) are estimated online and the plant parameter estimates are used to calculate the controller parameters from some design equation. For direct adaptive control, the plant parameters are parametrized implicitly in terms of a set of parameters of a nominal controller, which are estimated online (directly) by an adaptive law to implement an adaptive controller.

The principle behind the design of the direct and indirect adaptive control schemes well-developed in the literature is quite simple: using either the estimates (which are obtained from an adaptive law) of either the plant parameters to calculate the controller parameters for an indirect design, or the estimates of the controller parameters for a direct design. This is called the *certainty equivalence principle*. The key to the successful use of this principle is that the adaptive law for updating the parameter estimates should lead to an estimation error between a measured system signal and its estimated version constructed using the parameter estimates, which has some convergence (to zero) property. Such a convergence property in turn implies the desired closed-loop stability and output tracking, which are the main objectives of adaptive control. It is important to note that for stability and tracking with adaptive control it is sufficient to have some desired signal matching property (plus some system structural property), that is, it is not necessary to have exact parameter matching. However, for the choice of a nominal controller structure it is usually necessary to ensure parameter matching between the closed-loop system and a given desired system.

A summary of adaptive control schemes is given in the appendix. For adaptive control of linear systems, a popular and powerful direct adaptive control approach is model reference adaptive control (MRAC), whose basic design theory is presented in Section A.1. Model reference control uses dynamic feedback to make the closed-loop system dynamics match that of a chosen reference model system, which generically leads to cancellation of zeros of the controlled system. Therefore, a model reference controller is usually applicable only to systems with stable zeros. Model reference adaptive control employs the structure of a model reference controller constructed based on the knowledge of the system order, as well as an adaptive law to update its parameters using the information of the tracking error between the system output and reference model output. A reference model system is constructed with the relative degree knowledge of the controlled system.

A model reference adaptive controller can be designed in three ways: using state feedback for state tracking (see Section A.1.1); using state feedback for output tracking (see Section A.1.2); and using output feedback for output tracking (see Section A.1.3). When the controlled system has multiple inputs and multiple outputs, a model reference adaptive controller needs to be designed with special care to handle interactions between system inputs and outputs (see Section A.2: Multivariable MRAC).

For adaptive control of linear systems, another commonly used (indirect) adaptive control design is adaptive pole placement control, whose design procedure and conditions are given in Section A.3. A pole placement controller is applicable to systems whose zeros may be unstable, as the controller aims to place all the closed-loop poles at some given desired ones and no cancellation of zeros is needed. For pole placement control, the dynamic model of the controlled system should be minimal, that is, the state-space realization is both controllable and observable; in other words, for the chosen system modeling order (which is to be used to construct the controller structure), there are no zero-pole cancellations. For adaptive pole placement control, the system parameters are estimated using an adaptive law, and the parameter estimates are then used to calculate the controller parameters from a design equation. The key to the successful implementation of an adaptive pole placement controller is the existence of solution to such a design equation (the Diophantine equation), which can be reinforced by special modifications to parameter estimates from a standard adaptive law.

Adaptive control of nonlinear systems has made major progress in recent years, with the help of advances in the differential geometric theory of nonlinear feedback [60], [71], [84]. The well-known methodology *feedback linearization* converts many nonlinear control problems into simpler linear problems solvable by linear control approaches. A thorough investigation on feedback linearization in given by Isidori in [60]. A recursive design, *backstepping*, is powerful for adaptive nonlinear control. A complete treatment of adaptive backstepping designs is given by Krstić et al in [71].

The goal of this book is to show how the existing adaptive control designs can be further developed to handle both system and failure uncertainties.

1.3 Adaptive Actuator Failure Compensation

Adaptive control of systems with actuator failures is aimed at compensating for uncertain failures with adaptive tuning of controller parameters based on system response errors to achieve desired system performance. In this book, we will present solutions to the adaptive actuator failure compensation problems: for linear and nonlinear systems with uncertain actuator failures.

1.3.1 Failure Compensation for Linear Systems

We first consider linear time-invariant plants described by

$$\dot{x}(t) = Ax(t) + Bu(t), \; y(t) = Cx(t) \tag{1.1}$$

where $A \in R^{n \times n}$, $B = [b_1, \ldots, b_m] \in R^{n \times m}$, $C \in R^{q \times n}$ are unknown constant parameter matrices, $u(t) = [u_1, \ldots, u_m]^T \in R^m$ is the input vector whose components (actuators) may fail during system operation (such failures are also unknown), and $y(t) \in R^q$ is the plant output.

The type of actuator failures [19] considered in this book is characterized by some unknown inputs being stuck at some unknown fixed values that cannot be influenced by control action, which are modeled as

$$u_j(t) = \bar{u}_j, \; t \geq t_j, \; j \in \{1, 2, \ldots, m\} \tag{1.2}$$

where the constant value $\bar{u}_j$ and the failure time instant t_j are unknown. This is an important type of actuator failure that often occurs. A typical example is an aircraft control surface (such as rudder, elevator, or stabilizer) stuck at some unknown fixed value.

In the presence of actuator failures, $u(t)$ can be expressed as

$$u(t) = v(t) + \sigma(\bar{u} - v(t)) \tag{1.3}$$

where $v(t)$ is the applied control input to be designed, and

$$\bar{u} = [\bar{u}_1, \bar{u}_2, \ldots, \bar{u}_m]^T, \; \sigma = \text{diag}\{\sigma_1, \sigma_2, \ldots, \sigma_m\} \tag{1.4}$$

$$\sigma_j = \begin{cases} 1 & \text{if the } j\text{th actuator fails, i.e., } u_j = \bar{u}_j \\ 0 & \text{otherwise.} \end{cases} \tag{1.5}$$

More general types of actuator failures are also considered in this book. One is the parametrizable time-varying failures modeled as

$$u_j(t) = \bar{u}_j + \bar{d}_j(t), \; t \geq t_j \tag{1.6}$$

where $\bar{u}_j$ is an unknown constant, and

$$\bar{d}_j(t) = \sum_{l=1}^{n_d} \bar{d}_{jl} f_{jl}(t) \tag{1.7}$$

for some unknown scalar constants $\bar{d}_{jl}$ and known scalar signals $f_{jl}(t)$, $j = 1, \ldots, m$, $l = 1, \ldots, n_d$, $n_d \geq 1$.

The other is the unparametrizable time-varying failures modeled as

$$u_j(t) = \bar{u}_j + \bar{d}_j(t) + \bar{\delta}_j(t), \; t \geq t_j \tag{1.8}$$

where $\bar{u}_j$ and $\bar{d}_j(t)$ are defined in (1.2) and (1.7), respectively, and $\bar{\delta}_j(t)$ is an unknown and unparametrizable but bounded term. The actuator failure model (1.6) or (1.8) can be used to closely approximate a large class of practical failures, by a proper selection of these "basis" functions $f_{jl}(t)$, while parametrized by $\bar{d}_{jl}$ (a large n_d would make $\bar{\delta}_j$ smaller).

For the plant (1.1) with unknown parameters and uncertain actuator failures modeled in either (1.2), (1.6), or (1.8) whose failure time (when a failure occurs), pattern (which actuators fail), and values (how much the failure is) are all unknown, we will achieve the following control objectives:

State feedback for state tracking: Design a state feedback control $v(t)$ such that all closed-loop system signals are bounded and the plant state vector $x(t)$ tracks a reference state $x_m(t)$ asymptotically.

State feedback for output tracking: Design a state feedback control $v(t)$ such that all closed-loop system signals are bounded and the plant output $y(t)$ tracks a reference output $y_m(t)$ asymptotically.

Output feedback for output tracking: Design an output feedback control $v(t)$ such that all closed-loop system signals are bounded and the plant output $y(t)$ tracks a reference output $y_m(t)$ asymptotically.

For state tracking, it is a multi-input, multi-output control problem, while for output tracking, it is a multi-input multi-output control problem when $q > 1$ for $y(t) \in R^q$ in (1.1) and it is a multi-input single-output control problem when $q = 1$. Input redundancy is an issue to be solved.

Such an adaptive actuator failure compensation control system structure is shown in Figure 1.1, where the actuator failure indicators σ_i and failure values $\bar{u}_i$, $i = 1, 2, \ldots, m$, are all uncertain ($\sigma_i = 0$ or $\sigma_i = 1$ as it depends on the failure pattern, which is unknown), and $r(t)$ is a reference input signal. The control task is to make the system output $y(t)$ (or state variable $x(t)$) to track a reference output (or state) signal determined by $r(t)$, in the presence of both the system parameter and actuator failure uncertainties.

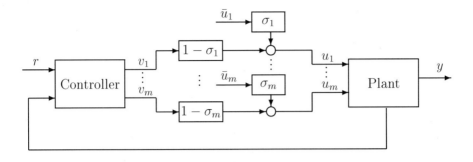

Fig. 1.1. Control system with actuator failures.

1.3.2 Failure Compensation for Nonlinear Systems

To investigate the adaptive actuator failure compensation control problems for nonlinear systems, we consider the nonlinear plant

$$\dot{x} = f(x) + \sum_{j=1}^{m} g_j(x)u_j$$

$$y = h(x) \tag{1.9}$$

where $x \in R^n$ is the state, $y \in R$ is the output, and $u_i \in R$ is the input component, which may fail during operation, and $f(x) \in R^n$, $g_j(x) \in R^n$, and $h(x) \in R$ are smooth (differentiable) functions.

For a plant with uncertain parameters, we consider its parametrized form

$$\dot{x} = f_0(x) + \sum_{i=1}^{l} \theta_i f_i(x) + \sum_{j=1}^{m} \mu_j g_j(x)u_j$$

$$y = h(x) \tag{1.10}$$

where $f_i(x) \in R^n$, $i = 0, 1, \ldots, l$, are smooth functions, and θ_i, $i = 1, 2, \ldots, l$, and μ_j, $j = 1, 2, \ldots, m$, are unknown constant parameters.

The actuator failures are modeled by (1.2), (1.6), or (1.8). The plant input expression (1.3) is valid for (1.2) (also for (1.6) or (1.8) if $\bar{u}_j$ is replaced by $\bar{u}_j + \bar{d}_j(t)$ or $\bar{u}_j + \bar{d}_j(t) + \bar{\delta}_j(t)$). With (1.3), the plant (1.9) is expressed as

$$\dot{x} = f(x) + g(x)\sigma\bar{u} + g(x)(I - \sigma)v$$

$$y = h(x) \tag{1.11}$$

where $g(x) = [g_1(x), g_2(x), \ldots, g_m(x)] \in R^{n \times m}$, for a failure pattern σ.

With the expression (1.3), we can also rewrite the plant (1.10) with the actuator failures (1.2) in the compact form

$$\begin{aligned} \dot{x} &= f_0(x) + F(x)\theta + g(x)\mu\sigma\bar{u} + g(x)\mu(I - \sigma)v \\ y &= h(x) \end{aligned} \qquad (1.12)$$

for a failure pattern σ, where

$$\begin{aligned} F(x) &= [f_1(x), f_2(x), \ldots, f_l(x)] \in R^{n \times l} \\ \theta &= [\theta_1, \theta_2, \ldots, \theta_l]^T \in R^l \\ g(x) &= [g_1(x), g_2(x), \ldots, g_m(x)] \in R^{n \times m} \\ \mu &= \mathrm{diag}\{\mu_1, \mu_2, \ldots, \mu_m\} \in R^{m \times m}. \end{aligned} \qquad (1.13)$$

From (1.11) and (1.12) it is clear that actuator failures not only introduce disturbance into systems, but also change the system structure.

The control objective is to use state feedback or output feedback control for the plant (1.12) or (1.13) in the presence of unknown actuator failures modeled as (1.2), (1.6), or (1.8), such that the closed-loop signal boundedness and asymptotic tracking of a reference signal $y_m(t)$ by the plant output $y(t)$ are ensured, despite the actuator failure uncertainties.

1.3.3 Basic Assumption

The goal of this research is to apply adaptive control techniques to design and analyze actuator failure compensation schemes based on theoretically characterizable and practically implementable conditions for desired closed-loop stability and asymptotic tracking despite actuator failure uncertainties.

The basic assumption on the plant (1.1), (1.9), or (1.10) for solving the actuator failure compensation problems in this book is stated as

(A1) The plant (1.1), (1.9), or (1.10) is so constructed that in the presence of up to any $m - q$ $(1 \le q \le m)$ actuator failures, the remaining functional actuators can still be used to implement control signals (which are designed using the knowledge of the plant parameters and that of the failure time, pattern, and parameters) to achieve a desired control objective.

This is the basic assumption for the existence of a nominal solution to the actuator failure compensation problem when the plant parameters and actuator failure pattern and parameters are all known. This basic assumption leads to characterizations of necessary and sufficient conditions for the existence of such a nominal solution, when applied to different types of plants as well as different types of control design methods.

The key task of adaptive actuator failure compensation is to adjust the remaining functional actuators to achieve the desired system performance when there are actuator failures whose failure time, pattern, and parameters are unknown, in addition to the unknown system parameters.

For example, suppose the control surfaces (e.g., aileron) of an aircraft are divided into several individually actuated segments. If some of the aileron segments stop moving and stay at fixed positions, the remaining ailerons may still be able to make a safe landing, but they need to be controlled in a proper manner. An adaptive controller is expected to automatically adjust the positions of the remaining ailerons to desired values for a safe landing, without knowing which and how many ailerons have failed, and at what fixed positions they have failed.

As will be shown in this book, the conditions (some of which appear restrictive but are necessary even for a nominal controller due to the demanding features of the actuator compensation problems) implied by the above basic assumption are also sufficient for the existence of adaptive solutions to the adaptive actuator failure compensation problems when the plant parameters and failure pattern and parameters are all unknown.

On the other hand, necessary and sufficient conditions derived from this basic assumption for different systems and different control schemes also represent the guidelines for constructing a dynamic system that has the ability of being compensated for its actuator failures.

1.3.4 Basic Actuation Scheme for Failure Compensation

Given that the system (1.1), (1.9), or (1.10) has m actuators to provide some redundancy needed for failure compensation, there are two different situations. The first is that all m actuators have similar physical characteristics; for example, they are segments of a multiple-segment rudder or elevator for an aircraft, or they are heating devices for an oven. The second situation is that the m actuators are divided into several groups and the actuators within each group have similar physical characteristics; for example, an aircraft has a group of four engines and a group of three rudder segments.

A meaningful design of actuation for the first situation is that all actuators have the same control signals, that is, the equal-actuation scheme

$$v_1(t) = v_2(t) = \cdots = v_m(t) \tag{1.14}$$

or the proportional-actuation scheme, that is,

$$v_1(t) = \alpha_2 v_2(t) = \cdots = \alpha_m v_m(t) \tag{1.15}$$

for some chosen constant $\alpha_i > 0$, $i = 2, 3, \ldots, m$. In other words, it is not realistic, for example, to make the two rudder segments move at opposite angles (in fact, in this case, it is desirable to use the actuation scheme (1.14) when $v_i(t)$ is the designed rudder segment angle for the ith rudder segment). In some other applications, the actuation scheme (1.15) may be more appropriate; for example, in temperature control, each heater has a different (proportional) effect on the temperature at a given point in an oven.

A meaningful design of actuation for the second situation is that all actuators in each group have the same or proportional control signals. The control signals for different groups should be designed based on a multivariable control scheme that takes into account the interactions between the outputs and the inputs from different actuator groups (see Chapter 5).

The control task is to design the feedback control signals $v_i(t)$, $i = 1, 2, \ldots, m$, without knowing which of the m actuators have failed (that is, it is not known whether $u_i(t) = v_i(t)$ or $u_i(t) = \bar{u}_i$), such that the plant output $y(t)$ (or state vector $x(t)$) can track a given reference model output $y_m(t)$ (or state vector $x_m(t)$) asymptotically, despite the uncertainties in the plant parameters and actuator failure parameters $\bar{u}_i$.

A further question is: Do actuators in different groups compensate for each other? The answer depends on the case of interest. For example, an elevator and a stabilizer may compensate for each other (this case will be illustrated by simulation results; see Chapter 7). Differential engine thrust may compensate for a failed rudder (this case is not considered in this book).

The basic assumption in Section 1.3.3 and the above basic actuation scheme will be used throughout this book (except for Chapter 5, where a multivariable design is given for different groups of actuators that may have different physical characteristics) for adaptive actuator failure compensation control designs for different systems with different control schemes. Without loss of generality, the equal-actuation scheme (1.14) will be used for many of the actuator failure compensation control designs. We should note that in Chapter 2, adaptive actuator failure compensation designs using actuation schemes more general than the above basic actuation scheme are also developed for the state feedback and state tracking case.

1.3.5 Redundancy and Failure Uncertainty

Redundancy and failure uncertainty are two basic issues that are unique for this adaptive failure compensation control problem.

Redundancy. It is not difficult to understand that for failure compensation, certain system redundancy such as actuator redundancy is necessary. System redundancy has been widely employed in the design of modern control systems such as aircraft flight control systems, rocket launch fairing systems, and other mission-critical systems. The existence of redundancy is not only common but also brings new challenges for feedback control designs, especially when the system failure pattern is unknown; for example, when the nominal system has five actuators and up to four of them may fail during system operation. Without knowing which of those up to four actuators have failed, control input signals have to be designed for all five actuators such that no matter which of up to four actuators fail, the remaining actuators can still control the system behavior. As given in (1.3), the actuation error $u - v = \sigma(\bar{u} - v)$, which is also dependent on v, is unknown to the design of the control input v. A desirable controller should be able to handle all possible failure patterns and values, with all possible degrees of redundancy. In this sense, redundancy is a part of the problem, not the proposed approach for solving the problem.

Failure Uncertainties. Systems failures introduce additional and large systems uncertainties. For example, when some actuators fail, the system structure from the active inputs (unfailed actuators) to the output experiences significant changes, and so do the system parameters. Furthermore, failures cause additional disturbances to systems, which influence a system's behavior at different actuation locations. Compensation of such disturbances alone can be a challenging problem (i.e., certain matching conditions have to be established by a proper controller design). In this sense, actuator failures do not just cause system gain changes, and they lead to system uncertainties not treated by existing adaptive control designs. To handle such uncertainties, existing design tools such as model reference adaptive control, adaptive pole placement control, and adaptive backstepping need to be further developed.

This book will present such new developments including plant-model matching, controller structures, error models, and adaptive laws (which are able to accommodate all possible failure patterns and values).

1.4 Book Outline

The book presents designs, analyses, and evaluations of adaptive actuator failure compensation control schemes for linear time-invariant systems with uncertain actuator failures in Chapters 2–8 and for some nonlinear systems with uncertain actuator failures in Chapters 9–11, illustrated by simulation results from aircraft flight control, with a summary in Chapter 12.

Chapter 2 presents state feedback for state tracking designs for linear time-invariant systems with up to $m - 1$ or up to $m - q$ ($1 \leq q \leq m$) uncertain failures of the total m actuators. Chapter 3 investigates the state feedback for output tracking problem for single-output linear time-invariant systems with unknown actuator failures. Chapter 4 develops an adaptive control scheme using output feedback for output tracking for linear time-invariant systems with unknown actuator failures. Chapter 5 deals with the output tracking problem for multi-output linear time-invariant systems using output feedback. Chapter 6 presents a study of an adaptive output feedback output tracking control design based on pole placement control for linear time-invariant systems with unknown actuator failures. Chapter 7 contains application of several adaptive control schemes developed in the previous chapters to a linearized longitudinal dynamic model of a transport aircraft model with unknown system parameters and failures. Chapter 8 considers adaptive actuator failure compensation for discrete-time systems and the associated issue of robust stability with respect to external disturbance and unmodeled dynamics. Chapter 9 develops adaptive actuator failure compensation control schemes for feedback linearizable systems with different structure conditions that characterize different classes of systems. Chapter 10 addresses the actuator failure compensation problem for nonlinear systems that can be transformed into parametric-strict-feedback form with zero dynamics. Chapter 11 presents an adaptive control scheme that achieves stability and output tracking with output feedback for nonlinear plants with unknown actuator failures. Chapter 12 concludes the book and presents some topics for further research on theory and applications.

The book has an appendix, which, to help the readers, presents the fundamentals of adaptive control, including model reference adaptive control using state feedback design for state tracking, state feedback design for output tracking, output feedback design for output tracking, as well as multivariable model reference adaptive control, and adaptive pole placement control.

Chapter 2

State Feedback Designs for State Tracking

In this chapter, we solve the adaptive actuator failure compensation problems for linear time-invariant systems with unknown actuator failures and unknown dynamics, using state feedback for state tracking. In Section 2.1, the plant-model state matching conditions, controller structure, and adaptive designs are presented for a linear time-invariant system with up to $m-1$ (where m is the total number of actuators) actuator failures characterized by some of the plant inputs being stuck at some unknown fixed or varying values that cannot be influenced by control action, for example, the "lock-in-place" type of actuator failures. Adaptive actuator failure designs for systems with up to $m-1$ parametrizable time-varying failures and up to $m-1$ unparametrizable time-varying failures are developed in Section 2.2 and Section 2.3, respectively. In Section 2.4, parametrization and design results such as plant-model matching and adaptive controller structure are extended to the case when the state reference model system has multiple inputs, which allows more freedom in characterizing desired system behavior. In Section 2.5, necessary and sufficient plant-model matching conditions, and adaptive control designs for systems with up to $m-q$, $1 \leq q \leq m-1$, lock-in-place actuator failures are derived and the effectiveness of adaptive compensation is verified by simulation results from Boeing 747 lateral control.

2.1 Fundamental Issues and Solutions

In this section, we formulate the basic control problems for adaptive actuator failure compensation for linear time-invariant systems with unknown actuator failures and unknown dynamics, using state feedback for state tracking, introduce some fundamental issues, and develop their solutions.

Consider a system described by the differential equation

$$\dot{x}(t) = Ax(t) + Bu(t) \tag{2.1}$$

where $A \in R^{n \times n}$, $B = [b_1, b_2, \ldots, b_m] \in R^{n \times m}$ are unknown constant parameter matrices, the state vector $x(t) \in R^n$ is available for measurement, and

$u(t) = [u_1, \ldots, u_m]^T \in R^m$ is the input vector whose components (actuators) may fail during system operation.

We start with the type of actuator failures modeled in (1.2):

$$u_j(t) = \bar{u}_j, \ t \geq t_j, \ j \in \{1, 2, \ldots, m\} \tag{2.2}$$

where the constant value $\bar{u}_j$ and the failure time instant t_j are unknown. This is an important type of failure that can occur in many applications, for example, when an aircraft control surface (such as the rudder or an aileron) is stuck at some unknown fixed value.

Recall the basic assumption (A1) (see Section 1.3.3), which may be restated for the control problem considered in this chapter as

(A2.0) For any up to $m - 1$ actuator failures, the remaining actuators (controls) can still achieve a desired control objective, when designed with the knowledge of the system parameters in (A, B) and the failure parameters j, $\bar{u}_j$, and t_j, for $j = j_1, j_2, \ldots, j_p$, with any $\{j_1, j_2, \ldots, j_p\} \subset \{1, 2, \ldots, m\}$ and $0 \leq p < m$.

This is the basic existence assumption for a nominal solution. The key task of actuator failure compensation control is to adaptively adjust the remaining controls to achieve the desired performance when there are up to $m - 1$ *unknown* actuator failures, *without* the knowledge of the system parameters.

Given that the plant dynamics matrices (A, B) are unknown, and so are the actuator failure time t_j, parameters $\bar{u}_j$, and pattern j, the control objective is to design a feedback control $v(t)$ such that all signals in the closed-loop system are bounded and the state vector $x(t)$ asymptotically tracks a given reference state vector $x_m(t)$ generated from the reference system

$$\dot{x}_m(t) = A_M x_m(t) + b_M r(t), \ x_m(t) \in R^n, \ r(t) \in R \tag{2.3}$$

where $A_M \in R^{n \times n}$, $b_M \in R^n$ are known constant matrices such that all the eigenvalues of A_M are in the open left-half complex plane, and $r(t)$ is bounded and piecewise continuous.

Remark 2.1.1. In the absence of actuator failures, such a state tracking problem was solved and well-understood when $m = 1$ in (2.1) [95]. Its solution employs a plant-model matching condition: $A + bk_1^{*T} = A_M$, $bk_2^* = b_M$, with $b = B$, for some $k_1^* \in R^n$ and $k_2^* \in R$. This condition characterizes the class of plants and models admissible to an adaptive state feedback and state tracking design. For issues in actuator failure compensation such as controller parametrization, matching, error model, adaptive law, and stability and tracking analysis, new solutions are developed next. □

2.1.1 Basic Plant-Model Matching Conditions

To develop adaptive control schemes for the system (2.1) with unknown parameters and unknown actuator failures, it is useful to understand the desirable controller structure and parameters for the case when the system parameters and actuator failure pattern and parameters are known.

Before addressing the control problem for arbitrary actuator failures, we first derive the controllers for the system (2.1) in two special cases: when all but one actuators fail, and when no actuator fails, to derive some basic plant-model matching conditions that are useful for controller parametrizations in fixed or adaptive designs for arbitrary actuator failures.

Denote the ith column of B as b_i, $i = 1, \ldots, m$, in view of the actuator failure model (2.2). For plant-model matching when the special actuator failure pattern $u_j(t) = \bar{u}_j = 0$, $j = 1, \ldots, i-1, i+1, \ldots, m$, $j \neq i$, occurs, Assumption (A2.0) implies that there exist constant vectors $k^*_{s1i} \in R^n$ and nonzero constant scalars $k^*_{s2i} \in R$, $i = 1, \ldots, m$, such that the following matching equations are satisfied:

$$A + b_i k^{*T}_{s1i} = A_M, \ b_i k^*_{s2i} = b_M. \tag{2.4}$$

In this case, the ith control input

$$u_i(t) = v_i(t) = v^*_i(t) = k^{*T}_{s1i} x(t) + k^*_{s2i} r(t) \tag{2.5}$$

leads to the closed-loop system

$$\dot{x}(t) = A_M x(t) + b_M r(t) \tag{2.6}$$

which matches the reference model so that the control objective is met.

On the other hand, when no actuator failure is present, that is, $Bu(t) = Bv(t)$, Assumption (A2.0) implies that there exist constant vectors $k^*_{1i} \in R^n$ and nonzero constant scalars $k^*_{2i} \in R$, $i = 1, \ldots, m$, such that for $K^*_1 = [k^*_{11}, \ldots, k^*_{1m}] \in R^{n \times m}$, $k^*_2 = [k^*_{21}, \ldots, k^*_{2m}]^T \in R^m$,

$$A + BK^{*T}_1 = A + \sum_{i=1}^{m} b_i k^{*T}_{1i} = A_M, \ Bk^*_2 = \sum_{i=1}^{m} b_i k^*_{2i} = b_M. \tag{2.7}$$

In this case, the feedback control law

$$u(t) = v(t) = v^*(t) = K^{*T}_1 x(t) + k^*_2 r(t) \tag{2.8}$$

also leads to the closed-loop system (2.6).

Now for a more general case, suppose there are p failed actuators, that is, $u_j(t) = \bar{u}_j$, $j = j_1, \ldots, j_p$, $1 \leq p \leq m-1$, but with $\bar{u}_j = 0$, $j = j_1, \ldots, j_p$. In this case, the desired matching conditions clearly are

$$A + \sum_{i \neq j_1, \ldots, j_p} b_i k_{1i}^{*T} = A_M, \quad \sum_{i \neq j_1, \ldots, j_p} b_i k_{2i}^* = b_M. \tag{2.9}$$

To characterize these conditions, using $b_i k_{s2i}^* = b_M$ in (2.4), $i = 1, \ldots, m$, we can write $\sum_{i \neq j_1, \ldots, j_p} b_i k_{2i}^* = b_M$ in (2.9) as

$$\sum_{i \neq j_1, \ldots, j_p} \frac{k_{2i}^*}{k_{s2i}^*} = 1. \tag{2.10}$$

Similarly, using $A + b_i k_{s1i}^{*T} = A_M$, $i = 1, \ldots, m$, and $A + \sum_{i \neq j_1, \ldots, j_p} b_i k_{1i}^{*T} = A_M$, we obtain

$$\sum_{i \neq j_1, \ldots, j_p} \frac{k_{1i}^*}{k_{s2i}^*} = \frac{k_{s1j}^*}{k_{s2j}^*}, \quad j = 1, \ldots, m. \tag{2.11}$$

Remark 2.1.2. The parameters k_{s1i}^* and k_{s2i}^* that satisfy (2.4) may be unique, while the parameters k_{1i}^* and k_{2i}^* in (2.9) are not unique and can be determined from (2.10) and (2.11) and their existence is ensured by the existence of k_{s1i}^* and k_{s2i}^*. The equalities (2.10) and (2.11) also hold for $p = 0$, that is, when there is no failure. The existence of such K_1^* and k_2^* is crucial for an adaptive control design that needs a nominal control law like that in (2.8) for a complete parametrization of an error model used for developing a suitable adaptive parameter update law. The condition (2.11) implies that $k_{s1j}^*/k_{s2j}^* = k_{s1i}^*/k_{s2i}^* = \bar{k}_{s1}^*$ for some $\bar{k}_{s1}^* \in R^n$ and any i, j. While they are non-uniquely defined, one choice of k_{1i}^* to satisfy (2.11) is $k_{1i}^* = \alpha_i \bar{k}_{s1}^*$, $i \neq j_1, \ldots, j_p$, for some $\alpha_i \in R$ such that $\sum_{i \neq j_1, \ldots, j_p} \alpha_i / k_{s2i}^* = 1$. □

Remark 2.1.3. It is clear that condition (2.4) implies condition (2.9), that is, for a special choice $k_{1i}^* = k_{s1i}^*$, $k_{2i}^* = k_{s2i}^*$, $k_{1j}^* = 0$, $k_{2j}^* = 0$, for all $j \neq i$. As indicated above, condition (2.4) is necessary for solving the formulated actuator failure compensation problems. As shown in the next sections, condition (2.4) is also sufficient for solving the adaptive actuator failure compensation problem. We should note that condition (2.4) is also the plant-model matching condition needed for a solution to the model reference control problem [95], without actuator failures and with or without plant parameter uncertainties when $m = 1$ in the plant (2.1) (see Remark 2.1.1). □

Condition (2.4) is used in Section 2.1.2 to derive a plant-model state matching controller in the presence of actuator failures modeled in (2.2) with known parameters, and in Sections 2.2, 2.3, and 2.4 for adaptive actuator failure compensation designs using parameter estimates for different failure models: constant, parametrizable, or unparametrizable varying failures. Some fundamental issues in actuator failure compensation control, including system modeling, controller structures, plant-model matching, error models, adaptive laws, stability analysis, which are different from those in the standard adaptive control cases without actuator failures, are to be addressed.

2.1.2 Actuator Failure Compensation

To solve the control problem with arbitrary actuator failures in a known system, we propose the following structure:

$$v(t) = v^*(t) = K_1^{*T} x(t) + k_2^* r(t) + k_3^* \tag{2.12}$$

where the parameters $K_1^* \in R^{n \times m}$ and $k_2^* \in R^m$ have been introduced in (2.7) and (2.8), and $k_3^* = [k_{31}^*, \ldots, k_{3m}^*]^T \in R^m$ is to be defined.

As given in Section 1.3.1, the input vector u to the plant (2.1) in the presence of actuator failures can be described as

$$u(t) = v(t) + \sigma(\bar{u} - v(t)) \tag{2.13}$$

where $v(t)$ is a designed control input, and

$$\bar{u} = [\bar{u}_1, \bar{u}_2, \ldots, \bar{u}_m]^T \tag{2.14}$$

$$\sigma = \text{diag}\{\sigma_1, \sigma_2, \ldots, \sigma_m\} \tag{2.15}$$

$$\sigma_i = \begin{cases} 1 & \text{if the } i\text{th actuator fails, i.e., } u_i = \bar{u}_i \\ 0 & \text{otherwise.} \end{cases} \tag{2.16}$$

Suppose there are p failed actuators at time t, that is, $u_j(t) = \bar{u}_j$, $j = j_1, \ldots, j_p$, $0 \le p \le m-1$. For arbitrary $\bar{u}_j$, $j = j_1, \ldots, j_p$, $0 \le p \le m-1$, the plant-model matching condition for K_1^* and k_2^* is also that in (2.9), that is,

$$A + B(I - \sigma)K_1^{*T} = A + \sum_{i \neq j_1, \ldots, j_p} b_i k_{1i}^{*T} = A_M$$

$$B(I - \sigma)k_2^* = \sum_{i \neq j_1, \ldots, j_p} b_i k_{2i}^* = b_M. \tag{2.17}$$

From (2.13) and (2.17), the controller (2.12) leads to the closed-loop system

$$\begin{aligned} \dot{x}(t) &= Ax(t) + Bv^*(t) + B\sigma(\bar{u} - v^*(t)) \\ &= A_M x(t) + b_M r(t) + B\sigma(K_1^{*T} x(t) + k_2^*(t)) + Bk_3^* + B\sigma(\bar{u} - v^*(t)) \\ &= A_M x(t) + b_M r(t) + B(I - \sigma)k_3^* + B\sigma\bar{u}. \end{aligned} \tag{2.18}$$

For this system to match the reference model system (2.3), we need to choose k_{3i}^*, $i \neq j_1, \ldots, j_p$, to make

$$B(I - \sigma)k_3^* + B\sigma\bar{u} = \sum_{i \neq j_1,\ldots,j_p} b_i k_{3i}^* + \sum_{j=j_1,\ldots,j_p} b_j \bar{u}_j = 0. \qquad (2.19)$$

In this case, the choice of k_{1j}^*, k_{2j}^*, and k_{3j}^*, $j = j_1, \ldots, j_p$, is not relevant because their actuators have failed (they are not present in the closed-loop system (2.18)). One nominal choice could be

$$k_{1j}^* = 0, \; k_{2j}^* = 0, \; k_{3j}^* = 0, \; j = j_1, \ldots, j_p. \qquad (2.20)$$

Under the condition (2.4), that is, $b_i = b_M/k_{s2i}^*$, from (2.19) we have the desired matching condition

$$\sum_{i \neq j_1,\ldots,j_p} \frac{k_{3i}^*}{k_{s2i}^*} + \sum_{j=j_1,\ldots,j_p} \frac{\bar{u}_j}{k_{s2j}^*} = 0 \qquad (2.21)$$

which determines the choice of the parameters k_{3i}^*, $i \neq j_1, \ldots, j_p$, for achieving the desired closed-loop system (2.6).

There are two special cases of (2.21). One is when only one actuator $u_j(t)$ fails, so that (2.21) becomes

$$\sum_{i \neq j} \frac{k_{3i}^*}{k_{s2i}^*} + \frac{\bar{u}_j}{k_{s2j}^*} = 0. \qquad (2.22)$$

Another case is when all $u_j(t)$ but $u_i(t)$ fail, that is, when $u_j(t) = \bar{u}_j$ for all $j \neq i$, and $u_i(t) = v_i(t)$. In this case, (2.21) leads to the unique choice

$$k_{3i}^* = -k_{s2i}^* \sum_{j \neq i} \frac{\bar{u}_j}{k_{s2j}^*}. \qquad (2.23)$$

For the case when no actuator fails, the parameters K_1^* and k_2^* are defined in Section 2.2, while the parameters in k_3^* are defined by

$$\sum_{i=1}^{m} \frac{k_{3i}^*}{k_{s2i}^*} = 0. \qquad (2.24)$$

One choice is $k_{3i}^* = 0$, $i = 1, \ldots, m$, as that in (2.8).

In all cases, the existence of K_1^*, k_2^*, and k_3^* in the controller (2.12) is ensured to make the closed-loop system (2.18) match the reference system (2.3), that is, resulting in the desired closed-loop system (2.6).

In summary, the plant-model matching conditions for solving the actuator failure compensation problem under Assumption (A2.0) are given as

Proposition 2.1.1. *For actuator failure compensation, it is necessary and sufficient that there exist constant vectors $k_{s1i}^* \in R^n$ and nonzero constant scalars $k_{s2i}^* \in R$, $i = 1, \ldots, m$, such that $A + b_i k_{s1i}^{*T} = A_M$, $b_i k_{s2i}^* = b_M$.*

The choice of the plant-model matching parameters K_1^*, k_2^*, and k_3^* is determined by (2.17) and (2.21), or by (2.4), (2.10), (2.11), and (2.21). Such desired controller parameters may not be unique. It is important to note that those parameters depend on the system parameters A and B as well as on the knowledge of the actuator failures. In system operations, whenever the actuator failure pattern changes, for example, two actuators fail first and another one fails later, the desired plant-model matching parameters K_1^*, k_2^*, and k_3^* may change according to the conditions in (2.17) and (2.21) or by (2.4), (2.10), (2.11), and (2.21). Therefore, the ideal matching parameters K_1^*, k_2^*, and k_3^* are actually piecewise constant time-varying parameters.

It will be shown next that the condition in Proposition 2.1.1 is also sufficient for constructing a stable adaptive control scheme for the case when both the system parameters and actuator failure parameters are unknown.

2.1.3 Adaptive Compensation Design

Now we develop an adaptive control scheme for the system (2.1) with unknown parameters A and B, and with unknown actuator failures (2.2). We propose the controller structure

$$v(t) = K_1^T(t)x(t) + k_2(t)r(t) + k_3(t) \qquad (2.25)$$

where $K_1(t) \in R^{n \times m}$, $k_2(t) \in R^m$, and $k_3(t) \in R^m$ are adaptive estimates of the unknown parameters K_1^*, k_2^* and, k_3^*. To design adaptive update laws for

$$K_1(t) = [k_{11}(t), \ldots, k_{1m}(t)], \; k_2(t) = [k_{21}(t), \ldots, k_{2m}(t)]^T$$
$$k_3(t) = [k_{31}(t), \ldots, k_{3m}(t)]^T \qquad (2.26)$$

we assume that

(A2.1) $\text{sign}[k_{s2i}^*]$, the sign of the parameter k_{s2i}^*, in (2.4) is known, for $i = 1, \ldots, m$.

Define the parameter errors

$$\tilde{k}_{1i}(t) = k_{1i}(t) - k_{1i}^*, \; \tilde{k}_{2i}(t) = k_{2i}(t) - k_{2i}^*, \; \tilde{k}_{3i}(t) = k_{3i}(t) - k_{3i}^* \qquad (2.27)$$

for $i = 1, \ldots, m$, and the tracking error $e(t) = x(t) - x_m(t)$.

Let (T_i, T_{i+1}), $i = 0, 1, \ldots, m_0$, with $T_0 = 0$, be the time intervals on which the actuator failure pattern is fixed, that is, actuators only fail at time T_i, $i = 1, \ldots, m_0$. Since there are m actuators, at least one of them does not fail, we have $m_0 < m$ and $T_{m_0+1} = \infty$. Then, at time T_j, $j = 1, \ldots, m_0$, the unknown plant-model matching parameters K_1^*, k_2^*, and k_3^* (that is, k_{1j}^*, k_{2j}^*, k_{3j}^*, $j = 1, \ldots, m$) change their values such that

$$k_{1j}^* = k_{1j(i)}^*, \ k_{2j}^* = k_{2j(i)}^*, \ k_{3j}^* = k_{3j(i)}^*, \ t \in (T_i, T_{i+1}) \qquad (2.28)$$

for $j = 1, \ldots, m$, $i = 0, 1, \ldots, m_0$.

Suppose there are p failed actuators, that is, $u_j(t) = \bar{u}_j$, $j = j_1, \ldots, j_p$, $1 \le p \le m - 1$, at time $t \in (T_i, T_{i+1})$ (with $i = i(p) \le p$ because there may be more than one actuator failing at the same time T_i). Using (2.1), (2.13), (2.17), (2.19), (2.25), and (2.26), we obtain

$$
\begin{aligned}
\dot{x}(t) &= Ax(t) + Bv(t) + B\sigma(\bar{u} - v(t)) \\
&= Ax(t) + B(I - \sigma)(K_1^{*T}x(t) + k_2^* r(t)) + B(I - \sigma)k_3^* + B\sigma\bar{u} \\
&\quad + B(I - \sigma)\left(\tilde{K}_1^T(t)x(t) + \tilde{k}_2(t)r(t) + \tilde{k}_3(t)\right) \\
&= A_M x(t) + b_M r(t) + B(I - \sigma)\left(\tilde{K}_1^T(t)x(t) + \tilde{k}_2(t)r(t) + \tilde{k}_3(t)\right) \\
&= A_M x(t) + b_M r(t) + \sum_{j \ne j_1, \ldots, j_p} b_j \tilde{k}_{1j}^T(t)x(t) + \sum_{j \ne j_1, \ldots, j_p} b_j \tilde{k}_{2j}(t)r(t) \\
&\quad + \sum_{j \ne j_1, \ldots, j_p} b_j \tilde{k}_{3j}(t) \\
&= A_M x(t) + b_M r(t) + b_M \\
&\quad \cdot \sum_{j \ne j_1, \ldots, j_p} \frac{1}{k_{s2j}^*}\left(\tilde{k}_{1j}^T(t)x(t) + \tilde{k}_{2j}(t)r(t) + \tilde{k}_{3j}(t)\right). \qquad (2.29)
\end{aligned}
$$

Then, using (2.3) and (2.29), we have the tracking error equation

$$\dot{e}(t) = A_M e(t) + b_M \sum_{j \ne j_1, \ldots, j_p} \frac{1}{k_{s2j}^*}\left(\tilde{k}_{1j}^T(t)x(t) + \tilde{k}_{2j}(t)r(t) + \tilde{k}_{3j}(t)\right). \quad (2.30)$$

Consider the positive definite function

$$
\begin{aligned}
&V_p(e, \tilde{k}_{1j}, \tilde{k}_{2j}, \tilde{k}_{3j}, j \ne j_1, \ldots, j_p) \\
&= e^T P e + \sum_{j \ne j_1, \ldots, j_p} \frac{1}{|k_{s2j}^*|}\left(\tilde{k}_{1j}^T \Gamma_{1j}^{-1}\tilde{k}_{1j} + \tilde{k}_{2j}^2 \gamma_{2j}^{-1} + \tilde{k}_{3j}^2 \gamma_{3j}^{-1}\right) \quad (2.31)
\end{aligned}
$$

where $k_{1j}^* = k_{1j(i)}^*$, $k_{2j}^* = k_{2j(i)}^*$, and $k_{3j}^* = k_{3j(i)}^*$ are defined in (2.28) (note that this $V_p(\cdot)$ is defined for all $t \in [0, \infty)$ with the matching parameters $k_{1j(i)}^*$, $k_{2j(i)}^*$, and $k_{3j(i)}^*$ for the case when there are p failed actuators over the interval (T_i, T_{i+1}), $i = i(p) \le p)$, $P \in R^{n \times n}$, $P = P^T > 0$ such that

$$PA_M + A_M^T P = -Q \qquad (2.32)$$

for any constant $Q \in R^{n \times n}$ such that $Q = Q^T > 0$, $\Gamma_{1j} \in R^{n \times n}$ is constant such that $\Gamma_{1j} = \Gamma_{1j}^T > 0$, $\gamma_{2j} > 0$, and $\gamma_{3j} > 0$ are constant, $j = 1, \ldots, m$.

Choose the adaptive laws for $k_{1j}(t)$, $k_{2j}(t)$, and $k_{3j}(t)$, $j = 1, \ldots, m$, as

$$\dot{k}_{1j}(t) = -\text{sign}[k_{s2j}^*]\Gamma_{1j}x(t)e^T(t)Pb_M \tag{2.33}$$

$$\dot{k}_{2j}(t) = -\text{sign}[k_{s2j}^*]\gamma_{2j}r(t)e^T(t)Pb_M \tag{2.34}$$

$$\dot{k}_{3j}(t) = -\text{sign}[k_{s2j}^*]\gamma_{3j}e^T(t)Pb_M. \tag{2.35}$$

Then, from (2.30)–(2.35) we have

$$\dot{V}_p = -e^T(t)Qe(t) \leq 0, \ t \in (T_{i(p)}, T_{i(p)+1}), \ i(p) \leq p \tag{2.36}$$

where $V_p(t) \overset{\triangle}{=} V_p(e(t), \tilde{k}_{1j}(t), \tilde{k}_{2j}(t), \tilde{k}_{3j}(t), j \neq j_1, \ldots, j_p)$.

Since there are only a finite number of time instants of actuator failures, that is, $i \leq m_0 < m$, and eventually there are $\bar{m}_0$ failed actuators, $m_0 \leq \bar{m}_0 < m$, from (2.36) it follows that $\dot{V}_{\bar{m}_0} = -e^T(t)Qe(t) \leq 0$, $t \in (T_{m_0}, \infty)$, so we can conclude that $V_{\bar{m}_0}(t)$ is bounded (it can be shown that $V_{\bar{m}_0}(T_{m_0}) \leq \alpha V_0(0) + \beta$ for some constant $\alpha > 0$ and $\beta > 0$; see the analysis example after Theorem 2.1.1) and so are $x(t)$, $k_{1j}(t)$, $k_{2j}(t)$, and $k_{3j}(t)$, $j \neq j_1, \ldots, j_{\bar{m}_0}$. To show the boundedness of other signals, from (2.33) we have

$$k_{1i}(t) = k_{1i}(0) - \text{sign}[k_{s2i}^*]\Gamma_{1i} \int_0^t x(\tau)e^T(\tau)Pb_M d\tau, \ i \neq j_1, \ldots, j_{\bar{m}_0} \tag{2.37}$$

$$k_{1j}(t) = k_{1j}(0) - \text{sign}[k_{s2j}^*]\Gamma_{1j} \int_0^t x(\tau)e^T(\tau)Pb_M d\tau, \ j = j_1, \ldots, j_{\bar{m}_0}. \tag{2.38}$$

From (2.37) we obtain that for $i \neq j_1, \ldots, j_{\bar{m}_0}$,

$$\int_0^t x(\tau)e^T(\tau)Pb_M d\tau = \text{sign}[k_{s2i}^*]\Gamma_{1i}^{-1}(k_{1i}(0) - k_{1i}(t)) \tag{2.39}$$

which, together with (2.38), implies

$$k_{1j}(t) = k_{1j}(0) - \text{sign}[k_{s2j}^*]\Gamma_{1j}\text{sign}[k_{s2i}^*]\Gamma_{1i}^{-1}(k_{1i}(0) - k_{1i}(t))$$

$$j = j_1, \ldots, j_{\bar{m}_0} \tag{2.40}$$

for any $i \neq j_1, \ldots, j_{\bar{m}_0}$. Since $k_{1i}(t)$ is bounded, $i \neq j_1, \ldots, j_{\bar{m}_0}$, it follows that $k_{1j}(t)$ is bounded, $j = j_1, \ldots, j_{\bar{m}_0}$. Similarly, $k_{2j}(t)$, and $k_{3j}(t)$ are bounded, $j = j_1, \ldots, j_{\bar{m}_0}$. From (2.25) it follows that $v(t)$ is bounded. Therefore, all signals in the closed-loop system are bounded.

Since at least one actuator does not fail, that is, $m_0 < m$ and $T_{m_0+1} = \infty$, over the time interval (T_{m_0}, ∞), the function $V_{\bar{m}_0}(t)$ has a finite initial value and its derivative is $\dot{V}_{\bar{m}_0} = -e^T(t)Qe(t) \leq 0$, which means that $e(t) \in L^2$.

From (2.30) and closed-loop signal boundedness, we have that $\dot{e}(t) \in L^\infty$ so that $\lim_{t\to\infty} e(t) = 0$. Furthermore, since $e(t) \in L^2 \cap L^\infty$ and $x(t) \in L^\infty$, we have from (2.33) that $\dot{k}_{1j}(t) \in L^2 \cap L^\infty$, $\dot{k}_{2j}(t) \in L^2 \cap L^\infty$, and $\dot{k}_{3j}(t) \in L^2 \cap L^\infty$, $j = 1, \ldots, m$. It can also be verified that $\ddot{k}_{1j}(t) \in L^\infty$, $\ddot{k}_{2j}(t) \in L^\infty$, and $\ddot{k}_{3j}(t) \in L^\infty$, $j = 1, \ldots, m$, so that $\lim_{t\to\infty} \dot{k}_{1j}(t) = 0$, $\lim_{t\to\infty} \dot{k}_{2j}(t) = 0$, and $\lim_{t\to\infty} \dot{k}_{3j}(t) = 0$, $j = 1, \ldots, m$.

In summary, we have the following main result.

Theorem 2.1.1. *The adaptive controller (2.25), with the adaptive laws (2.33)–(2.35), applied to the system (2.1) with actuator failures (2.2) guarantees that all closed-loop signals are bounded and $\lim_{t\to\infty}(x(t) - x_m(t)) = 0$.*

It should be noted that for $t \in (T_i, T_{i+1})$, $i = 0, 1, \ldots, m_0 - 1$, the developed adaptive actuator failure compensation control scheme ensures that $\dot{V}_p = -e^T(t)Qe(t) \leq 0$, a normal performance measure that can be achieved in an adaptive control system without actuator failures.

Boundedness of $V_p(t)$. As an example to illustrate the analysis of the boundedness of $V_p(t)$, we consider the case when $m = 3$, in which the actuator $u_3(t)$ fails at $t = T_1$, the actuator $u_2(t)$ fails at $t = T_2 > T_1$, and the actuator $u_1(t)$ does not fail (that is, $u_1(t) = v_1(t)$, $t \geq 0$). In this case, we have

$$V_0(e, \tilde{k}_{1i}, \tilde{k}_{2i}, \tilde{k}_{3i}, i = 1, 2, 3)$$
$$= e^T Pe + \sum_{i=1}^{3} \frac{1}{|k_{s2i}^*|} \left(\tilde{k}_{1i}^T \Gamma_{1i}^{-1} \tilde{k}_{1i} + \tilde{k}_{2i}^2 \gamma_{2i}^{-1} + \tilde{k}_{3i}^2 \gamma_{3i}^{-1} \right) \qquad (2.41)$$

$$V_1(e, \tilde{k}_{1i}, \tilde{k}_{2i}, \tilde{k}_{3i}, i = 1, 2)$$
$$= e^T Pe + \sum_{i=1}^{2} \frac{1}{|k_{s2i}^*|} \left(\tilde{k}_{1i}^T \Gamma_{1i}^{-1} \tilde{k}_{1i} + \tilde{k}_{2i}^2 \gamma_{2i}^{-1} + \tilde{k}_{3i}^2 \gamma_{3i}^{-1} \right) \qquad (2.42)$$

$$V_2(e, \tilde{k}_{11}, \tilde{k}_{21}, \tilde{k}_{31})$$
$$= e^T Pe + \frac{1}{|k_{s21}^*|} \left(\tilde{k}_{11}^T \Gamma_{11}^{-1} \tilde{k}_{11} + \tilde{k}_{21}^2 \gamma_{21}^{-1} + \tilde{k}_{31}^2 \gamma_{31}^{-1} \right) \qquad (2.43)$$

that is, $m_0 = \bar{m}_0 = 2$. The time derivatives of these functions are

$$\dot{V}_0 = -e^T Qe, \ t \in [0, T_1)$$
$$\dot{V}_1 = -e^T Qe, \ t \in (T_1, T_2)$$
$$\dot{V}_2 = -e^T Qe, \ t \in (T_2, \infty) \qquad (2.44)$$

For V_0, V_1, and V_2 as functions of t, it follows from (2.41)–(2.44) that

$$V_0(0) \geq V_0(T_1), \ V_1(T_1) \geq V_1(T_2), \ V_2(T_2) \geq V_2(t), \ t \geq T_2. \qquad (2.45)$$

For simplicity of presentation, let $\Gamma_{1i} = I$, $\gamma_{2i} = \gamma_{3i} = 1$, and $k_{s2i}^* = 1$ in (2.41) and (2.42). Then, we have

$$V_0 = e^T Pe + \sum_{i=1}^{3}(k_{1i}(t) - k_{1i(0)}^*)^T(k_{1i}(t) - k_{1i(0)}^*) + \sum_{i=1}^{3}(k_{2i}(t) - k_{2i(0)}^*)^2$$

$$+ \sum_{i=1}^{3}(k_{3i}(t) - k_{3i(0)}^*) \tag{2.46}$$

$$V_1 = e^T Pe + \sum_{i=1}^{2}(k_{1i}(t) - k_{1i(1)}^*)^T(k_{1i}(t) - k_{1i(1)}^*) + \sum_{i=1}^{2}(k_{2i}(t) - k_{2i(1)}^*)^2$$

$$+ \sum_{i=1}^{2}(k_{3i}(t) - k_{3i(1)}^*) \tag{2.47}$$

$$V_2 = e^T Pe + (k_{11}(t) - k_{11(2)}^*)^T(k_{11}(t) - k_{11(2)}^*) + (k_{21}(t) - k_{21(2)}^*)^2$$

$$+ (k_{31}(t) - k_{31(2)}^*)^2 \tag{2.48}$$

where $k_{1i(j)}^*$, $k_{2i(j)}^*$, and $k_{3i(j)}^*$ are defined in (2.28). Using the inequality

$$(k_{11}(t) - k_{11(2)}^*)^T(k_{11}(t) - k_{11(2)}^*)$$

$$\leq 2\Big((k_{11}(t) - k_{11(1)}^*)^T(k_{11}(t) - k_{11(1)}^*)$$

$$+ (k_{11(2)}^* - k_{11(1)}^*)^T(k_{11(2)}^* - k_{11(1)}^*)\Big) \tag{2.49}$$

and the similar ones for $(k_{21}(t) - k_{21(2)}^*)^2$ and $(k_{31}(t) - k_{31(2)}^*)^2$, we obtain

$$V_2 \leq 2V_1 + 2\Big((k_{11(2)}^* - k_{11(1)}^*)^T(k_{11(2)}^* - k_{11(1)}^*)$$

$$+ (k_{21(2)}^* - k_{21(1)}^*)^2 + (k_{31(2)}^* - k_{31(1)}^*)^2\Big) \tag{2.50}$$

$$V_1 \leq 2V_0 + 2\sum_{i=1}^{2}\Big((k_{1i(2)}^* - k_{1i(1)}^*)^T(k_{1i(2)}^* - k_{1i(1)}^*)$$

$$+ (k_{2i(2)}^* - k_{2i(1)}^*)^2 + (k_{3i(2)}^* - k_{3i(1)}^*)^2\Big). \tag{2.51}$$

Let ΔV_1 and ΔV_2 be the finite parameter variations in (2.50) and (2.51): $V_1 \leq 2V_0 + \Delta V_1$, $V_2 \leq 2V_2 + \Delta V_2$. Using (2.45), (2.50), and (2.51), we have

$$V_1(T_1) \leq 2V_0(T_1) + \Delta V_1 \leq 2V_0(0) + \Delta V_1 \tag{2.52}$$

$$V_2(T_2) \leq 2V_1(T_2) + \Delta V_2 \leq 2V_1(T_1) + \Delta V_2 \leq 4V_0(0) + 2\Delta V_1 + \Delta V_2. \tag{2.53}$$

Therefore, it follows from (2.45) and (2.53) that

$$V_2(t) \leq V_2(T_2) \leq 4V_0(0) + 2\Delta V_1 + \Delta V_2 \tag{2.54}$$

is bounded. Then, from (2.37)–(2.40), $V_0(t)$ and $V_1(t)$ are also bounded. $\square$

Remark 2.1.4. Suppose eventually q actuators are working, that is, $u_i(t)$ $= v_i(t)$, $t > T_{m_0}$, $i = i_1, \ldots, i_q$, $q \geq 1$. Assume that the corresponding parameters $k_{1i}(t)$ converge: $\lim_{t \to \infty} k_{1i}(t) = \bar{k}_{1i}$ for some $\bar{k}_{1i} \in R^n$, $i = i_1, \ldots, i_q$. A question of interest would be, can the matching condition

$$A + \sum_{i=i_1,\ldots,i_q} b_i \bar{k}_{1i}^T = A_M \tag{2.55}$$

be satisfied with such $\bar{k}_{1i}$ from the adaptive law (2.33) with arbitrary $k_{1i}(0)$?

In view of (2.4), the condition (2.55) is equivalent to either of

$$\sum_{i=i_1,\ldots,i_q} b_i \bar{k}_{1i}^T = b_j k_{s1j}^{*T}, \; j = 1,\ldots,m \tag{2.56}$$

$$\sum_{i=i_1,\ldots,i_q} \frac{\bar{k}_{1i}^T}{k_{s2i}^*} = \bar{k}_{s1}^{*T} \triangleq \frac{k_{s1j}^{*T}}{k_{s2j}^*}, \; j = 1,\ldots,m. \tag{2.57}$$

From (2.33) we have

$$k_{1i}(t) = k_{1i}(0) - \text{sign}[k_{s2i}^*]\Gamma_{1i} \int_0^t x(\tau)e^T(\tau)Pb_M d\tau, \; i = i_1,\ldots,i_q \tag{2.58}$$

$$\bar{k}_{1i} = \lim_{t \to \infty} k_{1i}(t) = k_{1i}(0) - \text{sign}[k_{s2i}^*]\Gamma_{1i}\bar{k}_{s1}, \; i = i_1,\ldots,i_q \tag{2.59}$$

where $\bar{k}_{s1} = \lim_{t \to \infty} \int_0^t x(\tau)e^T(\tau)Pb_M d\tau$. Hence, (2.56) becomes

$$\sum_{i=i_1,\ldots,i_q} \frac{k_{1i}(0) - \text{sign}[k_{s2i}^*]\Gamma_{1i}\bar{k}_{s1}}{k_{s2i}^*} = \bar{k}_{s1}^* \tag{2.60}$$

which is satisfied with

$$\bar{k}_{s1} = \left(\sum_{i=i_1,\ldots,i_q} \frac{1}{|k_{s2i}^*|}\Gamma_{1i} \right)^{-1} \left(\sum_{i=i_1,\ldots,i_q} \frac{k_{1i}(0)}{k_{s2i}^*} - \bar{k}_{s1}^* \right). \tag{2.61}$$

This analysis indicates that it is possible for the developed adaptive controller to meet the matching condition (2.55), even if arbitrary initial conditions $k_{1i}(0)$, $i = 1,\ldots,m$, are used for the adaptive laws (2.33) (it would be an interesting topic to study when this matching condition is met). □

Remark 2.1.5. If one wants to implement an equal-actuation scheme (1.14): $v_1(t) = v_2(t) = \cdots = v_m(t)$, or the proportional-actuation scheme (1.15): $v_1(t) = \alpha_2 v_2(t) = \cdots = \alpha_m v_m(t)$, one can choose $\Gamma_{1j} = \alpha_j \Gamma_{11}$, $\gamma_{2j} = \alpha_j \gamma_{21}$, $\gamma_j = \alpha_j \gamma_1$, $k_{1j}(0) = k_{11}(0)$, $k_{2j}(0) = k_{21}(0)$, $k_{3j}(0) = k_{31}(0)$, $j = 2,3,\ldots,m$, for some $\alpha_j > 0$, in the adaptive law (2.33)–(2.35), to make $v_1(t) = \alpha_2 v_2(t) = \cdots = \alpha_m v_m(t)$. This choice is desirable when all m actuators have the same physical characteristics with values proportional to each other. □

2.2 Designs for Parametrized Varying Failures

In this section, we consider a more general type of time-varying actuator failure. Such failures can occur, for example, due to hydraulics failures that can produce unintended movements in the control surfaces of an aircraft. Other examples include orbiting satellites subjected to periodic torques of unknown magnitude due to gravity gradient, Earth's magnetic field, or onboard rotating machinery with variable loads.

We consider the actuator failure model described by

$$u_j(t) = \bar{u}_j + \bar{d}_j(t), \ t \geq t_j, \ j \in \{1, \ldots, m\} \tag{2.62}$$

where $\bar{u}_j$ is an unknown constant, and

$$\bar{d}_j(t) = \sum_{l=1}^{n_d} \bar{d}_{jl} f_{jl}(t) \tag{2.63}$$

for some unknown constants $\bar{d}_{jl} \in R$ and known signals $f_{jl}(t) \in R$, $l = 1, \ldots, n_d$, $n_d \geq 1$ (for simplicity, a single index n_d is used for m actuators). Parametrized by $\bar{u}_j$ and $\bar{d}_{jl}$, $\bar{u}_j + \bar{d}_j(t)$ can approximate a practical time-varying failure, by increasing n_d, the number of "basis" functions $f_{jl}(t)$.

With such actuator failures, the actuation vector Bu can be described as

$$Bu(t) = Bv(t) + B\sigma(\bar{u} + \bar{d}(t) - v(t)) \tag{2.64}$$

where $v(t)$ is a designed control input, and

$$\bar{d}(t) = [\bar{d}_1(t), \ldots, \bar{d}_m(t)]^T. \tag{2.65}$$

2.2.1 Plant-Model Matching Control Design

We first develop a plant-model matching controller of the form

$$v(t) = v^*(t) = K_1^{*T} x(t) + k_2^* r(t) + k_3^* + g^*(t) \tag{2.66}$$

where $K_1^* = [k_{11}^*, \ldots, k_{1m}^*] \in R^{n \times m}$, $k_2^* = [k_{21}^*, \ldots, k_{2m}^*]^T \in R^m$, and $k_3^* = [k_{31}^*, \ldots, k_{3m}^*]^T \in R^m$ are parameters, and $g^*(t) = [g_1^*(t), \ldots, g_m^*(t)]^T \in R^m$ is a vector signal to be specified.

Suppose at time t there are p failed actuators, that is, $u_j(t) = \bar{u}_j + \bar{d}_j(t)$, $j = j_1, \ldots, j_p$, $1 \leq p \leq m - 1$. Then, the parameters K_1^*, k_2^*, and k_3^* are chosen to satisfy (2.17) and (2.19), plus (2.20). With the controller (2.66), using (2.63), and (2.66), the closed-loop system becomes

$$
\begin{aligned}
\dot{x}(t) &= Ax(t) + Bv^*(t) + B\sigma(\bar{u} - v^*(t)) \\
&= A_M x(t) + b_M r(t) + B\sigma(K_1^{*T} x(t) + k_2^* r(t)) + Bk_3^* \\
&\quad + Bg^*(t) + B\sigma(\bar{u} - v^*(t)) \\
&= A_M x(t) + b_M r(t) + B(I - \sigma)k_3^* + B\sigma\bar{u} + B(I - \sigma)g^*(t) + B\sigma\bar{d}(t) \\
&= A_M x(t) + b_M r(t) + B(I - \sigma)g^*(t) + B\sigma\bar{d}(t). \tag{2.67}
\end{aligned}
$$

To match this system to (2.3), the design task now is to choose $g^*(t)$ to make

$$
B(I - \sigma)g^*(t) + B\sigma\bar{d}(t) = \sum_{i\neq j_1,\ldots,j_p} b_i g_i^*(t) + \sum_{j=j_1,\ldots,j_p} b_j \bar{d}_j^*(t) = 0. \tag{2.68}
$$

Using (2.4), that is, $b_i = b_M/k_{s2i}^*$, from (2.68) we have the plant-model matching condition for $g_i^*(t)$, $i \neq j_1,\ldots,j_p$:

$$
\sum_{i\neq j_1,\ldots,j_p} \frac{g_i^*(t)}{k_{s2i}^*} + \sum_{j=j_1,\ldots,j_p} \frac{\bar{d}_j(t)}{k_{s2j}^*} = 0. \tag{2.69}
$$

Similar to (2.20), the terms $g_j^*(t)$, $j = j_1,\ldots,j_p$, are irrelevant for the closed-loop system, and they could be chosen as

$$
g_j^*(t) = 0, \ j = j_1,\ldots,j_p. \tag{2.70}
$$

However, the condition (2.69) does not give an explicit parametrized form of $g_i^*(t)$. To derive a desirable parametrized form of $g_i^*(t)$, we first consider the case when all $u_j(t)$ but $u_i(t)$ fail, that is, when $u_j(t) = \bar{u}_j + \bar{d}_j(t)$, $j = 1, 2, \ldots, m$, $j \neq i$, and $u_i(t) = v_i(t) = v_i^*(t)$. In this case, (2.69) becomes

$$
g_i^*(t) = -k_{s2i}^* \sum_{j\neq i} \frac{\bar{d}_j(t)}{k_{s2j}^*}. \tag{2.71}
$$

In view of (2.62), we rewrite $g_i^*(t)$ as

$$
g_i^*(t) = \sum_{j\neq i}^{m} \sum_{l=1}^{n_d} g_{ijl}^* f_{jl}(t) \tag{2.72}
$$

where the parameters g_{ijl}^* are

$$
g_{ijl}^* = -\frac{k_{s2i}^*}{k_{s2j}^*} d_{jl}^*, \ j = 1,\ldots,m, \ j \neq i, \ l = 1,\ldots,n_d. \tag{2.73}
$$

Although the parametrized form of $g_i^*(t)$ in (2.72) is derived from the special case when all $u_j(t)$ but $u_i(t)$ fail, it is the right form for the general case when there are p failed actuators, that is, $u_j(t) = \bar{u}_j + \bar{d}_j(t)$, $j = j_1,\ldots,j_p$.

Proposition 2.2.1. *The parametrized form (2.72) of $g_i^*(t)$, $i = 1, \ldots, m$, is necessary and sufficient for the matching condition (2.69).*

Proof: The necessity is clear from (2.70), when all $u_j(t)$ but $u_i(t)$ fail. To prove the sufficiency, we see that the term in (2.69), $\sum_{j=j_1,\ldots,j_p} \bar{d}_j(t)/k_{s2j}^*$, contains $f_{jl}(t)$, $j = j_1, \ldots, j_p$, $l = 1, \ldots, n_d$. Since $p < m$, in the term $\sum_{i \neq j_1,\ldots,j_p} g_i^*(t)/k_{s2i}^*$, there is at least a $g_{j_{p+1}}^*$ that from (2.72), has the form

$$g_{j_{p+1}}^*(t) = \sum_{j \neq j_{p+1}}^{m} \sum_{l=1}^{n_d} g_{j_{p+1}jl}^* f_{jl}(t). \tag{2.74}$$

This $g_{j_{p+1}}^*(t)$ does not contain $f_{j_{p+1}l}(t)$, $l = 1, \ldots, n_d$, but all other $f_{jl}(t)$, $j = 1, \ldots, m$, $j \neq j_{p+1}$, $l = 1, \ldots, n_d$, that is, $g_{j_{p+1}}^*(t)$ also contains $f_{jl}(t)$, $j = j_1, \ldots, j_p$, $l = 1, \ldots, n_d$. Therefore, the parametrized form (2.72) of the individual $g_i^*(t)$, $i = 1, \ldots, m$, is sufficient for parametrizing the set of $g_i^*(t)$, $i \neq j_1, \ldots, j_p$, to meet the matching condition (2.69) in which $\sum_{j=j_1,\ldots,j_p} \bar{d}_j(t)/k_{s2j}^*$, with $\bar{d}_j(t)$ in (2.62), is parametrizable by $f_{jl}(t)$, $j = j_1, \ldots, j_p$, $l = 1, \ldots, n_d$. It should be noted that when there are p failed actuators, that is, $u_j(t) = \bar{u}_j + \bar{d}_j(t)$, $j = j_1, \ldots, j_p$, $g_j^*(t) = 0$ (see (2.70)), is a special case of (2.72) with parameters $g_{jkl}^* = 0$. ∇

For the case when no actuator fails, the parameters K_1^* and k_2^* are defined in (2.7)–(2.11), while the parameters in k_3^* are defined in (2.24). The signals $g_i^*(t)$, $i = 1, \ldots, m$, are defined by

$$\sum_{i=1}^{m} \frac{g_i^*(t)}{k_{s2i}^*} = 0. \tag{2.75}$$

One choice is that $g_i^*(t) = 0$, $i = 1, \ldots, m$.

For the case when only one $u_j(t)$ fails, the matching condition (2.69) nonuniquely specifies the parameters of $g_i^*(t)$, $i \neq j$:

$$\sum_{i \neq j} \frac{g_i^*(t)}{k_{s2i}^*} + \frac{\bar{d}_j(t)}{k_{s2j}^*} = 0. \tag{2.76}$$

2.2.2 Adaptive Control Design

We now develop an adaptive version of the controller (2.66):

$$v(t) = K_1^T(t)x(t) + k_2(t)r(t) + k_3(t) + g(t) \tag{2.77}$$

where $K_1(t)$, $k_2(t)$, and $k_3(t)$ are the estimates of K_1^*, k_2^*, and k_3^*, and

$$g(t) = [g_1(t), \ldots, g_m(t)]^T \tag{2.78}$$

$$g_i(t) = \sum_{j \neq i}^{m} \sum_{l=1}^{n_d} g_{ijl}(t) f_{jl}(t) \tag{2.79}$$

with $g_{ijl}(t)$ being the estimate of g_{ijl}^* defined in (2.73).

Suppose at time t there are p failed actuators, that is, $u_j(t) = \bar{u}_j + \bar{d}_j(t)$, $j = j_1, \ldots, j_p$, $1 \leq p \leq m - 1$. Our task now is to develop adaptive laws to update the parameter estimates $K_1(t)$, $k_2(t)$, $k_3(t)$, and $g_{ijl}(t)$. We start with the closed-loop control system with (2.1), (2.13), (2.17), (2.19), (2.26), (2.68), and (2.77), and $\tilde{g}(t) = g(t) - g^*(t)$, that is,

$$
\begin{aligned}
\dot{x}(t) &= Ax(t) + Bv(t) + B\sigma(\bar{u} + \bar{d}(t) - v(t)) \\
&= Ax(t) + B(I - \sigma)(K_1^{*T}x(t) + k_2^*r(t)) + B(I - \sigma)k_3^* + B\sigma\bar{u} \\
&\quad + B(I - \sigma)\left(\tilde{K}_1^T(t)x(t) + \tilde{k}_2(t)r(t) + \tilde{k}_3(t) + \tilde{g}(t)\right) \\
&\quad + B(I - \sigma)g^*(t) + B\sigma\bar{d}(t) - B\sigma(v(t) - v^*(t)) \\
&= A_M x(t) + b_M r(t) + B(I - \sigma)\left(\tilde{K}_1^T(t)x(t) + \tilde{k}_2(t)r(t) + \tilde{k}_3(t) + \tilde{g}(t)\right) \\
&= A_M x(t) + b_M r(t) + \sum_{i \neq j_1, \ldots, j_p} b_i \tilde{k}_{1i}^T(t)x(t) + \sum_{i \neq j_1, \ldots, j_p} b_i \tilde{k}_{2i}(t)r(t) \\
&\quad + \sum_{i \neq j_1, \ldots, j_p} b_i \tilde{k}_{3i}(t) + \sum_{i \neq j_1, \ldots, j_p} b_i \sum_{j \neq i}^{m} \sum_{l=1}^{n_d} \tilde{g}_{ijl}(t) f_{jl}(t) \\
&= A_M x(t) + b_M r(t) + b_M \left(\sum_{i \neq j_1, \ldots, j_p} \frac{1}{k_{s2i}^*} \tilde{k}_{1i}^T(t)x(t) \right. \\
&\quad + \sum_{i \neq j_1, \ldots, j_p} \frac{1}{k_{s2i}^*} \tilde{k}_{2i}(t)r(t) + \sum_{i \neq j_1, \ldots, j_p} \frac{1}{k_{s2i}^*} \tilde{k}_{3i}(t) \\
&\quad \left. + \sum_{i \neq j_1, \ldots, j_p} \frac{1}{k_{s2i}^*} \sum_{j \neq i}^{m} \sum_{l=1}^{n_d} \tilde{g}_{ijl}(t) f_{jl}(t) \right). \tag{2.80}
\end{aligned}
$$

Then, the error equation for $e(t) = x(t) - x_m(t)$ is

$$
\begin{aligned}
\dot{e}(t) &= A_M e(t) + b_M \left(\sum_{i \neq j_1, \ldots, j_p} \frac{1}{k_{s2i}^*} \tilde{k}_{1i}^T(t)x(t) + \sum_{i \neq j_1, \ldots, j_p} \frac{1}{k_{s2i}^*} \tilde{k}_{2i}(t)r(t) \right. \\
&\quad \left. + \sum_{i \neq j_1, \ldots, j_p} \frac{1}{k_{s2i}^*} \tilde{k}_{3i}(t) + \sum_{i \neq j_1, \ldots, j_p} \frac{1}{k_{s2i}^*} \sum_{j \neq i}^{m} \sum_{l=1}^{n_d} \tilde{g}_{ijl}(t) f_{jl}(t) \right) \tag{2.81}
\end{aligned}
$$

as similar to (2.31) for $d(t) = g(t) = 0$, where $g_{ijl}(t) = g_{ijl}(t) - g_{ijl}^*$, and $x_m(t)$ is the reference system state vector as from in (2.3).

Consider the positive definite function

$$V_p(e, \tilde{k}_{1i}, \tilde{k}_{2i}, \tilde{k}_{3i}, \tilde{g}_{ijl}, i \neq j_1, \ldots, j_p)$$

$$= e^T P e + \sum_{i \neq j_1, \ldots, j_p} \frac{1}{|k^*_{s2i}|} \tilde{k}^T_{1i} \Gamma^{-1}_{1i} \tilde{k}_{1i} + \sum_{i \neq j_1, \ldots, j_p} \frac{1}{|k^*_{s2i}|} \tilde{k}^2_{2i} \gamma^{-1}_{2i}$$

$$+ \sum_{i \neq j_1, \ldots, j_p} \frac{1}{|k^*_{s2i}|} \tilde{k}^2_{3i} \gamma^{-1}_{3i} + \sum_{i \neq j_1, \ldots, j_p} \frac{1}{|k^*_{s2i}|} \sum_{j \neq i}^{m} \sum_{l=1}^{n_d} \tilde{g}^2_{ijl} \gamma^{-1}_{ijl} \quad (2.82)$$

where P is defined in (2.32), $\Gamma_{1i} \in R^{n \times n}$ such that $\Gamma_{1i} = \Gamma^T_{1i} > 0$, and $\gamma_{2i} > 0$, $\gamma_{3i} > 0$, and $\gamma_{ijl} > 0$, $i = 1, \ldots, m$, $j \neq i$, $l = 1, \ldots, n_d$.
Choose the adaptive laws for $g_{ijl}(t)$ as

$$\dot{g}_{ijl}(t) = -\text{sign}[k^*_{s2i}] \Gamma_{ijl} f_{jl}(t) e^T(t) P b_M \quad (2.83)$$

for $i = 1, \ldots, m$, $j \neq i$, $l = 1, \ldots, n_d$, and use the adaptive laws for $k_{1i}(t)$, $k_{2i}(t)$, and $k_{3i}(t)$ as those given in (2.33)–(2.35).

Then, similar to (2.36), we can obtain

$$\dot{V}_p = -e^T(t) Q e(t) \leq 0, \ t \in (T_{i(p)}, T_{i(p)+1}), \ i(p) \leq p \quad (2.84)$$

where, as in Section 2.1, $i = i(p) \leq p$ is such that there are p failed actuators for $t \in (T_i, T_{i+1})$, that is, actuators only fail at time T_i, and (T_i, T_{i+1}), $i = 0, 1, \ldots, m_0 < m$, are the time intervals on which the actuator failure pattern is fixed. Since eventually there are only $\bar{m}_0$ failed actuators, $m_0 \leq \bar{m}_0 < m$, it follows that $\dot{V}_{\bar{m}_0} = -e^T(t) Q e(t) \leq 0$, $t \in (T_{m_0}, \infty)$. Based on a similar analysis to that for Theorem 2.1.1, we have the following results.

Theorem 2.2.1. *The adaptive controller (2.77) (via (2.13)), with the adaptive laws (2.33)–(2.35) and (2.83), applied to the system (2.1), guarantees that all closed-loop signals are bounded and $\lim_{t \to \infty} (x(t) - x_m(t)) = 0$.*

2.3 Designs for Unparametrizable Failures

In this section, we consider the even more general actuator failure model:

$$u_j(t) = \bar{u}_j + \bar{d}_j(t) + \bar{\delta}_j(t), \ t \geq t_j, \ j \in \{1, 2, \ldots, m\} \quad (2.85)$$

where $\bar{u}_j$ and $\bar{d}_j(t)$ are defined in (2.2) and (2.62), respectively, and $\bar{\delta}_j(t)$ is an unknown and unparametrizable but bounded term that may represent some practical failures. In this case, the actuation vector Bu is

$$Bu(t) = Bv(t) + B\sigma(\bar{u} + \bar{d}(t) + \bar{\delta}(t) - v(t)) \quad (2.86)$$

where $v(t)$ is a designed control input, and

$$\bar{\delta}(t) = [\bar{\delta}_1(t), \ldots, \bar{\delta}_m(t)]^T. \tag{2.87}$$

The control objective is to design a feedback control $v(t)$ such that all closed-loop signals are bounded and the state vector $x(t)$ of the system (2.1) tracks the state vector $x_m(t)$ of the reference model (2.1) as closely as possible (due to the unparametrizable $\bar{\delta}_j(t)$, asymptotic tracking may not be possible).

2.3.1 Stabilizing Control

We first develop a stabilizing controller of the form

$$v(t) = v^*(t) = K_1^{*T}x(t) + k_2^*r(t) + k_3^* + g^*(t) + v_s(t) \tag{2.88}$$

where $K_1^* = [k_{11}^*, \ldots, k_{1m}^*] \in R^{n \times m}$, $k_2^* = [k_{21}^*, \ldots, k_{2m}^*]^T \in R^m$, and $k_3^* = [k_{31}^*, \ldots, k_{3m}^*]^T \in R^m$ are parameters (see Section 2.3), $g^*(t) = [g_1^*(t), \ldots, g_m^*(t)]^T \in R^m$ is a design signal (see Section 2.2.1), and

$$v_s(t) = [v_{s1}(t), \ldots, v_{sm}(t)]^T \tag{2.89}$$

is to be designed to compensate the unknown $\bar{\delta}(t)$ for stability and tracking.

Suppose at time t there are p failed actuators, that is, $u_j(t) = \bar{u}_j + \bar{d}_j(t) + \bar{\delta}_j(t)$, $j = j_1, \ldots, j_p$, $1 \le p \le m - 1$. Similar to (2.19), (2.68) for k_3^*, $g^*(t)$, an ideal choice of $v_s(t)$ would be such that

$$B(I - \sigma)v_s(t) + B\sigma\bar{\delta}(t) = \sum_{i \neq j_1, \ldots, j_p} b_i v_{si}(t) + \sum_{j=j_1, \ldots, j_p} b_j \bar{\delta}_j(t) = 0 \tag{2.90}$$

which, in view of (2.4), would imply

$$\sum_{i \neq j_1, \ldots, j_p} \frac{v_{si}(t)}{k_{s2i}^*} + \sum_{j=j_1, \ldots, j_p} \frac{\bar{\delta}_j(t)}{k_{s2j}^*} = 0 \tag{2.91}$$

which, however, cannot be parametrized because $\bar{\delta}_j(t)$ is not parametrizable and thus is not useful for deriving a desirable adaptive control design for $v_{si}(t)$ in the case when $\bar{\delta}_j(t)$ is unknown.

With the choice of K_1^*, k_2^*, k_3^*, and $g^*(t)$ in (2.17), (2.19), (2.20), (2.69), and (2.70), we consider the resulting closed-loop system

$$\dot{x}(t) = A_M x(t) + b_M r(t) + B(I - \sigma)v_s(t) + B\sigma\bar{\delta}(t) \tag{2.92}$$

which leads to the error system for $e(t) = x(t) - x_m(t)$ as

$$
\begin{aligned}
\dot{e}(t) &= A_M e(t) + B(I - \sigma)v_s(t) + B\sigma\bar{\delta}(t) \\
&= A_M e(t) + \sum_{i \neq j_1,\dots,j_p} b_i v_{si}(t) + \sum_{j=j_1,\dots,j_p} b_j \bar{\delta}_j(t) \\
&= A_M e(t) + b_M \left(\sum_{i \neq j_1,\dots,j_p} \frac{v_{si}(t)}{k^*_{s2i}} + \sum_{j=j_1,\dots,j_p} \frac{\bar{\delta}_j(t)}{k^*_{s2j}} \right).
\end{aligned} \tag{2.93}
$$

We choose the design signal $v_{si}(t)$ as

$$
v_{si}(t) = -k^0_{2i}\delta^0 \mathrm{sgn}[e^T(t)Pb_M]\mathrm{sign}[k^*_{s2i}], \; i \neq j_1, \dots, j_p \tag{2.94}
$$

where k^0_{2i} is a known upper bound on $|k^*_{s2i}|$: $k^0_{2i} > |k^*_{s2i}|$,

$$
\delta^0 \geq \max\{\delta_1, \dots, \delta_{m-1}\} \tag{2.95}
$$

$$
\delta_p = \max_{j_1,\dots,j_p \in \{1,\dots,m\}} \Big| \sum_{j=j_1,\dots,j_p} \frac{\bar{\delta}_j(t)}{k^*_{s2j}} \Big|, \; 1 \leq p \leq m-1 \tag{2.96}
$$

$P = P^T > 0$ satisfying (2.32), and $\mathrm{sgn}[\chi] = \begin{cases} 1 & \text{if } \chi > 0 \\ 0 & \text{if } \chi = 0 \\ -1 & \text{if } \chi < 0. \end{cases}$

Consider the positive definite function

$$
W(e) = e^T P e \tag{2.97}
$$

and its time derivative

$$
\begin{aligned}
\dot{W} &= -e^T Q e - 2|e^T P b_M|\delta^0 \sum_{i \neq j_1,\dots,j_p} \frac{k^0_{2i}}{|k^*_{s2i}|} + 2e^T P b_M \sum_{j=j_1,\dots,j_p} \frac{\bar{\delta}_j(t)}{k^*_{s2j}} \\
&\leq -e^T Q e - 2|e^T P b_M|\delta^0 \sum_{i \neq j_1,\dots,j_p} \frac{k^0_{2i}}{|k^*_{s2i}|} + 2|e^T P b_M|\delta_p \\
&\leq -e^T Q e.
\end{aligned} \tag{2.98}
$$

Since the control signal $v_{si}(t)$ in (2.94) is not continuous, although $\dot{W} \leq -e^T(t)Qe(t)$, the closed-loop system is a variable structure system and its solution should be understood in the Filippov sense [33], [134].

To avoid system chattering caused by such discontinuous control laws, one can use the following common approximations for the sgn[·] function:

$$
\mathrm{sgn}[\chi] \approx \sigma_{s1}[\chi] = \frac{\chi}{|\chi| + \epsilon}, \; \epsilon > 0 \tag{2.99}
$$

$$
\mathrm{sgn}[\chi] \approx \sigma_{s2}[\chi] = \begin{cases} \chi/|\chi| & \text{if } |\chi| \geq \epsilon \\ \chi/\epsilon & \text{if } |\chi| < \epsilon \end{cases} \; \epsilon > 0. \tag{2.100}
$$

With sgn[·] in (2.94) replaced by $\sigma_{s1}[\cdot]$ in (2.99), we obtain

$$\dot{W} = -e^T Q e - 2\frac{|e^T P b_M|^2}{|e^T P b_M| + \epsilon} \delta^0 \sum_{i \neq j_1, \dots, j_p} \frac{k_{2i}^0}{|k_{s2i}^*|} + 2e^T P b_M \sum_{j=j_1, \dots, j_p} \frac{\bar{\delta}_j(t)}{k_{s2j}^*}$$

$$\leq -e^T Q e - 2|e^T P b_M| \left(\frac{|e^T P b_M|}{|e^T P b_M| + \epsilon} \delta^0 \sum_{i \neq j_1, \dots, j_p} \frac{k_{2i}^0}{|k_{s2i}^*|} - \delta_p \right)$$

$$\leq -e^T Q e + 2\delta_0 \frac{|e^T P b_M| \epsilon}{|e^T P b_M| + \epsilon} \sum_{i \neq j_1, \dots, j_p} \frac{k_{2i}^0}{|k_{s2i}^*|}$$

$$\leq -e^T Q e + 2k_0 \epsilon \qquad\qquad (2.101)$$

where $k_0 = \delta_0 \sum_{i \neq j_1, \dots, j_p} k_{2i}^0 / |k_{s2i}^*|$ is a constant. From (2.101) we have

$$\dot{W} \leq -\|e\|^2 q_1 + 2k_0 \epsilon \qquad\qquad (2.102)$$

where q_1 is the minimum eigenvalue of Q. Hence, $\dot{W} < 0$ whenever $\|e\|^2 > 2k_0\epsilon/q_1$, i.e., $\|e\| > \sqrt{2k_0\epsilon/q_1}$, which means that $\|e\|$ decreases to a lower bound proportional to $\sqrt{\epsilon}$. If $\epsilon > 0$ is chosen to be small, the tracking error $e(t)$ will converge to a set of a small size.

With $\mathrm{sgn}[\cdot]$ in (2.94) replaced by $\sigma_{s2}[\cdot]$ in (2.100), we obtain

$$\dot{W} \leq \begin{cases} -e^T Q e & \text{if } |e^T P b_M| \geq \epsilon \\ -e^T Q e - 2|e^T P b_M| & \\ \cdot \left(\frac{|e^T P b_M|}{\epsilon} \delta^0 \sum_{i \neq -j_1, \dots, j_p} \frac{k_{2i}^0}{|k_{s2i}^*|} - \delta_p \right) & \text{if } |e^T P b_M| < \epsilon. \end{cases} \qquad (2.103)$$

When $|e^T P b_M| < \epsilon$, we have

$$\dot{W} \leq -e^T Q e - 2|e^T P b_M| \left(\left(\frac{|e^T P b_M|}{\epsilon} - 1 \right) \delta^0 \sum_{i \neq j_1, \dots, j_p} \frac{k_{2i}^0}{|k_{s2i}^*|} \right)$$

$$\leq -e^T Q e + 2k_0 \epsilon \qquad\qquad (2.104)$$

for the same k_0 in (2.101), which means that $\|e\|$ decreases to a lower bound proportional to $\sqrt{\epsilon}$ that may be chosen to be small.

2.3.2 Adaptive Control Design

For the case of unknown parameters, the adaptive controller structure is

$$v(t) = K_1^T(t)x(t) + k_2(t)r(t) + k_3(t) + g(t) + v_s(t) \qquad (2.105)$$

where $K_1(t)$, $k_2(t)$, $k_3(t)$, and $g(t)$ are adaptive estimates of K_1^*, k_2^*, k_3^*, and $g^*(t)$, respectively, and $v_s(t)$ is given in (2.94).

Suppose there are p failed actuators, that is, $u_j(t) = \bar{u}_j + \bar{d}_j(t) + \bar{\delta}_j(t)$, $j = j_1, \dots, j_p$, $1 \leq p \leq m - 1$, at time $t \in (T_i, T_{i+1})$, $i = i(p) < p$. To develop adaptive update laws for $K_1(t)$, $k_2(t)$, $k_3(t)$, and $g(t)$, in view of (2.83) and (2.93), and similar to (2.31) and (2.83), with the controller (2.105), we have the tracking error equation as

$$
\begin{aligned}
\dot{e}(t) = A_M e(t) + b_M \Bigg(&\sum_{i \neq j_1,\ldots,j_p} \frac{1}{k^*_{s2i}} \tilde{k}^T_{1i}(t) x(t) + \sum_{i \neq j_1,\ldots,j_p} \frac{1}{k^*_{s2i}} \tilde{k}_{2i}(t) r(t) \\
&+ \sum_{i \neq j_1,\ldots,j_p} \frac{1}{k^*_{s2i}} \tilde{k}_{3i}(t) + \sum_{i \neq j_1,\ldots,j_p} \frac{1}{k^*_{s2i}} \sum_{j \neq i}^{m} \sum_{l=1}^{n_d} \tilde{g}_{ijl}(t) f_{jl}(t) \\
&+ \sum_{i \neq j_1,\ldots,j_p} \frac{v_{si}(t)}{k^*_{s2i}} + \sum_{j=j_1,\ldots,j_p} \frac{\bar{\delta}_j(t)}{k^*_{s2j}} \Bigg).
\end{aligned}
\tag{2.106}
$$

With the adaptive laws (2.33)–(2.35), (2.83) for $K_1(t)$, $k_2(t)$, $k_3(t)$, and $g(t)$, respectively, and the design signal $v_s(t)$ in (2.94), we also have the time derivative of $V_p(e, \tilde{k}_{1i}, \tilde{k}_{2i}, \tilde{k}_{3i}, \tilde{g}_{ijl}, i \neq j_1, \ldots, j_p)$ in (2.82) as that given in (2.84), while, due to discontinuities in $v_s(t)$, the system solution should be understood in the Filippov sense [33], [134].

When the approximations in (2.99) and (2.100) are used for the sgn[·] function in (2.94), the adaptive laws (2.33)–(2.35), (2.83) ensure

$$
\dot{V}_p \leq -e^T Q e + 2k_0 \epsilon, \ t \in (T_{i(p)}, T_{i(p)+1}), \ i(p) \leq p
\tag{2.107}
$$

for a design parameter $\epsilon > 0$ (which can be made small) and some constant $k_0 > 0$ independent of ϵ. The inequality (2.107) implies that, with V_p in (2.82), the tracking error $e(t)$ is bounded. However, it does not mean that the parameters $K_1(t)$, $k_2(t)$, $k_3(t)$, and $g(t)$ are bounded (that is, parameter drifts [55] can occur). To enhance the robustness of the adaptive control system, we need to modify the individual adaptive law (2.33), (2.34), (2.35), or (2.83) with an additional design signal [55], [95], [125]. A standard parameter projection [125] or a soft parameter projection (the switching-σ modification [55]) for such a design signal (see Section 2.3.3) ensures the boundedness of the parameter estimates $K_1(t)$, $k_2(t)$, $k_3(t)$, and $g(t)$ in addition to the desired inequality (2.107), which also implies that $e(t)$ is bounded. Therefore, all closed-loop signals are bounded. Then, using (2.107), we obtain

$$
\int_{t_1}^{t_2} \|e(t)\|^2 dt \leq \frac{V_p(t_1) - V_p(t_2)}{q_1} + \frac{2k_0 \epsilon}{q_1}(t_2 - t_1)
$$
$$
\forall t_2 > t_1, \ t_2, t_1 \in (T_{i(p)}, T_{i(p)+1})
\tag{2.108}
$$

where q_1 is the minimum eigenvalue of Q. Since $V_p(t)$ is bounded, we have

$$\frac{1}{t_2 - t_1} \int_{t_1}^{t_2} \|e(t)\|^2 dt \le \frac{c_0}{t_2 - t_1} + \frac{2k_0\epsilon}{q_1}$$

$$\forall t_2 > t_1,\ t_2, t_1 \in (T_{i(p)}, T_{i(p)+1}) \qquad (2.109)$$

for some constant $c_0 > 0$. The last inequality indicates that the mean value of the error $\|e(t)\|^2$ can be made small by a choice of small ϵ.

When $t_2 > t_1$ but $t_1 \in [T_i, T_{i+1}]$ and $t_2 \in [T_j, T_{j+1}]$, $j > i$, the error $e(t)$ is bounded for $t \in [t_1, t_2]$ so that the mean value of $\|e(t)\|^2$ is also bounded. After a finite number of actuator failures, $T_{m_0+1} = \infty$ for some m_0 but T_{m_0} is finite. Then, for any $t_2 > t_1$, $t_2, t_1 \in (T_{m_0}, \infty)$, we have (2.108).

In summary we have the following result.

Theorem 2.3.1. *The adaptive controller (2.105) with the design signal $v_s(t)$ in (2.94) approximated with (2.99) or (2.100), updated by the adaptive laws (2.33)–(2.35) and (2.83) modified by parameter projection, applied to the system (2.1), guarantees that all closed-loop signals are bounded and the tracking error $e(t)$ satisfies (2.108).*

2.3.3 Robust Adaptation

Robust parameter adaptation is important for adaptive control in the presence of modeling errors that cannot be parametrized such as system structural changes, parameter variations, external disturbance, and unmodeled dynamics. A standard adaptive law $\dot{\theta}(t) = -\Gamma\phi(t)$ needs to be modified to ensure robust stability (parameter boundedness and mean error smallness) (see Theorem 2.3.1). Common robustness modifications are the dead-zone modification [95], σ-modification [53], switching-σ modification [55], [56], ϵ-modification [95], and parameter projection [44], [125]. Parametrizable actuator failures cause system structural change and parameter variation, which, under our compensation control designs, only lead to some transition error responses (which are asymptotically decaying to zero) of the closed-loop system (see Sections 2.1 and 2.2). Unparametrizable actuator failures cause transition errors that may not decay to zero, for which robust adaptation is needed. In this subsection, we present two of them: a parameter projection modification and a switching-σ modification. For the case of external disturbances and unmodeled dynamics, robust adaptive control theory [55], [95] can be applied to design robust adaptive laws to ensure robust stability of the closed-loop system. In Chapter 8, we develop such a robust adaptive actuator failure compensation control scheme for discrete-time systems with unknown actuator failures, to show the robust stability of such an adaptive approach. In Chapter 12, we have further discussion on the robustness issue for adaptive actuator failure compensation.

Parameter Projection Modification. To design a projection signal $f(t) = [f_1, f_2, \ldots, f_{n_\theta}]^T$ for an adaptive law

$$\dot{\theta}(t) = -\Gamma\phi(t) + f(t) \tag{2.110}$$

where $\theta(t) \in R^{n_\theta}$ is the estimate of θ^*, we choose Γ as

$$\Gamma = \text{diag}\{\gamma_1, \gamma_2, \ldots, \gamma_{n_\theta}\}, \ \gamma_i > 0, \ i = 1, 2, \ldots, n_\theta. \tag{2.111}$$

Denote $\theta_j(t)$, $f_j(t)$, and $p_j(t)$ as the jth components of $\theta(t)$, $f(t)$, and

$$p(t) = -\Gamma\phi(t) \tag{2.112}$$

respectively, for $j = 1, 2, \ldots, n_\theta$, and choose the initial estimates as $\theta_j(0) \in [\theta_j^a, \theta_j^b]$ such that $\theta_j^* \in [\theta_j^a, \theta_j^b]$, $j = 1, 2, \ldots, n_\theta$, for $\theta^* = [\theta_1^*, \theta_2^*, \ldots, \theta_{n_\theta}^*]^T$, and set the projection function components as

$$f_j(t) = \begin{cases} 0 & \text{if } \theta_j(t) \in (\theta_j^a, \theta_j^b), \text{ or} \\ & \text{if } \theta_j(t) = \theta_j^a, \ p_j(t) \geq 0, \text{ or} \\ & \text{if } \theta_j(t) = \theta_j^b, \ p_j(t) \leq 0 \\ -p_j(t) & \text{otherwise.} \end{cases} \tag{2.113}$$

This $f(t)$ ensures that $\theta_j(t) \in [\theta_j^a, \theta_j^b]$, $i = 1, 2, \ldots, n_\theta$, that is, $\theta(t)$ is bounded, and that $(\theta_i(t) - \theta_i^*)f_i(t) \leq 0$, $i = 1, 2, \ldots, n_\theta$ so that (2.109) is ensured.

Switching σ-Modification. A switching σ-modification $f(t)$ is given as

$$f(t) = -\sigma(t)\Gamma\theta(t)$$

$$\sigma(t) = \begin{cases} 0 & \text{if } \|\theta(t)\|_2 < M_\theta \\ \sigma_0(\frac{\|\theta(t)\|_2}{M_\theta} - 1) & \text{if } M_\theta \leq \|\theta(t)\|_2 < 2M_\theta \\ \sigma_0 & \text{if } \|\theta(t)\|_2 \geq 2M_\theta \end{cases} \tag{2.114}$$

where $M_\theta \geq \|\theta^*\|_2$, the Euclidean norm of θ^*, and $\sigma_0 > 0$ is a design parameter. With this choice of $f(t)$, it follows that

$$\tilde{\theta}^T(t)\Gamma^{-1}f(t) = -\sigma(t)(\theta^T(t)\theta(t) - \theta^{*T}\theta(t))$$

$$\leq -\sigma(t)\|\theta(t)\|_2(\|\theta(t)\|_2 - \|\theta^*\|_2) \leq 0 \tag{2.115}$$

which also leads to (2.109). Since $\tilde{\theta}^T(t)\Gamma^{-1}f(t)$ goes to $-\infty$ as $\theta(t)$ goes to ∞, $\dot{V}_p$ becomes negative for large $\|\theta(t)\|_2$ so that $\theta(t)$ is bounded.

2.4 Designs for Multigroup Actuators

In this section, we address the adaptive actuator failure compensation control problem when the controlled system has multiple groups of actuators and whose behavior is expected to match that of a multiple-input reference model. For simplicity, we consider the case of two-group actuators.

The system to be controlled is

$$\dot{x} = Ax + Bu \tag{2.116}$$

where $A \in R^{n \times n}$, $B \in R^{n \times m}$, and $u = [u_1, \ldots, u_m]$. The reference model is

$$\dot{x}_m(t) = A_M x_m(t) + B_M r(t) \tag{2.117}$$

where $x_m \in R^n$, $A_M \in R^{n \times n}$, $r(t) \in R^2$, and $B_M = [b_{M1}, b_{M2}] \in R^{n \times 2}$.

We solve the adaptive actuator failure compensation control problem in which each u_i in (2.116) may fail, under the assumption

(A2.0m) The columns of B are separated into two groups, i.e.,

$$B = [b_{11}, \ldots, b_{1m_1}, b_{21}, \ldots, b_{2m_2}], \quad m_1 + m_2 = m \tag{2.118}$$

the corresponding inputs are $u = [u_{11}, \ldots, u_{1m_1}, u_{21}, \ldots, u_{2m_2}]^T$. The system is so designed that for any up to $m_1 - 1$ actuator failures in group one and for any up to $m_2 - 1$ actuator failures in group two, the remaining actuators can still be used to achieve a control objective (in this case, it means signal boundedness and tracking of the two-input reference model state x_m by the system state x).

If the system has only one actuator working in each group, such as u_{1i} and u_{2j}, from Assumption (A2.0m), there exist constant vectors k^*_{s11i} and k^*_{s12j} and nonzero constant scalars k^*_{s21i} and k^*_{s22j}, $1 \leq i \leq m_1$, $1 \leq j \leq m_2$, such that the following matching equations are satisfied:

$$A + [b_{1i}, b_{2j}][k^*_{s11i}, k^*_{s12j}]^T = A_M \tag{2.119}$$

$$[b_{1i}, b_{2j}] \begin{bmatrix} k^*_{s21i} & 0 \\ 0 & k^*_{s22j} \end{bmatrix} = B_M. \tag{2.120}$$

Since $B_M = [b_{M1}, b_{M2}]$, Eq. (2.120) can be rewritten as

$$b_{1i} k^*_{s21i} = b_{M1}, \quad b_{2j} k^*_{s22j} = b_{M2}. \tag{2.121}$$

On the other hand, if there is no actuator failure in the system, there should also exist matrices K_1^* and K_2^* such that

$$A + BK_1^{*T} = A_M, \quad BK_2^* = B_M \tag{2.122}$$

$$K_1^* = [k_{111}^*, \ldots, k_{11m_1}^* k_{121}^*, \ldots, k_{12m_2}^*] \in R^{n \times m} \tag{2.123}$$

$$K_2^* = \begin{bmatrix} k_{211}^*, & \cdots, & k_{21m_1}^*, & 0, & \cdots, & 0 \\ 0, & \cdots, & 0, & k_{221}^*, & \cdots, & k_{22m_2}^* \end{bmatrix}^T \in R^{m \times 2} \tag{2.124}$$

From (2.119)–(2.124) we can derive that the components of K_1^* and K_2^* can be obtained from the following equations:

$$\sum_{i=1}^{m_1} b_{1i} k_{11i}^{*T} + \sum_{j=1}^{m_2} b_{2j} k_{12j}^{*T} = b_{1i} k_{s11i}^* + b_{2j} k_{s12j}^* \tag{2.125}$$

$$\left[\sum_{i=1}^{m_1} b_{1i} k_{21i}^*, \sum_{j=1}^{m_2} b_{2j} k_{22j}^* \right] = [b_{M1}, b_{M2}]. \tag{2.126}$$

Following the steps in previous sections, we can develop adaptive control schemes for the system (2.116) with unknown actuator failures. To show such a design, we consider the failure pattern (2.2), i.e., the failed inputs are constants. The plant-model matching controller structure is

$$v(t) = v^*(t) = K_1^{*T} x(t) + K_2^* r(t) + k_3^* \tag{2.127}$$

where $k_3^* = [k_{311}^*, \ldots, k_{31m_1}^*, k_{321}^*, \ldots, k_{32m_2}^*]^T \in R^m$ is the term for compensating actuator failures in the plant (2.118).

Suppose at time t there are p_1 failed actuators in group one and p_2 failed actuators in group two, $1 \le p_1 \le m_1 - 1$, $1 \le p_2 \le m_2 - 1$, i.e., $u_{1i}(t) = \bar{u}_{1i}, i = i_1, i_2, \ldots, i_{p_1}$, and $u_{2j}(t) = \bar{u}_{2j}, j = j_1, j_2, \ldots, j_{p_2}$. For this two-group case, the desired plant-model matching conditions on k_{11i}^*, k_{11j}^*, k_{21i}^*, and k_{22j}^*, $i \ne i_1, i_2, \ldots, i_{p_1}$, $j \ne j_1, j_2, \ldots, j_{p_2}$, as similar to that in (2.17) for the single-group case, become

$$A + \sum_{i \ne i_1, \ldots, i_{p_1}} b_{1i} k_{11i}^{*T} + \sum_{j \ne j_1, \ldots, j_{p_2}} b_{2j} k_{12j}^{*T} = A_M$$

$$\left[\sum_{i \ne i_1, \ldots, i_{p_1}} b_{1i} k_{21i}^*, \sum_{j \ne j_1, \ldots, j_{p_2}} b_{2j} k_{22j}^* \right] = B_M. \tag{2.128}$$

Similar to that in (2.18), the closed-loop system is

$$\dot{x}(t) = A_M x(t) + B_M r(t) + \sum_{i \ne i_1, \ldots, i_{p_1}} b_{1i} k_{1i}^* + \sum_{j \ne j_1, \ldots, j_{p_2}} b_{2j} k_{2j}^*$$

$$+ \sum_{i = i_1, \ldots, i_{p_1}} b_{1i} \bar{u}_{1i} + \sum_{j = j_1, \ldots, j_{p_2}} b_{2j} \bar{u}_{2j}. \tag{2.129}$$

Then, the desired actuator failure compensation condition on k_{31i}^* and k_{32j}^*, $i \neq i_1, \ldots, i_{p_1}, j \neq j_1, \ldots, j_{p_2}$, is

$$\sum_{i \neq i_1, \ldots, i_{p_1}} b_{1i} k_{1i}^* + \sum_{j \neq j_1, \ldots, j_{p_2}} b_{2j} k_{2j}^* + \sum_{i = i_1, \ldots, i_{p_1}} b_{1i} \bar{u}_{1i} + \sum_{j = j_1, \ldots, j_{p_2}} b_{2j} \bar{u}_{2j} = 0. \quad (2.130)$$

In view of (2.126) for B_M, we have the desired compensation conditions as

$$\sum_{i \neq i_1, \ldots, i_{p_1}} \frac{k_{31i}^*}{k_{s21i}^*} + \sum_{i = i_1, \ldots, i_{p_1}} \frac{\bar{u}_{1i}}{k_{s21i}^*} = 0 \quad (2.131)$$

$$\sum_{j \neq j_1, \ldots, j_{p_2}} \frac{k_{32j}^*}{k_{s22j}^*} + \sum_{j = j_1, \ldots, j_{p_2}} \frac{\bar{u}_{2j}}{k_{s22j}^*} = 0 \quad (2.132)$$

which determine the choice of k_{31i}^* and k_{32j}^*, $i \neq 1, \ldots, i_{p_1}, j \neq j_1, \ldots, j_{p_2}$, to achieve the desired closed-loop system

$$\dot{x}(t) = A_M x(t) + B_M r(t) \quad (2.133)$$

which matches the reference model system (2.117).

The choice of k_{11i}^*, k_{11j}^*, k_{21i}^*, k_{22j}^*, k_{31i}^*, and k_{32j}^*, $i = i_1, i_2, \ldots, i_{p_1}, j = j_1, j_2, \ldots, j_{p_2}$, is irrelevant because their actuators have failed.

The adaptive version of the controller (2.127) is

$$v(t) = K_1^T(t) x(t) + K_2(t) r(t) + k_3(t) \quad (2.134)$$

where the controller parameters

$$K_1 = [k_{111}, \ldots, k_{11m_1}, k_{121}, \ldots, k_{12m_2}] \in R^{n \times m}$$

$$K_2 = \begin{bmatrix} k_{211}, & \ldots, & k_{21m_1}, & 0, & \ldots, & 0 \\ 0, & \ldots, & 0, & k_{221}, & \ldots, & k_{22m_2} \end{bmatrix}^T \in R^{m \times 2}$$

$$k = [k_{311}, \ldots, k_{31m_1}, k_{321}, \ldots, k_{32m_2}]^T \in R^m \quad (2.135)$$

are the estimates of K_1^*, K_2^*, and k_3^*, respectively.

For adaptive control, we assume that

(A2.1m) $\text{sign}[k_{s21i}^*]$ and $\text{sign}[k_{s22j}^*]$, $i = 1, \ldots, m_1, j = 1, \ldots, m_2$, are all known.

Defining the parameter errors

$$\begin{aligned}
\tilde{k}_{11i}(t) &= k_{11i}(t) - k_{11i}^*, & \tilde{k}_{12j}(t) &= k_{12j}(t) - k_{12j}^* \\
\tilde{k}_{21i}(t) &= k_{21i}(t) - k_{21i}^*, & \tilde{k}_{22j}(t) &= k_{22j}(t) - k_{22j}^* \\
\tilde{k}_{31i}(t) &= k_{31i}(t) - k_{31i}^*, & \tilde{k}_{32j}(t) &= k_{32j}(t) - k_{32j}^*
\end{aligned} \quad (2.136)$$

for $i = 1, \ldots, m_1$, $j = 1, \ldots, m_2$, and using (2.116), (2.122), (2.128), similar to (2.29) for the single actuator group case, we obtain

$$
\dot{x} = A_M x + B_M r + \left(\sum_{i \neq i_1, \ldots, i_{p_1}} b_{1i} \tilde{k}_{11i}^T x + \sum_{j \neq j_1, \ldots, j_{p_2}} b_{2j} \tilde{k}_{12j}^T x \right)
$$

$$
+ \left(\sum_{i \neq i_1, \ldots, i_{p_1}} b_{1i} \tilde{k}_{21i} r_1 + \sum_{j \neq j_1, \ldots, j_{p_2}} b_{2j} \tilde{k}_{22j} r_2 \right)
$$

$$
+ \left(\sum_{i \neq i_1, \ldots, i_{p_1}} b_{1i} \tilde{k}_{31i} + \sum_{j \neq j_1, \ldots, j_{p_2}} b_{2j} \tilde{k}_{32j} \right)
$$

$$
= A_M x + B_M r + \sum_{i \neq i_1, \ldots, i_{p_1}} \frac{b_1}{k_{s21i}^*} \left(\tilde{k}_{11i}^T x + \tilde{k}_{21i} r_1 + \tilde{k}_{31i} \right)
$$

$$
+ \sum_{j \neq j_1, \ldots, j_{p_2}} \frac{b_2}{k_{s22j}^*} \left(\tilde{k}_{12j}^T x + \tilde{k}_{22j} r_2 + \tilde{k}_{32j} \right) \tag{2.137}
$$

Define the positive definite function

$$
V_{p_1 p_2} \left(e, \tilde{k}_{11i}, \tilde{k}_{12j}, \tilde{k}_{21i}, \tilde{k}_{22j}, \tilde{k}_{31i}, \tilde{k}_{32j}, i \neq i_1, \ldots, i_{p_1}, j \neq j_1, \ldots, j_{p_2} \right)
$$

$$
= e^T P e + \sum_{i \neq i_1, \ldots, i_{p_1}} \frac{1}{|k_{s21i}^*|} \left(\tilde{k}_{11i}^T \Gamma_{11i}^{-1} \tilde{k}_{11i} + \tilde{k}_{21i}^2 \gamma_{21i}^{-1} + \tilde{k}_{31i}^2 \gamma_{31i}^{-1} \right)
$$

$$
+ \sum_{j \neq j_1, \ldots, j_{p_2}} \frac{1}{|k_{s22j}^*|} \left(\tilde{k}_{12j}^T \Gamma_{12j}^{-1} \tilde{k}_{12j} + \tilde{k}_{22j}^2 \gamma_{22j}^{-1} + \tilde{k}_{32j}^2 \gamma_{32j}^{-1} \right) \tag{2.138}
$$

where $P = P^T > 0$ and $Q = Q^T > 0$, satisfying the Lyapunov equation $P A_M + A_M^T P = -Q$, and $\Gamma_{11i} = \Gamma_{11i}^T > 0, \Gamma_{12j} = \Gamma_{12j}^T > 0$, $\gamma_{21i}, \gamma_{22j}, \gamma_{31i}$, $\gamma_{32j} > 0$, $i = 1, 2, \ldots, m_1$, $j = 1, 2, \ldots, m_2$.

We choose the adaptive laws as

$$
\dot{k}_{11i} = -\text{sign}[k_{s21i}^*] \Gamma_{11i} x(t) e^T P b_{M1} \tag{2.139}
$$

$$
\dot{k}_{12j} = -\text{sign}[k_{s22j}^*] \Gamma_{12j} x(t) e^T P b_{M2} \tag{2.140}
$$

$$
\dot{k}_{21i} = -\text{sign}[k_{s21i}^*] \gamma_{21i} r_1(t) e^T P b_{M1} \tag{2.141}
$$

$$
\dot{k}_{22j} = -\text{sign}[k_{s22j}^*] \gamma_{22j} r_2(t) e^T P b_{M2} \tag{2.142}
$$

$$
\dot{k}_{31i} = -\text{sign}[k_{s21i}^*] \gamma_{31i} e^T P b_{M1} \tag{2.143}
$$

$$
\dot{k}_{32j} = -\text{sign}[k_{s22j}^*] \gamma_{32j} e^T P b_{M2} \tag{2.144}
$$

for $i = 1, \ldots, m_1$, $j = 1, \ldots, m_2$.

Similar to the single actuator group case in Section 2.1, we let T_i be the time when actuators fail, $i = 1, 2, \ldots, m_0$, $m_0 \leq m - 2$ (note that m_0

may be equal to $m_1 - 1$ if those $m_1 - 1$ actuators in the first group of $u = [u_{11}, \ldots, u_{1m_1}, u_{21}, \ldots, u_{2m_2}]^T$ fail one by one first). Over the intervals (T_i, T_{i+1}), $i = 0, 1, \ldots, m_0$, with $T_0 = 0$ and $T_{m_0+1} = \infty$, the plant-model matching parameters K_1^*, K_2^*, and k_3^* are constant. Therefore, using (2.139)–(2.144), we have

$$\dot{V}_{p_1 p_2} = -e^T(t)Qe(t) \leq 0, \ t \in (T_{i(p)}, T_{i(p)+1}), \ i(p) \leq p = p_1 + p_2. \quad (2.145)$$

Since eventually there are $\bar{m}_{10}$ failed actuators in group one and $\bar{m}_{20}$ failed actuators in group two (see (2.118)), in particular (2.145) holds for $(p_1, p_2) = (\bar{m}_{10}, \bar{m}_{20})$ for some $\bar{m}_{10} < m_1$ and $\bar{m}_{20} < m_2$ such that $m_0 \leq \bar{m}_{10} + \bar{m}_{20}$, and $t \in (T_{m_0}, \infty)$ for some $T_{m_0} < \infty$. Therefore, we conclude that $V_{\bar{m}_{10} \bar{m}_{20}}(t)$ is bounded, and so are $x(t)$, $k_{11i}(t)$, $k_{12j}(t)$, $k_{21i}(t)$, $k_{22j}(t)$, $k_{31i}(t)$, and $k_{32j}(t)$, $i \neq i_1, \ldots, \bar{m}_{10}$, $j \neq j_1, \ldots, \bar{m}_{10}$, and that $e(t) \in L^2$. The boundedness of $k_{11i}(t)$, $k_{12j}(t)$, $k_{21i}(t)$, $k_{22j}(t)$, $k_{31i}(t)$, and $k_{32j}(t)$, $i = i_1, \ldots, \bar{m}_{10}$, $j = j_1, \ldots, \bar{m}_{20}$, can be also proved (similar to that in (2.37)–(2.40) for the single actuator group case). Then, we have the boundedness of $v(t)$ in (2.130), and also $\dot{e}(t) \in L^\infty$ so that $\lim_{t \to \infty} e(t) = 0$.

In summary, we have the following result.

Theorem 2.4.1. *The adaptive controller (2.133), with adaptive law (2.139)–(2.144), applied to the system (2.116), guarantees that all closed-loop signals are bounded and $\lim_{t \to \infty}(x(t) - x_m(t)) = 0$.*

So far we have derived the adaptive actuator failure compensation scheme for systems with two actuator groups to follow a two-input reference system and proved the stability and asymptotic tracking of the adaptive control system. For the more general case of more than two actuator groups, the same idea can be used to design an adaptive control scheme for tracking a reference system with more than two inputs.

2.5 Design for up to $m - q$ Actuator Failures

In Section 2.1, adaptive state feedback control schemes are developed for state tracking for linear time-invariant plants with up to $m - 1$ actuator failures, where m is the number of actuators. For a given plant $\dot{x}(t) = Ax(t) + Bu(t)$, where $u \in R^m$, and a chosen reference model $\dot{x}_m(t) = A_M x_m(t) + B_m r(t)$, where $r \in R$, to achieve desired plant-model state dynamics matching when there are up to $m - 1$ actuator failures in $u \in R^m$, a necessary condition is that the columns of B are parallel to each other. For many applications, this condition may be restrictive, and in many practical cases, there can be more than one actuator remaining fully functional, so it is useful and desirable to study the design conditions and adaptive control designs for state tracking when more than one actuators are active.

In this section, we study the more general case of adaptive actuator failure compensation for linear plants with up to $m - q$ ($1 \leq q \leq m - 1$) actuator failures (note that the case of up to $m - 1$ actuator failures is a special case of this, and that the no-failure case, that is, when $q = m$, can also be included). The goals of this section are to derive the necessary and sufficient conditions for achieving desired plant-model state dynamics matching when there are up to $m - q$ unknown actuator failures, and to present the adaptive control scheme for systems with unknown parameters and unknown failures, achieving closed-loop system stability and asymptotic state tracking.

2.5.1 Problem Statement

Consider a linear time-invariant plant

$$\dot{x}(t) = Ax(t) + Bu(t) \tag{2.146}$$

where $A \in R^{n \times n}$, $B \in R^{n \times m}$ are constant parameter matrices, the state vector $x(t) \in R^n$ is available for measurement, $u(t) = [u_1, \ldots, u_m]^T \in R^m$ is the input vector whose components (actuators) may fail during system operation. In this section, we consider the case that there are up to $m - q$ ($1 \leq q \leq m - 1$) unknown actuator failures, that is, any $m - q$ of the m actuators may fail during operation, and if an actuator fails, its failure value, location, and time are unknown. We assume that the system (2.146) is so constructed that for any up to $m - q$ actuator failures, the remaining actuators can still achieve a desired control objective.

We consider the same type of actuator failure as that in Section 2.1,

$$u_j(t) = \bar{u}_j, \ t \geq t_j, \ j \in \{1, 2, \ldots, m\} \tag{2.147}$$

where the constant value $\bar{u}_j$ and the failure time instant t_j are unknown.

As in (2.3), in the presence of actuator failures, $u(t)$ can be expressed as

$$u(t) = v(t) + \sigma(\bar{u} - v(t)) \tag{2.148}$$

where $v(t)$ is an applied control input to be designed, and

$$\bar{u} = [\bar{u}_1, \bar{u}_2, \ldots, \bar{u}_m]^T \tag{2.149}$$

$$\sigma = \text{diag}\{\sigma_1, \sigma_2, \ldots, \sigma_m\} \tag{2.150}$$

$$\sigma_i = \begin{cases} 1 & \text{if the } i\text{th actuator fails, i.e., } u_i = \bar{u}_i \\ 0 & \text{otherwise.} \end{cases} \tag{2.151}$$

The control objective is to design a feedback control $v(t)$ for the plant (2.146) with up to $m - q$ actuator failures such that despite the uncertain control error $u - v = \sigma(\bar{u} - v)$ caused by actuator failures, all closed-loop signals are bounded and the plant state vector $x(t)$ asymptotically tracks a given reference state vector $x_m(t)$ generated from the reference model

$$\dot{x}_m(t) = A_M x_m(t) + B_M r(t) \tag{2.152}$$

where $A_M \in R^{n \times n}$, $B_M \in R^{n \times l}$ are known constant matrices such that all the eigenvalues of A_M are in the left-half complex plane, all columns of B_M are independent, and $r(t) \in R^l$ is bounded.

2.5.2 Plant-Model Matching Control

Before developing the adaptive control schemes for plant (2.146) with un-known actuator failures, it is useful to understand the desirable controller structure and parameters when both the plant and actuator failure parameters are known. For the control objective and the failure model (2.147), we consider the controller structure

$$v(t) = v^*(t) = K_1^{*T} x(t) + K_2^{*T} r(t) + k_3^* \tag{2.153}$$

where $K_1^* = [k_{11}^*, \ldots, k_{1m}^*] \in R^{n \times m}$, $K_2^* = [k_{21}^*, \ldots, k_{2m}^*] \in R^{l \times m}$ to be defined for plant-model matching, and $k_3^* = [k_{31}^*, \ldots, k_{3m}^*]^T \in R^m$ is a vector to be chosen for compensation of the actuation error $Bu - Bv = B\sigma(\bar{u} - v)$.

To derive controller parametrizations for plant-model matching, denote

$$B = [b_1, \ldots, b_m], \ b_i \in R^n, \ i = 1, \ldots, m. \tag{2.154}$$

Plant-Model Matching with p Actuator Failures. When the system has p failed actuators at time t, that is, $u_j(t) = \bar{u}_j$, $j = j_1, \ldots, j_p$, $1 \le p \le m - q$, we have the closed-loop system as

$$
\begin{aligned}
\dot{x}(t) &= \left(A + BK_1^{*T}\right) x(t) + BK_2^{*T} r(t) + Bk_3^* + B\sigma(\bar{u} - v(t)) \\
&= \left(A + \sum_{j \ne j_1, \ldots, j_p} b_j k_{1j}^{*T}\right) x(t) + \sum_{j \ne j_1, \ldots, j_p} b_j k_{2j}^{*T} r(t) + \sum_{j \ne j_1, \ldots, j_p} b_j k_{3j}^* \\
&\quad + \sum_{j = j_1, \ldots, j_p} b_j \bar{u}_j \tag{2.155}
\end{aligned}
$$

We collect the $m - p$ columns of B, $b_j, j \neq j_1, \ldots, j_p$, to construct a matrix $B_a \in R^{n \times (m-p)}$, and the p columns $b_j, j = j_1, \ldots, j_p$, to construct a matrix $B_f \in R^{n \times p}$. Similarly, by grouping the columns and entries k_{1j}^*, k_{2j}^*, and $k_{3j}^*, j \neq j_1, \ldots, j_p$, from K_1^*, K_2^*, and k_3^*, we can construct matrices $K_{1a}^* \in R^{n \times (m-p)}$ and $K_{2a}^* \in R^{l \times (m-p)}$, and a constant vector $k_{3a}^* \in R^{m-p}$. Also, we define the failure value vector as

$$\bar{u}_f = [\bar{u}_{j_1}, \ldots, \bar{u}_{j_p}]^T. \tag{2.156}$$

Then, we rewrite the closed-loop plant as

$$\dot{x}(t) = (A + B_a K_{1a}^{*T}) x(t) + B_a K_{2a}^{*T} r(t) + B_a k_{3a}^* + B_f \bar{u}_f \tag{2.157}$$

For this plant to match the desired closed-loop system (2.152), we need to choose K_{1a}^*, K_{2a}^*, and k_{3a}^* to satisfy the conditions

$$B_a K_{1a}^{*T} = A_M - A \tag{2.158}$$

$$B_a K_{2a}^{*T} = B_M \tag{2.159}$$

$$B_a k_{3a}^* = -B_f \bar{u}_f \tag{2.160}$$

for plant-model state matching.

Based on these conditions, we have the following result.

Proposition 2.5.1. *Necessary and sufficient conditions for the existence of constant matrices K_{1a}^* and K_{2a}^* and a constant vector k_{3a}^* to satisfy the matching conditions (2.158)–(2.160) are*

$$\text{rank}(B_a) = \text{rank}([B_a | A_M - A]) \tag{2.161}$$

$$\text{rank}(B_a) = \text{rank}([B_a | B_M]) \tag{2.162}$$

$$\text{rank}(B_a) = \text{rank}(B) \tag{2.163}$$

where $B_a \in R^{n \times (m-p)}$ consists of $m - p$ columns of $B, b_j, j \neq j_1, \ldots, j_p$.

Proof: From linear equation theory, it is clear that conditions (2.161) and (2.162) are necessary and sufficient for the existence of K_{1a}^* and K_{2a}^* to satisfy the matching conditions (2.158) and (2.159).

Since B_a consists of only $b_j, j \neq j_1, \ldots, j_p$, $\text{rank}(B_a) < \text{rank}(B)$ means that there is at least one of $b_j, j = j_1, \ldots, j_p$ (i.e., a column of B_f), that cannot be expressed as a linear combination of the columns of B_a. Since $\bar{u}_j, j = j_1, \ldots, j_p$, can be any value, $B_f \bar{u}_f$ cannot be a linear combination of the columns of B_a either, which means there exists no such k_{3a}^* that can satisfy the matching condition (2.160). Therefore, the condition (2.163) is necessary for the matching condition (2.160).

To prove the sufficiency, we see that for any $\bar{u}_f$,

$$\text{rank}(B_a) \leq \text{rank}([B_a|B_f\bar{u}_f]) \leq \text{rank}(B) \tag{2.164}$$

so that the condition (2.163) means

$$\text{rank}(B_a) = \text{rank}([B_a|B_f\bar{u}_f]). \tag{2.165}$$

Hence, there exists a k_{3a}^* to satisfy (2.160). ∇

Plant-Model Matching with up to $m - q$ Actuator Failures. Now we consider the case of up to $m-q$ actuator failures. Clearly, for the state vector $x(t)$ of plant (2.146) to track the reference state vector $x_m(t)$ generated from the reference model (2.152) in the presence of up to $m - q$ actuator failures, it is required that for every $B_a \in R^{n \times q}$ consisting of q columns of B, there exist $K_{1a}^* \in R^{n \times q}, K_{2a}^* \in R^{l \times q}$, and $k_{3a}^* \in R^q$ such that the matching conditions (2.158)–(2.160) are satisfied. Hence, we have the following conditions for plant-model matching in the presence of up to $m - q$ actuator failures.

Proposition 2.5.2. *In the presence of up to $m - q$ actuator failures, the plant-model matching can be achieved if and only if for every $B_a \in R^{n \times q}$ consisting of q columns of B,*

$$\text{rank}(B_a) = \text{rank}([B_a|A_M - A]) \tag{2.166}$$

$$\text{rank}(B_a) = \text{rank}([B_a|B_M]) \tag{2.167}$$

$$\text{rank}(B_a) = \text{rank}(B). \tag{2.168}$$

Proposition 2.5.2 follows from Proposition 2.5.1, as conditions (2.166)–(2.168) imply (2.161)–(2.163).

Remark 2.5.1. Condition (2.168) is crucial for (2.166)–(2.168). If condition (2.168) holds for every B_a, then the following conditions

$$\text{rank}(B) = \text{rank}([B|A_M - A]) \tag{2.169}$$

$$\text{rank}(B) = \text{rank}([B|B_M]) \tag{2.170}$$

are equivalent to conditions (2.166) and (2.167). Using conditions (2.169) and (2.170) instead of (2.166) and (2.167) may considerably reduce the number of times the matching conditions need to be checked.

If condition (2.168) holds for every B_a, and (2.166) and (2.167) hold for one B_a, then (2.166) and (2.167) hold for every B_a. This property means that we can check conditions (2.166) and (2.167) only once.

Condition (2.168) also means that to achieve state tracking by state feedback for arbitrary failed values $\bar{u}_j$, it is necessary that at least $q \geq \text{rank}(B)$ actuators remain active.   □

Remark 2.5.2. For the actuator failure compensation problem considered in Section 2.2 where there are up to $m - 1$ actuator failures, that is, $q = 1$, conditions (2.166)–(2.168) become

$$\text{rank}(b_i) = \text{rank}([b_i|A_M - A]) \tag{2.171}$$

$$\text{rank}(b_i) = \text{rank}([b_i|B_M]) \tag{2.172}$$

$$\text{rank}(b_i) = \text{rank}(B) \tag{2.173}$$

for any $i = 1, 2, \ldots, m$. Condition (2.173) implies that $b_i, i = 1, 2, \ldots, m$, are parallel to each other. This is the conclusion we made in Section 2.1. Condition (2.172) implies that the columns of B_M and B are parallel to each other, which further implies that for the reference model system (2.152), one input, that is, $l = 1$, is the nontrivial choice. □

Remark 2.5.3. An interesting problem is how to find the largest integer l (the dimension of $r(t)$) to fit any up to $m - q$ actuator failures. It requires that matching condition (2.168) holds for every B_a. As stated in Remark 2.5.1, if (2.168) holds for every B_a, then $\text{rank}(B) = \text{rank}([B|B_M])$ implies that $\text{rank}(B_a) = \text{rank}([B_a|B_m])$, which means that plant-model matching can still be achieved, and $l = \text{rank}(B)$ is the solution. □

2.5.3 Adaptive Control Design

When the parameters of actuator failures (including the failed actuators, the failure values, and the failure time instants) are unknown, the matching parameters K_1^*, K_2^*, and k_3^* are unknown, and adaptive control schemes are needed for actuator failure compensation. In this section, we first develop an adaptive actuator failure compensation scheme for the case when B is known, and then we discuss some issues for B unknown. In our adaptive designs, the matrix A is not needed to be known. We should note that the matching conditions (2.166)–(2.168) are also sufficient for an adaptive design, and we assume that they are satisfied in the adaptive control case.

Adaptive Control Design with Known B. As the first adaptive actuator failure compensation scheme based on the matching conditions (2.166)–(2.168), we now develop an adaptive design for the case when the matrix B is known. The assumption that B is known does not make the control problem trivial as the essential system uncertainties are the unknown actuator failures in the sense that it is uncertain which of the columns of B correspond to failed actuators $u_j = \bar{u}_j$, $j = j_1, j_2, \ldots, j_p \in \{1, 2, \ldots, m\}$ with $0 \le p < m$, that is, the indices $j_1, j_2, \ldots, j_p$ are unknown. The control task is to generate the feedback control signals $v_i(t)$, $i = 1, 2, \ldots, m$, such that in the presence of actuator failure uncertainties those actuating signals $v_i(t)$, $i \neq j_1, j_2, \ldots, j_p$, can stabilize the closed-loop system and achieve asymptotic state tracking despite the uncertain failure values $\bar{u}_j$, $j = j_1, j_2, \ldots, j_p$.

To fulfill this task, we use the controller structure

$$v(t) = K_1^T(t)x(t) + K_2^T(t)r(t) + k_3(t) \tag{2.174}$$

where K_1, K_2, and k_3 are the estimates of K_1^*, K_2^*, and k_3^*. The resulting closed-loop system is

$$\dot{x}(t) = (A + B_a K_{1a}^T)x(t) + B_a K_{2a}^T r(t) + B_a k_{3a} + B_f \bar{u}_f \tag{2.175}$$

where K_{1a}, K_{2a}, and k_{3a} are the estimates of K_{1a}^*, K_{2a}^*, and k_{3a}^*, and are composed of the corresponding columns and entries of K_1, K_2, and k_3.

We define the parameter errors

$$\tilde{K}_{1a} = K_{1a} - K_{1a}^*, \ \tilde{K}_{2a} = K_{2a} - K_{2a}^*, \ \tilde{k}_{3a} = k_{3a} - k_{3a}^* \tag{2.176}$$

and the tracking error

$$e(t) = x(t) - x_m(t). \tag{2.177}$$

As in Section 2.1 for up to $m - 1$ actuator failures, we let $(T_i, T_{i+1}), i = 0, 1, \ldots, m_0$, with $T_0 = 0$ be the time intervals on which the actuator failure pattern is fixed, that is, actuators only fail at unknown time instants $T_i, i = 1, \ldots, m_0$. Since at least q actuators do not fail, we have $m_0 < m$ and $T_{m_0+1} = \infty$. Then, at time $T_i, i = 1, \ldots, m_0$, the unknown plant-model matching parameters K_1^*, K_2^*, and k_3^* change their values such that

$$k_{1j}^* = k_{1j(i)}^*, \ k_{2j}^* = k_{2j(i)}^*, \ k_{3j}^* = k_{3j(i)}^*, \ t \in (T_i, T_{i+1}) \tag{2.178}$$

for $j = 1, \ldots, m, i = 1, \ldots, m_0$. Suppose there are p failed actuators, that is, $u_j(t) = \bar{u}_j, j = j_1, \ldots, j_p, 1 \leq p \leq m_0$, at time $t \in (T_i, T_{i+1})$ (with $i = i(p) \leq p$ because more than one actuator may fail at the same time T_i). From (2.152) we have the tracking error equation as

$$
\begin{aligned}
\dot{e}(t) &= \dot{x}(t) - \dot{x}_m(t) \\
&= (A + B_a K_{1a}^T)x(t) + B_a K_{2a}^T r(t) + B_a k_{3a} + B_f \bar{u}_f - A_M x_m(t) - B_M r(t) \\
&= A_M x(t) + B_M r(t) + B_a \left(\tilde{K}_{1a}^T x(t) + \tilde{K}_{2a}^T r(t) + \tilde{k}_{3a} \right) \\
&\quad - A_M x_m(t) - B_M r(t) \\
&= A_M e(t) + B_a \left(\tilde{K}_{1a}^T x(t) + \tilde{K}_{2a}^T r(t) + \tilde{k}_{3a} \right).
\end{aligned}
\tag{2.179}
$$

To derive adaptive laws for k_{1j}, k_{2j}, k_{3j}, define the positive definite function

$$V_p \left(e, \tilde{k}_{1j}, \tilde{k}_{2j}, \tilde{k}_{3j}, j \neq j_1, \dots, j_p \right)$$
$$= e^T P e + \sum_{j \neq j_1, \dots, j_p} \left(\tilde{k}_{1j}^T \Gamma_{1j}^{-1} \tilde{k}_{1j} + \tilde{k}_{2j}^T \Gamma_{2j}^{-1} \tilde{k}_{2j} + \tilde{k}_{3j}^2 \gamma_{3j}^{-1} \right) \quad (2.180)$$

where $k_{1j}^* = k_{1j(i)}^*$, $k_{2j}^* = k_{2j(i)}^*$, and $k_{3j}^* = k_{3j(i)}^*$ are defined in (2.178) (note that V_p is defined for all $t \in [0, \infty)$ with the matching parameters k_{1j}^*, k_{2j}^* and k_{3j}^* for the case when there are p failed actuators over (T_i, T_{i+1}), $i = i(p) \leq p$), $P \in R^{n \times n}$, $P = P^T > 0$ such that

$$PA_M + A_M^T P = -Q \quad (2.181)$$

for any constant $Q \in R^{n \times n}$, $Q = Q^T > 0$, $\Gamma_{1j} \in R^{n \times n}$, and $\Gamma_{2j} \in R^{l \times l}$ are constant such that $\Gamma_{1j} = \Gamma_{1j}^T > 0$, $\Gamma_{2j} = \Gamma_{2j}^T > 0$, and $\gamma_{3j} > 0$ is constant, $j = 1, \dots, m$. Then, the time derivative of V_p is

$$\dot{V}_p = 2e^T P \dot{e} + 2 \sum_{j \neq j_1, \dots, j_p} \left(\tilde{k}_{1j}^T \Gamma_{1j}^{-1} \dot{\tilde{k}}_{1j} + \tilde{k}_{2j}^T \Gamma_{2j}^{-1} \dot{\tilde{k}}_{2j} + \tilde{k}_{3j} \gamma_{3j}^{-1} \dot{\tilde{k}}_{3j} \right). \quad (2.182)$$

For $j = 1, \dots, m$, we choose the adaptive laws as

$$\dot{k}_{1j}(t) = -\Gamma_{1j} x(t) e^T(t) P b_j \quad (2.183)$$
$$\dot{k}_{2j}(t) = -\Gamma_{2j} r(t) e^T(t) P b_j \quad (2.184)$$
$$\dot{k}_{3j}(t) = -\gamma_{3j} e^T(t) P b_j. \quad (2.185)$$

Then, from (2.179), (2.181)–(2.185) we have

$$\dot{V}_p = -e^T(t) Q e(t) \leq 0, \ t \in (T_{i(p)}, T_{i(p+1)}), \ i(p) \leq p. \quad (2.186)$$

Since there are only a finite number of time instants of actuators failures, that is, $i \leq m_0 < m$, and eventually there are $\bar{m}_0$ failed actuators such that $m_0 \leq \bar{m}_0 < m$, from (2.186) it follows that

$$\dot{V}_{\bar{m}_0} = -e^T(t) Q e(t) \leq 0, \ t \in (T_{m_0}, \infty) \quad (2.187)$$

so we can conclude that $V_{\bar{m}_0}(t)$ is bounded, and so are $x(t), k_{1j}(t), k_{2j}(t)$, and $k_{3j}(t), j \neq j_1, \dots, j_{\bar{m}_0}$.

Noting that any $b_j, j = j_1, \dots, j_{\bar{m}_0}$, is a linear combination of $b_i, i \neq j_1, \dots, j_{\bar{m}_0}$, that is,

$$b_j = \sum_{i \neq j_1, \dots, j_{\bar{m}_0}} \beta_{ij} b_i, \ j = j_1, \dots, j_{\bar{m}_0} \quad (2.188)$$

for some constant $\beta_{ij}, i \neq j_1, \dots, j_{\bar{m}_0}, j = j_1, \dots, j_{\bar{m}_0}$, from (2.183), we have

$$\dot{k}_{1j}(t) = -\Gamma_{1j}x(t)e^T(t)P \sum_{i\neq j_1,\ldots,j_{\bar{m}_0}} \beta_{ij}b_i = \Gamma_{1j} \sum_{i\neq j_1,\ldots,j_{\bar{m}_0}} \beta_{ij}\Gamma_{1i}^{-1}\dot{k}_{1i}(t)$$

which implies that for $j = j_1, \ldots, j_{\bar{m}_0}$,

$$k_{1j}(t) = k_{1j}(0) - \Gamma_{1j} \sum_{i\neq j_1,\ldots,j_{\bar{m}_0}} \beta_{ij}\Gamma_{1i}^{-1}(k_{1i}(0) - k_{1i}(t)) \qquad (2.189)$$

Since $k_{1i}(t), i \neq j_1, \ldots, j_{\bar{m}_0}$, is bounded, it follows that $k_{1j}(t), j = j_1, \ldots, j_{\bar{m}_0}$, is bounded. Similarly, $k_{2j}(t)$ and $k_{3j}(t), j = j_1, \ldots, j_{\bar{m}_0}$, are bounded, and so are all other closed-loop signals.

Since $m_0 < m$ and $T_{m_0+1} = \infty$, over the time interval (T_{m_0}, ∞), the function $V_{\bar{m}_0}$ has a finite initial value and its derivative is $\dot{V}_{\bar{m}_0} = -e^T(t)Qe(t) \leq 0$, which means that $e(t) \in L^2$. From (2.179) and signal boundedness, we have $\dot{e}(t) \in L^\infty$, and with $e(t) \in L^2$, we have $\lim_{t\to\infty} e(t) = 0$.

In summary, we have the following result.

Theorem 2.5.1. *The adaptive controller (2.174), with adaptive law (2.183)–(2.185), applied to the system (2.146) with actuator failure (2.147), guarantees that all closed-loop signals are bounded and* $\lim_{t\to\infty}(x(t) - x_m(t)) = 0$.

Issues When B Is Unknown. When the matrix B is not exactly known, some modifications of the scheme developed above are necessary.

Assume there are some parameter uncertainties in B such that

$$B = B_0 + \Delta B \qquad (2.190)$$

where $B = [b_1, \ldots, b_m] \in R^{n\times m}$ is the unknown true parameter, $B_0 = [b_{01}, \ldots, b_{0m}] \in R^{n\times m}$ is the known nominal parameter, and $\Delta B = [\Delta b_1, \ldots, \Delta b_m] \in R^{n\times m}$ is the parameter uncertainty.

Since B is unknown, we cannot use b_j in the adaptive laws (2.183)–(2.185). Instead, we use its nominal value b_{0j}. With this change, the time derivative of V_p defined in (2.180) becomes

$$\dot{V}_p = -e^T(t)Qe(t) + 2e^T(t)P \sum_{j\neq j_1,\ldots,j_p} \Delta b_j[\tilde{k}_{1j}^T x(t) + \tilde{k}_{2j}^T r(t) + \tilde{k}_{3j}]. \qquad (2.191)$$

If $\tilde{k}_{1j}, \tilde{k}_{2j}, \tilde{k}_{3j}, j = 1, \ldots, m$, are bounded, and $\|\Delta b_j\|, j = 1, \ldots, m$, are small, it can be established that $V_p \in L^\infty$, and so are $e(t)$ and $x(t)$. This analysis applies to all the failure patterns.

However, the boundedness of $\tilde{k}_{1j}, \tilde{k}_{2j}, \tilde{k}_{3j}$ (that is, the boundedness of k_{1j}, k_{2j}, and k_{3j}), $j = 1, \ldots, m$, cannot be guaranteed if b_{0j} instead of the unknown true parameter b_j is used in the adaptive laws (2.183)–(2.185). To ensure the boundedness of k_{1j}, k_{2j}, and k_{3j}, the adaptive laws need to be modified. Although ΔB, A, and $\bar{u}_f$ are unknown, it is reasonable to assume that we know the range of ΔB, A, and $\bar{u}_f$. Since in total there are m actuators and at least q actuators remain active during system operation, the total number of possible actuator failure patterns is countable, and the matching parameters k_{1j}^*, k_{2j}^*, and k_{3j}^* lie in a known convex bounded set S such that

$$m_{1ji} \leq k_{1ji}^* \leq M_{1ji}, \ m_{2ji} \leq k_{2ji}^* \leq M_{2ji}, \ m_{3j} \leq k_{3j}^* \leq M_{3j} \qquad (2.192)$$

where $k_{1ji}, i = 1, \ldots, n$, $k_{2ji}, i = 1, \ldots, l$, are the ith elements of k_{1j} and k_{2j}, respectively. We can then modify the adaptive laws (2.183)–(2.185) with the projection algorithm described in Section 2.3.3 to ensure the boundedness of k_{1j}, k_{2j}, and k_{3j}, and, provided that $\|\Delta b_j\|, j = 1, \ldots, m$, are small, the boundedness of the closed-loop signals is guaranteed.

2.5.4 Boeing 747 Lateral Control Simulation

As an illustrative example, we consider the linearized lateral dynamic model of a Boeing 747 airplane [37] as the controlled plant with three augmented rudder actuation vectors $b_2 u_2$, $b_3 u_3$, and $b_4 u_4$ for the study of adaptive actuator failure compensation:

$$\dot{x}(t) = Ax(t) + Bu(t)$$
$$x(t) = [\beta, r, p, \phi]^T, \ B = [b_1, b_2, b_3, b_4] \qquad (2.193)$$

where β is the side-slip angle, r is the yaw rate, p is the roll rate, ϕ is the roll angle, and u contains four control signals $u = [u_1, u_2, u_3, u_4]^T$ representing four rudder servos: δ_{r1}, δ_{r2}, δ_{r3}, δ_{r4}, from a four-piece rudder for achieving compensation in the presence of actuator failures.

From the data provided in [37] (p. 686), in horizontal flight at 40,000 ft and nominal forward speed 774 ft/sec (Mach 0.8), the Boeing 747 lateral-perturbation dynamics matrices are

$$A = \begin{bmatrix} -0.0558 & -0.9968 & 0.0802 & 0.0415 \\ 0.598 & -0.115 & -0.0318 & 0 \\ -3.05 & 0.388 & -0.465 & 0 \\ 0 & 0.0805 & 1 & 0 \end{bmatrix} \qquad (2.194)$$

$$b_1 = \begin{bmatrix} 0.0073 \\ -0.473 \\ 0.153 \\ 0 \end{bmatrix}, \ b_2 = \begin{bmatrix} 0.02 \\ -0.8 \\ 0.32 \\ 0 \end{bmatrix}$$

$$b_3 = \begin{bmatrix} 0.015 \\ -0.7 \\ 0.26 \\ 0 \end{bmatrix}, \; b_4 = \begin{bmatrix} 0.01 \\ -0.5 \\ 0.18 \\ 0 \end{bmatrix} \tag{2.195}$$

where b_1 is an original actuation vector, and b_2, b_3, and b_4 are added for studying actuator failure compensation. The reference model matrices are

$$A_M = \begin{bmatrix} -0.0638 & -0.9744 & 0.0784 & 0.0406 \\ 0.9180 & -1.0120 & -0.0407 & 0.0350 \\ -3.1780 & 0.7466 & -0.4939 & -0.0142 \\ 0 & 0.0805 & 1 & 0 \end{bmatrix}$$

$$B_M = \begin{bmatrix} 0.0075 & 0.005 \\ -0.4 & -0.3 \\ 0.14 & 0.1 \\ 0 & 0 \end{bmatrix}. \tag{2.196}$$

With $\mathrm{rank}(B) = 2$, we can specify two independent reference inputs. The reference inputs are given as $r_1(t) = 0.005\sin(0.1t)$ rad, $r_2(t) = 0.005$ rad. It can be verified that the matching conditions (2.166)–(2.168) hold for every B_a consisting of two columns of B, that is, two actuator failures can be compensated by our adaptive control design. In this study, we consider the following two actuator failures: u_4 fails at $t = 30$ sec with $\bar{u}_4 = 0.15$ rad, and then u_3 fails at $t = 60$ sec with $\bar{u}_3 = 0.1$ rad.

For initial conditions $x_0 = [0.1, 0.01, -0.01, 0.5]^T$, $x_m(0) = [0, 0, 0, 0.6]^T$, $\Gamma_{1j} = 5I$, $\Gamma_{2j} = I$, $j = 1, \ldots, 4$, and $\gamma_{3j} = 1$, $j = 1, \ldots, 4$, the simulation results are shown in Figures 2.1 and 2.2 for state tracking errors, actuating input $u(t)$, and the evolution of controller parameter $k_{3j}, j = 1, \ldots, 4$.

The system responses are as expected: Signal boundedness and asymptotic state tracking are achieved by the adaptive compensation control scheme. At the time instant when one actuator fails, there is a transient response in the state tracking errors, as the system actuation structure changes, for which the controller parameters (for example, $k_3(t)$ in Figure 2.2) also change to compensate for the actuator failure.

2.6 Concluding Remarks

In this chapter, we address some fundamental issues in adaptive control of systems with unknown actuator failures and present several solutions to the problem of adaptive state feedback control for state tracking in the presence of unknown actuator failures that are constant, or parametrizable time-varying,

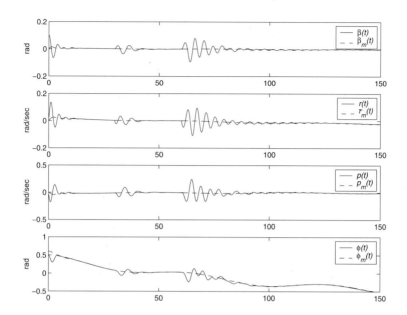

Fig. 2.1. System states $x = [\beta, r, p, \phi]^T$ and $x_m = [\beta_m, r_m, p_m, \phi_m]^T$.

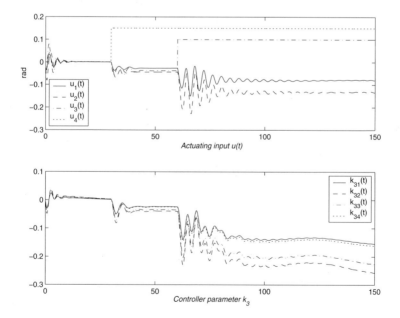

Fig. 2.2. Actuating input $u(t)$ and controller parameter k_3.

or nonparametrizable time-varying. We derive necessary and sufficient conditions for the existence of a nominal actuator failure compensation controller based on the knowledge of the system and failure parameters and we develop the controller structures, parametrizations, and error models. We show that such conditions are also sufficient for solving the adaptive control problem when the knowledge of the system and failure parameters is not available, and we develop the adaptive laws for controller parameter adaptation. We analyze the system stability and tracking performance and demonstrate these properties by simulation results.

The adaptive actuator failure compensation schemes are applicable to systems with up to $m - 1$ (or $m - q$, $1 \leq q \leq m - 1$) failures of the total m actuators, without knowing how many and which actuators have failed, in what manner, and at what time. A complete solution is provided for the adaptive actuator failure compensation problem when both the system parameters and failure parameters are unknown. A novel Lyapunov stability analysis is used to prove the desired closed-loop system properties, in particular to ensure the boundedness of the control signals that are applied to all actuators (including the failed ones).

The ideas and concepts presented in this chapter for the formulation and solutions of the problems are instrumental in developing solutions to other adaptive actuator failure compensation problems using state feedback or output feedback for output tracking, based on relaxed matching conditions, and involving nonlinear dynamics, as reported in the chapters to follow, with additional solutions to some new issues.

Chapter 3

State Feedback Designs for Output Tracking

In Chapter 2, we developed adaptive *state* tracking control schemes for linear time-invariant plants with actuator failures. For a controlled plant $\dot{x}(t) = Ax(t) + Bu(t)$ whose input $u(t)$ may have failed components, to achieve desired plant-model *state* dynamics matching in the presence of actuator failures, it is necessary that there exist constant vectors $k_{s1i}^* \in R^m$ and nonzero constant scalars $k_{s2i}^* \in R$, $i = 1, \ldots, m$, such that $A + b_i k_{s1i}^{*T} = A_M$, $b_i k_{s2i}^* = b_M$, where b_i, $i = 1, \ldots, m$, is the ith column of B, and A_M and b_M are a pair of reference model matrices independent of A and B.

For many applications when only output tracking is the desired control objective, less restrictive conditions on the system matrices A and B are needed for achieving the desired system performance. It is the goal of this chapter to present adaptive control schemes for systems with unknown parameters and unknown failures, achieving closed-loop system stability and asymptotic *output* tracking, with both of the above conditions relaxed. In Section 3.1, an adaptive state feedback compensation control scheme is developed to accommodate actuator failures where some unknown plant inputs are stuck at some fixed values, such as a hydraulic failure, referred to as "lock-in-place" actuator failures. In Section 3.2, the more general problem of adaptive state feedback compensation of both unknown "lock-in-place" and time-varying actuator failures is addressed, based on a theoretical framework for adaptive output rejection of unmatched input disturbances. In both cases, numerous simulation results from Boeing 747 lateral control are presented to illustrate the effectiveness of adaptive actuator failure compensation.

3.1 Designs for Lock-in-Place Failures

In this section, we formulate the actuator failure compensation control problem for plant-model output matching, choose a suitable controller structure, derive the needed design conditions, and develop an adaptive design for plants with unknown dynamics and actuator failure parameters.

3.1.1 Problem Statement

Consider a linear time-invariant plant

$$\dot{x}(t) = Ax(t) + Bu(t), \; y(t) = Cx(t) \tag{3.1}$$

where $A \in R^{n \times n}$, $B \in R^{n \times m}$, $C \in R^{1 \times n}$ are unknown constant parameter matrices, the state vector $x(t) \in R^n$ is available for measurement, $u(t) = [u_1, \ldots, u_m]^T \in R^m$ is the input vector whose components may fail during system operation, and $y(t) \in R$ is the plant output.

We first consider the following type of actuator failures (see (1.2)):

$$u_j(t) = \bar{u}_j, \; t \geq t_j, \; j \in \{1, 2, \ldots, m\} \tag{3.2}$$

where the constant value $\bar{u}_j$ and the failure time instant t_j are unknown.

The basic assumption for the actuator failure compensation problem is

(A3.0) The system (3.1) is so designed that for any up to $m - 1$ actuator failures, the remaining actuators can still achieve a desired control objective, with the knowledge of plant and failure parameters.

Adaptive control task is to adjust the remaining controls (actuators) to achieve the desired system performance when the actuator failure parameters, together with plant parameters, are unknown.

The control objective is to design a feedback control $v(t)$ for the plant (3.1) with actuator failures described by (3.2), such that despite the control error, all closed-loop signals are bounded and the plant output $y(t)$ asymptotically tracks a given reference output $y_m(t)$ from

$$y_m(t) = W_m(s)[r](t), \; W_m(s) = \frac{1}{P_m(s)} \tag{3.3}$$

where $P_m(s)$ is a stable monic polynomial of degree n^*, and $r(t)$ is a bounded and piecewise continuous reference input.

3.1.2 A Plant-Model Output Matching Controller

For state feedback control, we consider the controller structure (2.12):

$$v(t) = v^*(t) = K_1^{*T} x(t) + k_2^* r(t) + k_3^* \tag{3.4}$$

with $K_1^* = [k_{11}^*, \ldots, k_{1m}^*] \in R^{n \times m}$, $k_2^* = [k_{21}^*, \ldots, k_{2m}^*]^T \in R^m$ to be defined for plant-model output matching, and $k_3^* = [k_{31}^*, \ldots, k_{3m}^*]^T \in R^m$ for compensation of the actuation error. If the controlled plant has p failed actuators, that is, $u_j(t) = \bar{u}_j, \; j = j_1, \ldots, j_p$, we have the closed-loop system as

$$\dot{x}(t) = (A + \sum_{j \neq j_1, \ldots, j_p} b_j k_{1j}^{*T}) x(t) + \sum_{j \neq j_1, \ldots, j_p} b_j k_{2j}^* r(t)$$

$$+ \sum_{j \neq j_1, \ldots, j_p} b_j k_{3j}^* + \sum_{j = j_1, \ldots, j_p} b_j \bar{u}_j$$

$$y(t) = C x(t) \tag{3.5}$$

where b_j is the jth column of B,

$$B = [b_1, \ldots, b_m], \ b_j \in R^n, j = 1, \ldots, m. \tag{3.6}$$

For the system (3.5) to match the reference system (3.3), the following matching conditions need to be satisfied by K_1^*, k_2^*, and k_3^*:

$$C(sI - A - \sum_{j \neq j_1, \ldots, j_p} b_j k_{1j}^{*T})^{-1} \sum_{j \neq j_1, \ldots, j_p} b_j k_{2j}^* = W_m(s) \tag{3.7}$$

$$C(sI - A - \sum_{j \neq j_1, \ldots, j_p} b_j k_{1j}^{*T})^{-1} (\sum_{j \neq j_1, \ldots, j_p} b_j k_{3j}^* + \sum_{j = j_1, \ldots, j_p} b_j \bar{u}_j) = 0. \tag{3.8}$$

If the columns b_i, $i = 1, \ldots, m$, of B are parallel to each other, the above conditions can be easily met by some nonunique K_1^*, k_2^*, and k_3^*. For the control problem considered in this chapter, we are interested in the case when b_i, $i = 1, \ldots, m$, are not parallel to each other. In this case, there may be different parametrization schemes suitable for different actuator failure compensation designs. As discussed in Section 1.3.4, if we assume that the redundant actuators in a group have similar physical effects on system dynamic behavior, an equal-actuation (or proportional-actuation) scheme may be used for generating the components of the control signal $v(t)$. Such an actuation scheme also leads to a simple actuator failure compensation control design. With this consideration, we choose the following controller structure:

$$v_1^*(t) = \cdots = v_m^*(t) = k_{11}^{*T} x(t) + k_{21}^* r(t) + k_{31}^* \tag{3.9}$$

for $v^*(t) = [v_1^*(t), \ldots, v_m^*(t)]^T$ in (3.4). This choice implies that the controller parameters in (3.4) have the special forms $k_{1i}^* = k_{11}^*$, $k_{2i}^* = k_{21}^*$, and $k_{3i}^* = k_{31}^*$, for $i = 2, 3, \ldots, m$. Such a controller structure is suitable for plant-model output matching with known parameters as well as for adaptive control for closed-loop stability and asymptotic output tracking with adaptive estimates of the unknown plant and failure parameters.

For this equal-actuation scheme to deal with any up to $m - 1$ failures of the total m actuators, the following assumptions are needed for model reference control based adaptive control designs:

(A3.1) $(A, \sum_{j \neq j_1, \dots, j_p} b_j)$, $\forall p \in \{0, 1, \dots, m - 1\}$, are controllable;

(A3.2) the transfer functions of the systems $(C, A, \sum_{j \neq j_1, \dots, j_p} b_j)$, $\forall p \in \{0, 1, \dots, m - 1\}$, have the same relative degree n^*;

(A3.3) $(C, A, \sum_{j \neq j_1, \dots, j_p} b_j)$, $\forall p \in \{0, 1, \dots, m - 1\}$, are minimum phase; and

(A3.4) for any $p \in \{0, 1, \dots, m - 1\}$, the high frequency gains $CA^{n^*-1} \sum_{j \neq j_1, \dots, j_p} b_j$ have the same sign, which is known:

$$\text{sign}[k_{21}^*] = \text{sign}[CA^{n^*-1} \sum_{j \neq j_1, \dots, j_p} b_j] = \text{constant}. \qquad (3.10)$$

Under Assumptions (A3.1)–(A3.3), for each actuator failure pattern (including the no-failure case), $u_j(t) = \bar{u}_j$, $j = j_1, \dots, j_p$, the matching condition (3.7) can be satisfied by

$$C(sI - A - \sum_{j \neq j_1, \dots, j_p} b_j k_{11}^{*T})^{-1} \sum_{j \neq j_1, \dots, j_p} b_j k_{21}^* = W_m(s) \qquad (3.11)$$

for some $k_{11}^* \in R^n$ and $k_{21}^* \in R$, that is, under (A3.1) and (A3.2), the vector gain k_{11}^* places the closed-loop poles at the poles of $W_m(s)$ and the zeros of the open-loop system $(C, A, \sum_{j \neq j_1, \dots, j_p} b_j)$, which are stable by (A3.3), and the scalar gain $k_{21}^* = k_p^{-1}$, where $k_p = CA^{n^*-1} \sum_{j \neq j_1, \dots, j_p} b_j$.

For the matching condition (3.8), treating $k_{31}^* \in R$ as an input and $\bar{u}_j$ as disturbances, we have

$$f_p(t) \triangleq C(sI - A - \sum_{j \neq j_1, \dots, j_p} b_j k_{11}^{*T})^{-1} [\sum_{j \neq j_1, \dots, j_p} b_j k_{31}^* + \sum_{j = j_1, \dots, j_p} b_j \bar{u}_j](t)$$

$$= C(sI - A - \sum_{j \neq j_1, \dots, j_p} b_j k_{11}^{*T})^{-1} [\sum_{j = j_1, \dots, j_p} b_j \bar{u}_j](t)$$

$$+ W_m(s) \begin{bmatrix} k_{31}^* \\ k_{21}^* \end{bmatrix}(t) \qquad (3.12)$$

whose s-domain expression is

$$F_p(s) = C(sI - A - \sum_{j \neq j_1, \dots, j_p} b_j k_{11}^{*T})^{-1} \left(\sum_{j \neq j_1, \dots, j_p} b_j \frac{k_{31}^*}{s} + \sum_{j = j_1, \dots, j_p} b_j \frac{\bar{u}_j}{s} \right)$$

$$= C(sI - A - \sum_{j \neq j_1, \dots, j_p} b_j k_{11}^{*T})^{-1} \sum_{j = j_1, \dots, j_p} b_j \frac{\bar{u}_j}{s} + W_m(s) \frac{k_{31}^*}{k_{21}^* s}. \qquad (3.13)$$

Since all zeros of $\det(sI - A - \sum_{j \neq j_1, \dots, j_p} b_j k_{11}^{*T})$ are stable, there exists a k_{31}^* such that $\lim_{s \to 0} sF_p(s) = 0$, that is, in the time domain, $\lim_{t \to \infty} f_p(t) = 0$ exponentially. This asymptotic property is crucial for the parametrization of an actuator failure compensation design for the control law (3.9), which ensures that $\lim_{t\to\infty}(y(t) - y_m(t)) = 0$, as we just verified.

As in Section 2.1.3, we let (T_i, T_{i+1}), $i = 0, 1, \ldots, m_0$, with $T_0 = 0$, be the time intervals on which the actuator failure pattern is fixed, that is, actuators only fail at time T_i, $i = 1, \ldots, m_0$. Since there are m actuators and at least one of them does not fail, we have $m_0 < m$ and $T_{m_0+1} = \infty$. Then, at time T_j, $j = 1, \ldots, m_0$, the unknown plant-model matching parameters k_{11}^*, k_{21}^*, and k_{31}^* change their values such that

$$k_{11}^* = k_{11(i)}^*, \; k_{21}^* = k_{21(i)}^*, \; k_{31}^* = k_{31(i)}^*, \; t \in (T_i, T_{i+1}) \qquad (3.14)$$

for $i = 0, 1, \ldots, m_0$, that is, the plant-model matching parameters k_{11}^*, k_{21}^*, and k_{31}^* are all piecewise constant, because the plant has different characterizations under different failure conditions so that the plant-model matching parameters are also different. At each time when the piecewise constant parameters change because of the actuator failure pattern change, a transient component occurs in the system response and the effect of such a transient response (there are a finite number of them during system operation) converges to zero exponentially, which follows from the analysis in [132].

Remark 3.1.1. When there are p failed actuators, the remaining $m - p$ actuators need to deliver desired controls to the controlled plant for output matching or tracking as well as for failure compensation. The design of the $m - p$ control signals is crucial for this task, given that in the adaptive control case it is not known which $m - p$ actuators of the total m actuators are functional and how much the failure values are. The equal-actuation based control design in (3.9) is a chosen design that is able to fulfill the task. For the design (3.4), Assumptions (A3.1)–(A3.4) are sufficient for a nonadaptive plant-model output matching control scheme, as well as for an adaptive plant-model output tracking control scheme, in the presence of up to $m - 1$ actuator failures, as shown in this chapter. These assumptions are, in fact, also necessary for desired system matching performance, under the controller structure (3.9), even in the case when the actuator failure pattern (when, how many, and how much failure) is known; that is, in order to satisfy the desired output matching and failure compensation equations (3.7) and (3.8), Assumptions (A3.1)–(A3.3) have to be satisfied. □

3.1.3 Adaptive Control Design

Now we develop an adaptive control scheme for the system (3.1) with unknown parameters A and B, and with unknown actuator failures (3.2). In this case, the parameters k_{11}^*, k_{21}^*, and k_{31}^* in (3.9) are unknown. As an adaptive version of (3.9), we use the controller structure

$$v_1(t) = v_2(t) = \cdots = v_m(t) = k_{11}^T(t)x(t) + k_{21}(t)r(t) + k_{31}(t) \qquad (3.15)$$

where $k_{11}(t) \in R^n$, $k_{21}(t) \in R$, and $k_{31}(t) \in R$ are the estimates of the unknown parameters k_{11}^*, k_{21}^*, and k_{31}^*. The resulting closed-loop system is

$$\begin{aligned}
\dot{x}(t) = {}& (A + \sum_{j \neq j_1,\ldots,j_p} b_j k_{11}^{*T})x(t) + \sum_{j \neq j_1,\ldots,j_p} b_j(k_{21}^* r(t) + k_{31}^*) \\
& + \sum_{j=j_1,\ldots,j_p} b_j \bar{u}_j + \sum_{j \neq j_1,\ldots,j_p} b_j(\tilde{k}_{11}^T x(t) + \tilde{k}_{21} r(t) + \tilde{k}_{31}) \quad (3.16)
\end{aligned}$$

where $\tilde{k}_{11} = k_{11} - k_{11}^*$, $\tilde{k}_{21} = k_{21} - k_{21}^*$, and $\tilde{k}_{31} = k_{31} - k_{31}^*$. In view of (3.5), (3.11), (3.12), and (3.14), the closed-loop system output can be expressed as

$$\begin{aligned}
y(t) = {}& y_m(t) + W_m(s)\left[\frac{1}{k_{21}^*}\left(\tilde{k}_{11}^T x + \tilde{k}_{21} r + \tilde{k}_{31}\right)\right](t) \\
& + f_p(t) + \epsilon_0(t) + \epsilon_t(t) \qquad (3.17)
\end{aligned}$$

where $f_p(t)$ is defined in (3.12) such that $\lim_{t\to\infty} f_p(t) = 0$ exponentially, $\epsilon_0(t)$ is related to the system initial conditions, and $\epsilon_t(t)$ is related to the transient system response equivalent to the effect of the piecewise constant parameter variations. Based on the analysis of [132], it can be shown that $\lim_{t\to\infty} \epsilon_0(t) = 0$ and $\lim_{t\to\infty} \epsilon_t(t) = 0$, both exponentially.

Ignoring these exponentially decaying terms, we have the tracking error

$$e(t) = y(t) - y_m(t) = W_m(s)\frac{1}{k_{21}^*}[\tilde{\theta}^T \omega](t) \qquad (3.18)$$

where $\tilde{\theta}(t) = \theta(t) - \theta^*$ with $\theta(t) = [k_{11}^T, k_{21}, k_{31}]^T$ being the estimate of $\theta^* = [k_{11}^{*T}, k_{21}^*, k_{31}^*]^T$, and $\omega = [x^T, r, 1]^T$.

Introducing the auxiliary signals

$$\begin{aligned}
\zeta(t) &= W_m[\omega](t) & (3.19) \\
\xi(t) &= \theta^T(t)\zeta(t) - W_m(s)[\theta^T \omega](t) & (3.20) \\
\epsilon(t) &= e(t) + \rho(t)\xi(t) & (3.21)
\end{aligned}$$

where $\rho(t)$ is the estimate of $\rho^* = 1/k_{21}^*$, we choose the adaptive laws as

$$\begin{aligned}
\dot{\theta}(t) &= -\frac{\text{sign}[k_{21}^*]\Gamma\zeta(t)\epsilon(t)}{1 + \zeta^T\zeta + \xi^2}, \quad \Gamma = \Gamma^T > 0 & (3.22) \\
\dot{\rho}(t) &= -\frac{\gamma\xi(t)\epsilon(t)}{1 + \zeta^T\zeta + \xi^2}, \quad \gamma > 0. & (3.23)
\end{aligned}$$

Note that k_{21}^* is the scalar gain defined in (3.11), and by Assumption (A3.4), the sign of k_{21}^* is fixed for different failure patterns and is known.

To analyze the stability and tracking performance of the adaptive control system, we define the positive definite function

$$V(\tilde{\theta}, \tilde{\rho}) = \frac{1}{2}\left(|\rho^*|\tilde{\theta}^T \Gamma^{-1}\tilde{\theta} + \gamma^{-1}\tilde{\rho}^2\right), \; \tilde{\theta} = \theta - \theta^*, \; \tilde{\rho} = \rho - \rho^*. \quad (3.24)$$

From (3.18)–(3.21), we obtain

$$\epsilon(t) = \rho^*\tilde{\theta}^T(t)\zeta(t) + \tilde{\rho}(t)\xi(t) + \epsilon_p(t), \; t \in (T_i, T_{i+1}), \; i = 0, 1, \ldots, m_0 \quad (3.25)$$

where

$$\epsilon_p(t) = \epsilon_{p1}(t) + \epsilon_{p2}(t) \quad (3.26)$$

$$\epsilon_{p1}(t) = \rho^*(\theta^{*T}(t)\zeta(t) - W_m(s)[\theta^{*T}\omega](t)) \quad (3.27)$$

$$\epsilon_{p2}(t) = W_m(s)\rho^*[\tilde{\theta}^T\omega](t) - \rho^*W_m(s)[\tilde{\theta}^T\omega](t). \quad (3.28)$$

Before evaluating the time derivative of V, we first show that

$$\lim_{t\to\infty} \epsilon_p(t) = 0 \text{ exponentially.} \quad (3.29)$$

Note that $p = m_0$ for $t \geq T_{m_0}$, that is, there are m_0 failed actuators for $t \in (T_{m0}, \infty)$. Let the impulse response function of $W_m(s)$ be $w_m(t)$. Then, for $t \in (T_i, T_{i+1})$, we can express (3.27) as

$$\begin{aligned}
\epsilon_{p_1}(t) &= \rho^*\left(\theta^{*T}(t)\int_0^t w_m(t-\tau)\omega(\tau)d\tau \right.\\
&\quad \left. - \int_0^t w_m(t-\tau)\theta^{*T}(\tau)\omega(\tau)d\tau\right)\\
&= \rho^*\left(\theta^{*T}(t)\int_0^{T_i} w_m(t-\tau)\omega(\tau)d\tau \right.\\
&\quad \left. - \int_0^{T_i} w_m(t-\tau)\theta^{*T}(\tau)\omega(\tau)d\tau\right).
\end{aligned} \quad (3.30)$$

For the second equality of (3.30), we used the fact that $\theta^*(t)$ is constant for $t \in (T_i, T_{i+1})$, that is,

$$\rho^*\left(\theta^{*T}(t)\int_{T_i}^t w_m(t-\tau)\omega(\tau)d\tau - \int_{T_i}^t w_m(t-\tau)\theta^{*T}(\tau)\omega(\tau)d\tau\right) = 0. \quad (3.31)$$

Therefore, for $t > T_{m_0}$, we have

$$\epsilon_{p1}(t) = \rho^*\int_0^{T_{m_0}} w_m(t-\tau)(\theta^*_{(m_0)} - \theta^*(\tau))^T\omega(\tau)d\tau. \quad (3.32)$$

Since the reference model system $W_m(s)$ is stable, we have $|w_m(t - \tau)| \leq \beta e^{-\alpha(t-\tau)}$ for some $\alpha > 0$, $\beta > 0$, so that

$$|\epsilon_{p1}(t)| \leq \beta|\rho^*|e^{-\alpha t} \int_0^{T_{m0}} e^{\alpha \tau} |(\theta^*_{(m_0)} - \theta^*(\tau))^T \omega(\tau)| d\tau. \tag{3.33}$$

As $\theta^*(t) = \theta^*_{(m_0)}$ is constant for $t > T_{m_0}$, $\theta^*(t)$ is piecewise constant in $(0, T_{m_0})$, T_{m_0} is finite, and $\omega(t)$ is bounded in $(0, T_{m_0})$, there exists a constant $a_1 > 0$ such that

$$\int_0^{T_{m0}} e^{\alpha \tau} |(\theta^*_{(m_0)} - \theta^*(\tau))^T \omega(\tau)| d\tau \leq a_1 \tag{3.34}$$

and $|\epsilon_{p1}(t)| \leq a_1 \beta |\rho^*| e^{-\alpha t}$.

Similarly, we have

$$
\begin{aligned}
\epsilon_{p2}(t) &= \int_0^{T_{m0}} w_m(t - \tau)(\rho^*(\tau) - \rho^*_{(m_0)})\tilde{\theta}^T(\tau)\omega(\tau) d\tau \\
&\leq \beta e^{-\alpha t} \int_0^{T_{m0}} e^{\alpha \tau} |(\rho^*_{(m_0)} - \rho^*(\tau))\tilde{\theta}^T(\tau)\omega(\tau)| d\tau. \tag{3.35}
\end{aligned}
$$

As $\rho^*(t) = \rho^*_{(m_0)}$ is constant for $t > T_{m_0}$, $\rho^*(t)$ is piecewise constant in $(0, T_{m_0})$, T_{m_0} is finite, and $\tilde{\theta}^T(t)\omega(t)$ is bounded in $(0, T_{m_0})$, there exists a constant a_2 such that

$$\int_0^{T_{m0}} e^{\alpha \tau} |(\rho^*_{(m_0)} - \rho^*(\tau))\tilde{\theta}^T(\tau)\omega(\tau)| d\tau \leq a_2 \tag{3.36}$$

and $|\epsilon_{p2}(t)| \leq a_2 \beta e^{-\alpha t}$. Therefore, $\lim_{t \to \infty} \epsilon_p(t) = 0$ exponentially.

With (3.19)–(3.23) and (3.25), ignoring the exponentially decaying term $\epsilon_p(t)$, which does not destabilize the gradient adaptive laws (3.22) and (3.23), the time derivative of V along (3.22) and (3.23) is

$$\dot{V} = -\frac{\epsilon^2(t)}{1 + \zeta^T(t)\zeta(t) + \xi^2(t)} \leq 0, \ t \in (T_i, T_{i+1}), \ i = 0, 1, \ldots, m_0. \tag{3.37}$$

It is important to note that $V(\cdot)$ as a function of t is not continuous because $\theta^* = [k_{11}^{*T}, k_{21}^*, k_{31}^*]^T = [k_{11(i)}^{*T}, k_{21(i)}^*, k_{1(i)}^*]^T$ is a piecewise constant parameter vector as described in (3.14). Since there are only a finite number of failures in the system, it follows that T_{m_0} is finite and

$$\dot{V} = -\frac{\epsilon^2(t)}{1 + \zeta^T(t)\zeta(t) + \xi^2(t)} \leq 0, \ t \in (T_{m_0}, \infty) \tag{3.38}$$

which implies that $\theta(t)$, $\rho(t) \in L^\infty$, $\epsilon(t)/\sqrt{1 + \zeta^T(t)\zeta(t) + \xi^2(t)} \in L^2 \cap L^\infty$, $\dot{\theta}(t) \in L^2 \cap L^\infty$, and $\rho(t) \in L^2 \cap L^\infty$. Based on this, closed-loop stability and asymptotic tracking can be proved as follows.

Theorem 3.1.1. *The adaptive controller (3.15), with the adaptive laws (3.22) and (3.23), applied to the system (3.1) with actuator failures (3.2) guarantees that all closed-loop signals are bounded and the tracking error $e(t) = y(t) - y_m(t)$ goes to zero as t goes to infinity.*

Proof (outline): The standard properties of the adaptive laws: $\theta(t), \rho(t) \in L^\infty$, $\epsilon(t)/\sqrt{1 + \zeta^T(t)\zeta(t) + \xi^2(t)} \in L^2 \cap L^\infty$, and $\dot{\theta}(t) \in L^2 \cap L^\infty$, ensure that the closed-loop system has a small loop gain so that it is stable in the sense that all signals are bounded [118]. Then, it follows that $\epsilon(t) \in L^2 \cap L^\infty$ and that $\xi(t) \in L^2 \cap L^\infty$ as $\dot{\theta}(t) \in L^2 \cap L^\infty$, so that $e(t) = \epsilon(t) - \rho(t)\xi(t) \in L^2 \cap L^\infty$. Finally, it follows from (3.18) that $\dot{e}(t)$ is bounded so that $\lim_{t \to \infty} e(t) = 0$.

A key feature of the adaptive control system is that during any time interval when the actuator failure pattern does not change, the system is completely parametrized so that based on the adaptive law properties the tracking error converges (to zero over the last time interval where no failure occurs). The details of the proof are given in [121].   ▽

3.1.4 Boeing 747 Lateral Control Simulation I

As an application example, we use the linearized lateral dynamic model of a Boeing 747 airplane ([37], p. 686) as the controlled plant. The original aircraft dynamic model is augmented with two actuation vectors $b_2 u_2$ and $b_3 u_3$ representing two additional rudder pieces for the study of actuator (rudder) failure compensation.

The plant with augmented actuation vectors is described by

$$\dot{x}(t) = Ax(t) + Bu(t), \ y(t) = Cx(t) \tag{3.39}$$

$$A = \begin{bmatrix} -0.0558 & -0.9968 & 0.0802 & 0.0415 \\ 0.598 & -0.115 & -0.0318 & 0 \\ -3.05 & 0.388 & -0.465 & 0 \\ 0 & 0.0805 & 1 & 0 \end{bmatrix} \tag{3.40}$$

$$B = [b_1, b_2, b_3], \ C = \begin{bmatrix} 0 & 1 & 0 & 0 \end{bmatrix} \tag{3.41}$$

$$b_1 = \begin{bmatrix} 0.00729 \\ -0.475 \\ 0.153 \\ 0 \end{bmatrix}, \ b_2 = \begin{bmatrix} 0.01 \\ -0.5 \\ 0.2 \\ 0 \end{bmatrix}, \ b_3 = \begin{bmatrix} 0.005 \\ -0.3 \\ 0.1 \\ 0 \end{bmatrix} \tag{3.42}$$

where $x(t) = [\beta, y_r, p, \phi]^T$, β is the side-slip angle, y_r is the yaw rate, p is the roll rate, ϕ is the roll angle, y is the plant output, which is the yaw rate y_r in this case, and u is the control input vector that contains three control signals $u = [u_1, u_2, u_3]^T$ to represent three rudder servos from a three-piece rudder for achieving compensation in the presence of actuator failures.

We simulated the case of two actuator failures, one fails at $t = 50$ sec, and one fails at $t = 100$ sec. We considered two failure patterns: (i) When an actuator fails, it jumps to another position, that is, the failure value is different from the control value as the failure occurs. In this simulation study, the failure pattern is modeled as

$$u_2(t) = -0.02 \, \text{rad}, \quad t \geq 50 \, \text{sec} \tag{3.43}$$

$$u_3(t) = 0.03 \, \text{rad}, \qquad t \geq 100 \, \text{sec}. \tag{3.44}$$

(ii) When an actuator fails, it is stuck at the position where it fails. In this simulation, the failures are modeled as

$$u_2(t) = u_2(50), \quad t \geq 50 \, \text{sec} \tag{3.45}$$

$$u_3(t) = u_3(100), \quad t \geq 100 \, \text{sec}. \tag{3.46}$$

Both patterns may happen in real-life systems, and we may expect that a failure as in pattern (i) may introduce a bigger transient behavior than that in failure pattern (ii). For both actuator failure patterns, the failures are assumed to be unknown to the adaptive controller.

The reference model was chosen as the $W_m(s) = 1/s + 3$, and we did simulation for two different reference inputs: $r(t) = 0.03$, and $r(t) = 0.03 \sin(0.1t)$. The adaptive gains used in the simulations are $\Gamma = 20I$, $\gamma = 2$. The initial conditions are $y_m(0) = 0$, $y(0) = -0.015$, $\theta(0) = 0.8\theta^*$ (θ^* is the matching parameter vector when there is no actuator failure), and $\rho(0) = 0$.

The simulation results for all the four cases are shown in Figures 3.1–3.4, including the plant output $y(t)$, reference output $y_m(t)$, tracking error $e(t)$, and the three control input signals $u_1(t), u_2(t)$, and $u_3(t)$.

The simulation results verify the effectiveness of the desired closed-loop system performance. As shown from the simulation results (Figures 3.1–3.4), at the beginning of simulation, there is a tracking error due to the initial conditions, and this tracking error goes to zero as time elapses. At each time instant when a failure occurs, there is a transient behavior resulting from the failure and the mismatch of the controller parameters. Such transient behavior decays very fast as the controller parameters are adjusted online, and asymptotic output tracking is achieved. We may also observe from the figures that a failure as in pattern (i) introduces a much larger transient than that in pattern (ii), which is what we expected.

Thus far, we have developed an adaptive control scheme for systems with unknown actuator failures. This scheme, when implemented with the true parameters, ensures desired plant-model matching in the presence of actuator

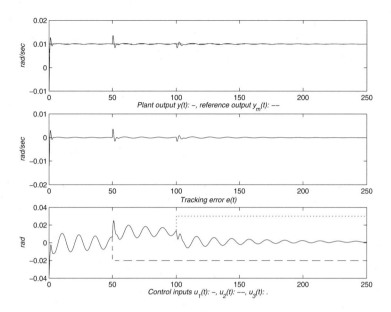

Fig. 3.1. System response for failure pattern (i), $r(t) = 0.03$.

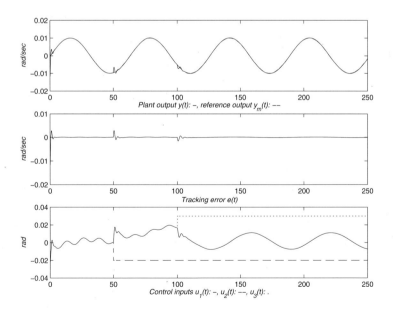

Fig. 3.2. System response for failure pattern (i), $r(t) = 0.03\sin(0.1t)$.

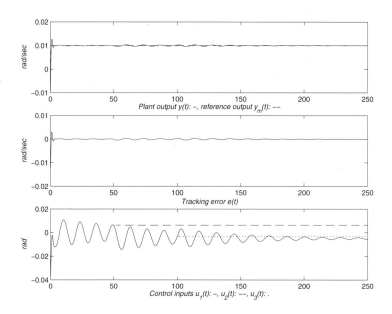

Fig. 3.3. System response for failure pattern (ii), $r(t) = 0.03$.

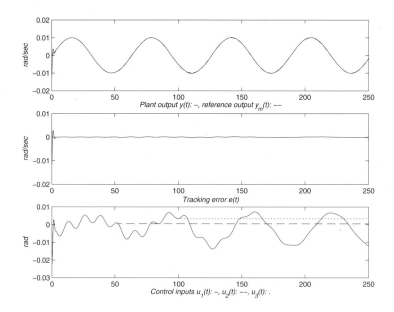

Fig. 3.4. System response for failure pattern (ii), $r(t) = 0.03\sin(0.1t)$.

failures, and, when implemented with adaptive parameter estimates, ensures asymptotic output tracking in the presence of unknown plant parameters and unknown actuator failure parameters. The design conditions on the plant matrices (A, B) are relaxed as compared with a state tracking design, so that adaptive actuator failure compensation is applicable to a larger class of systems. Simulation results verified the desired performance of the developed adaptive actuator failure compensation design.

3.2 Designs for Varying Failures

Actuator failures cause input disturbances to the controlled system in addition to structural changes. To adaptively reject the effect of certain unmatched input disturbances on the output of a linear time-invariant system, a transfer function matching condition is needed. In this section, a lemma that presents a basic property of linear systems is first derived to characterize system conditions for such transfer function matching. An adaptive disturbance rejection control scheme is then developed for such systems with uncertain dynamics parameters and disturbance parameters. This adaptive control technique is applicable to control of systems with possibly time-varying actuator failures whose failure values, failure time instants, and failure patterns are unknown. An adaptive actuator failure compensation scheme is developed, which ensures closed-loop stability and asymptotic output tracking, in the presence of any up to $m - 1$ uncertain failures of the total m actuators. Desired adaptive system performance is verified by simulation results.

3.2.1 Problem Statement

Consider the linear time-invariant plant

$$\dot{x}(t) = Ax(t) + b_i u_i(t) + b_j u_j(t), \; y(t) = cx(t) \tag{3.47}$$

where $A \in R^{n \times n}$, $b_i, b_j \in R^n$, $c \in R^{1 \times n}$ are constant parameter matrices, the state vector $x(t) \in R^n$ is available for measurement, $u_i(t)$ and $u_j(t) \in R$ are two actuating inputs, and $y(t) \in R$ is the plant output. The basic problem is to design one actuating (control) input $u_i(t)$ to cancel the effect of the other actuating (disturbance) input $u_j(t)$ on the plant output $y(t)$.

As in Section 3.1, let the desired behavior of $y(t)$ be given by the output $y_m(t)$ of the reference model system

$$y_m(t) = W_m(s)[r](t), \ W_m(s) = \frac{1}{P_m(s)} \tag{3.48}$$

where $P_m(s)$ is a stable monic polynomial of degree n^*, and $r(t)$ is bounded. For the case of known plant parameters, the ideal controller structure is

$$u_i(t) = u_i^*(t) = k_{1i}^{*T} x(t) + k_{2i}^* r(t) + f_i^*(t) \tag{3.49}$$

with $k_{1i}^* \in R^n$, k_{2i}, $f_i^*(t) \in R$, which leads to the closed-loop system

$$\dot{x}(t) = (A + b_i k_{1i}^{*T})x(t) + b_i k_{2i}^* r(t) + b_i f_i^*(t) + b_j u_j(t), \ y(t) = cx(t). \tag{3.50}$$

If (A, b_i) is controllable and all zeros of (c, A, b_i) are stable, there exist k_{1i}^* and k_{2i}^* such that

$$c(sI - A - b_i k_{1i}^{*T})^{-1} b_i k_{2i}^* = W_m(s) \tag{3.51}$$

with stable zero-pole cancellations, that is, $A + b_i k_{1i}^{*T}$ has all stable eigenvalues. With this choice of k_{1i}^* and k_{2i}^*, the closed-loop system (3.50) has the frequency-domain representation

$$y(s) = y_m(s) + c(sI - A - b_i k_{1i}^{*T})^{-1} x(0) + \frac{1}{k_{2i}^*} W_m(s) f_i^*(s)$$

$$+ c(sI - A - b_i k_{1i}^{*T})^{-1} b_j u_j(s). \tag{3.52}$$

Based on (3.52), rejection of three classes of disturbances may be studied:

(i) $u_j(t) = \bar{u}_j \in R$ is a constant (that is, $u_j(s) = \bar{u}_j/s$);
(ii) $u_j(t) = \sum_{l=1}^{q}(\alpha_l \sin \omega_l t + \beta_l \cos \omega_l t)$ for some constants α_l, β_l, and ω_l (that is, $u_j(s) = \sum_{l=1}^{q}(\alpha_l \omega_l/(s^2 + \omega_l^2) + \beta_l s/(s^2 + \omega_l^2)))$; and
(iii) $u_j(t) = \sum_{l=1}^{q} \gamma_l g_l(t)$, where $g_l(t)$, $l = 1, 2, \ldots, q$, are some known, bounded, and piecewise continuous functions, and γ_l, $l = 1, 2 \ldots, q$, are constant parameters.

For class (i) disturbances, with the choice of $f_i^*(t) = \lambda^*$ (that is, $f_i^*(s) = \lambda^*/s$), for any $b_j \in R^n$ and $\bar{u}_j \in R$ (that is, $u_j(s) = \bar{u}_j/s$), there exists a constant λ^* such that

$$\lim_{s \to 0} s \left(\frac{1}{k_{2i}^*} W_m(s) f_i^*(s) + c(sI - A - b_i k_{1i}^{*T})^{-1} b_j u_j(s) \right) = 0. \tag{3.53}$$

For class (ii) disturbances, with the choice of $f_i^*(t) = \sum_{l=1}^{q}(c_l \sin \omega_l t + d_l \cos \omega_l t)$ (that is, $f_i^*(s) = \sum_{l=1}^{q}(c_l \omega_l/(s^2 + \omega_l^2) + d_l s/(s^2 + \omega_l^2)))$, for any b_j and $u_j(t)$ in the specified form, there exist constants c_l and d_l, $l = 1, 2, \ldots, q$, such that (3.53) also holds. In both cases the function $s(W_m(s)f_i^*(s)/k_{2i}^*$ ⏐

$c(sI - A - b_i k_{1i}^{*T})^{-1} b_j u_j(s))$ does not have any pole in $Re[s] \geq 0$. There-fore, from the final value theorem of the Laplace transform, the time do-main function of $W_m(s) f_i^*(s)/k_{2i}^* + c(sI - A - b_i k_{1i}^{*T})^{-1} b_j u_j(s)$ converges to zero exponentially fast as time t goes to infinity, which means that $\lim_{t \to \infty}(y(t) - y_m(t)) = 0$ exponentially fast, that is, output rejection of the *unmatched* input disturbance $u_j(t)$ of class (i) and class (ii) can be achieved. When the plant parameters A, b_i, c and the disturbance param-eters $\bar{u}_j$ or α_l and β_l, $l = 1, 2, \ldots, q$, are unknown, adaptive estimates of the controller parameters in (3.49) can be generated from some stable adap-tive laws such that closed-loop stability is ensured and asymptotic tracking $\lim_{t \to \infty}(y(t) - y_m(t)) = 0$ is achieved.

For class (iii) disturbances, since $g_l(t)$, $l = 1, 2, \ldots, q$, may be arbitrary, the existence of an $f_i^*(t)$ of the same components $g_l(t)$ to meet the above final value theorem condition may be impossible unless the following transfer function matching is ensured:

$$\frac{1}{k_{2i}^*} W_m(s) k_{3i}^* + c(sI - A - b_i k_{1i}^{*T})^{-1} b_j = 0 \qquad (3.54)$$

for some constant $k_{3i}^* \in R$. Under this condition, the choice of $f_i^*(t) = k_{3i}^* u_j(t)$ in (3.49) (that is, $f_i^*(s) = k_{3i}^* u_j(s)$) ensures that the plant-model output matching $y(s) = y_m(s)$ is achieved for $x(0) = 0$, and $\lim_{t \to \infty}(y(t) - y_m(t)) = 0$ exponentially, for $x(0) \neq 0$. Clearly, if $b_j = \alpha_{ji} b_i$ for some constant $\alpha_{ji} \in R$ (that is, the case of *matched* input disturbances), then k_{3i}^* exists to satisfy (3.54), which is crucial for adaptive rejection of the class (iii) disturbances $u_j(t)$ when the parameters γ_l, $l = 1, 2 \ldots, q$, are unknown.

What if b_i and b_j are not parallel? This is the first question to be answered in this section: What class of (c, A, b_i, b_j) can satisfy the matching equations (3.51) and (3.54) simultaneously? In Section 3.2.2, we derive a necessary and sufficient condition for meeting (3.51) and (3.54) simultaneously. Based on this condition we develop an adaptive scheme in Section 3.2.3 for rejecting the class (iii) disturbances $u_j(t)$ with unknown parameters γ_l, $l = 1, 2, \ldots, q$. In Section 3.2.4, we use this adaptive disturbance rejection result to derive a solution to the actuator failure compensation problem: Both $u_i(t)$ and $u_j(t)$ are subject to a failure model described by

$$u_k(t) = \bar{u}_k + \sum_{l=1}^{q} \bar{d}_{kl} f_{kl}(t), \ t \geq t_k, \ k \in \{i, j\} \qquad (3.55)$$

such that either $u_i(t)$ or $u_j(t)$ fails as described by (3.55) or both $u_i(t)$ and $u_j(t)$ do not fail, where $\bar{u}_k \in R$ is an unknown constant, $\bar{d}_{kl} \in R$ are some unknown constants, and $f_{kl}(t) \in R$ are some known bounded signals, $k = i, j$,

$l = 1, 2, \ldots, q$. The control objective of failure compensation is to design an adaptive control scheme for the plant (3.47), which ensure closed-loop stability and output tracking, for all three possible unknown situations: (a) $u_i(t) = v_i(t)$ and $u_j(t) = \bar{u}_j + \sum_{l=1}^{q} \bar{d}_{jl} f_{jl}(t)$; (b) $u_j(t) = v_j(t)$ and $u_i(t) = \bar{u}_i + \sum_{l=1}^{q} \bar{d}_{il} f_{il}(t)$; and (c) $u_i(t) = v_i(t)$ and $u_j(t) = v_j(t)$, where $v_i(t)$ and $v_j(t)$ are the applied control signals generated from an adaptive feedback design. In Section 3.2.5, we discuss extensions of these results.

3.2.2 A Lemma for Output Matching

We first derive the following lemma, which gives a characterization of the set (c, A, b_i, b_j) for the existence of $k_{1i}^* \in R^n$ and $k_{2i}^*, k_{3i}^* \in R$ to satisfy the matching equations (3.51) and (3.54).

Lemma 3.2.1. *Assume that (A, b_i) is controllable. There exist constant $k_{1i}^* \in R^n$, $k_{2i}^*, k_{3i}^* \in R$ such that*

$$c(sI - A - b_i k_{1i}^{*T})^{-1} b_i k_{2i}^* = W_m(s) = \frac{1}{P_m(s)} \tag{3.56}$$

$$\frac{1}{k_{2i}^*} W_m(s) k_{3i}^* + c(sI - A - b_i k_{1i}^{*T})^{-1} b_j = 0 \tag{3.57}$$

where $P_m(s)$ is a monic polynomial of degree n^*, if and only if the two systems (c, A, b_i) and (c, A, b_j) have the same relative degree n^*.

Proof: *Sufficiency*:   Given that (A, b_i) is controllable, and (c, A, b_i) and $W_m(s) = 1/P_m(s)$ have the same relative degree n^*, from pole placement theory, there exist constant $k_{1i}^* \in R^n$ and $k_{2i}^* = 1/(cA^{n^*-1} b_i)$ such that (3.56) is satisfied. The main task now is to use the condition

$$c(sI - A - b_i k_{1i}^{*T})^{-1} b_i = \alpha_{ii} W_m(s), \quad \alpha_{ii} = cA^{n^*-1} b_i, \quad W_m(s) = \frac{1}{P_m(s)} \tag{3.58}$$

to show the existence of a $k_{3i}^* \in R$ such that (3.57) is met, i.e., for some α_{ij},

$$c(sI - A - b_i k_{1i}^{*T})^{-1} b_j = \alpha_{ij} W_m(s) \tag{3.59}$$

such that k_{3i}^* can be chosen as $k_{3i}^* = -\alpha_{ij}/\alpha_{ii}$.
   For $\bar{A} = A + b_i k_{1i}^{*T} \in R^{n \times n}$, $c \in R^{1 \times n}$, and $b \in R^{n \times 1}$, the identity

$$c(sI - \bar{A})^{-1} b = \frac{N(s)}{\det(sI - \bar{A})} \tag{3.60}$$

with polynomials $\det(sI - \bar{A})$ and $N(s)$ given by

$$\det(sI - \bar{A}) = s^n + a_n s^{n-1} + \cdots + a_2 s + a_1 \tag{3.61}$$

$$N(s) = s^{n-1}cb + s^{n-2}(a_n cb + c\bar{A}b) + \cdots$$
$$+ s^{n-n^*}(a_{n-n^*+2}cb + a_{n-n^*+3}c\bar{A}b + \cdots + a_n c\bar{A}^{n^*-2}b + c\bar{A}^{n^*-1}b)$$
$$+ \cdots + s(a_3 cb + a_4 c\bar{A}b + \cdots + a_n c\bar{A}^{n-3}b + c\bar{A}^{n-2}b)$$
$$+ a_2 cb + a_3 c\bar{A}b + \cdots + a_n c\bar{A}^{n-2}b + c\bar{A}^{n-1}b \qquad (3.62)$$

which follows from the resolvent formula [34], will be used to simplify some polynomial equations in deriving the desired result (3.59).

Since $c(sI-A)^{-1}b_i$ and $c(sI-A)^{-1}b_j$ have the same relative degree n^*, we have that $cA^k b_i = 0$, $cA^k b_j = 0$, $k = 0,\ldots,n^*-2$, and $cA^{n^*-1}b_i \neq 0$, $cA^{n^*-1}b_j \neq 0$. Without loss of generality, we let $cA^{n^*-1}b_i = 1$ (which implies that $\alpha_{ii} = 1$ in (3.58)), and set $\bar{A}$, c, and b_i as

$$\bar{A} = A + b_i k_{1i}^{*T} = \begin{bmatrix} 0 & 1 & 0 & 0 & \cdots & 0 \\ 0 & 0 & 1 & 0 & \cdots & 0 \\ & & \cdots & \cdots & & \\ 0 & 0 & \cdots & 0 & 0 & 1 \\ -a_1 & -a_2 & \cdots & -a_{n-2} & -a_{n-1} & -a_n \end{bmatrix} \in R^{n\times n}$$
$$c = [c_1, c_2, \ldots, c_{n-n^*}, 1, 0, \ldots, 0] \in R^{1\times n}$$
$$b_i = [0,\ldots,0,1]^T \in R^{n\times 1} \qquad (3.63)$$

because (A, b_i) is controllable. From (3.58) with $\alpha_{ii} = 1$, we obtain

$$c(sI-\bar{A})^{-1}b_i P_m(s) = 1 \qquad (3.64)$$

where $P_m(s)$ has the form

$$P_m(s) = s^{n^*} + a_{n^*}^* s^{n^*-1} + a_{n^*-1}^* s^{n^*-2} + \cdots + a_2^* s + a_1^*. \qquad (3.65)$$

Hence, we have the explicit expressions

$$c = [c_1, c_2, \ldots, c_{n-n^*}, 1, 0, \ldots, 0]$$
$$c\bar{A} = [0, c_1, c_2, \ldots, c_{n-n^*}, 1, 0, \ldots, 0]$$
$$\cdots \cdots$$
$$c\bar{A}^{n^*-1} = [0, 0, \ldots, 0, c_1, c_2, \ldots, c_{n-n^*}, 1]$$
$$c\bar{A}^{n^*} = [-a_1, -a_2, \ldots, -a_{n^*}, c_1 - a_{n^*+1},$$
$$c_2 - a_{n^*+2}, \ldots, c_{n-n^*} - a_n]. \qquad (3.66)$$

From the fact that $c(sI-\bar{A})^{-1}b_i$ and $c(sI-A)^{-1}b_i$ have the same zero polynomial $s^{n-n^*} + c_{n-n^*}s^{n-n^*-1} + \cdots + c_2 s + c_1$, we have

$$c(sI-\bar{A})^{-1}b_i = \frac{s^{n-n^*} + c_{n-n^*}s^{n-n^*-1} + \cdots + c_2 s + c_1}{s^n + a_n s^{n-1} + \cdots + a_2 s + a_1}. \qquad (3.67)$$

Using this expression in (3.64), we get

$$[a_1, a_2, \ldots, a_n, 1, 0, \ldots, 0]$$
$$= [0, \ldots, 0, c] + [0, \ldots, 0, c, 0]a_{n^*}^* + [0, \ldots, 0, c, 0, 0]a_{n^*-1}^* + \cdots$$
$$+ [c, 0, \ldots, 0]a_1^* \in R^{1 \times (n+n^*)}. \tag{3.68}$$

Combining (3.66) and (3.68), we conclude that

$$c(\bar{A}^{n^*} + a_{n^*}^* \bar{A}^{n^*-1} + a_{n^*-1}^* \bar{A}^{n^*-2} + \cdots + a_2^* \bar{A} + a_1^* I) = 0. \tag{3.69}$$

It follows from (3.69) that

$$\begin{aligned} c\bar{A}^k b_i &= -a_{n^*}^* c\bar{A}^{k-1} b_i - a_{n^*-1}^* c\bar{A}^{k-2} b_i - \cdots \\ &\quad -a_2^* c\bar{A}^{k-n^*+1} b_i - a_1^* c\bar{A}^{k-n^*} b_i, \end{aligned} \tag{3.70}$$

$$\begin{aligned} c\bar{A}^k b_j &= -a_{n^*}^* c\bar{A}^{k-1} b_j - a_{n^*-1}^* c\bar{A}^{k-2} b_j - \cdots \\ &\quad -a_2^* c\bar{A}^{k-n^*+1} b_j - a_1^* c\bar{A}^{k-n^*} b_j \end{aligned} \tag{3.71}$$

for $k = n^*, n^* + 1, \ldots, n - 1$. From the condition that $cA^k b_i = 0$, $cA^k b_j = 0$, $k = 0, 1, \ldots, n^* - 2$, we get

$$c\bar{A}^k b_i = \frac{1}{\alpha_{ij}} c\bar{A}^k b_j = 0, \quad k = 0, 1, \ldots, n^* - 2 \tag{3.72}$$

independent of α_{ij}. Choosing the parameter α_{ij} as

$$\alpha_{ij} = \frac{c\bar{A}^{n^*-1} b_j}{c\bar{A}^{n^*-1} b_i} = c\bar{A}^{n^*-1} b_j, \quad c\bar{A}^{n^*-1} b_i = 1 \tag{3.73}$$

we obtain

$$c\bar{A}^{n^*-1} b_i = \frac{1}{\alpha_{ij}} c\bar{A}^{n^*-1} b_j. \tag{3.74}$$

Finally from (3.70), (3.71), (3.72), and (3.74), we have

$$c\bar{A}^k b_i = \frac{1}{\alpha_{ij}} c\bar{A}^k b_j \tag{3.75}$$

from $k = n*$ to $k = n - 1$. In summary, we have

$$c\bar{A}^k \left(b_i - \frac{1}{\alpha_{ij}} b_j \right) = 0, k = 0, 1, \ldots, n - 1. \tag{3.76}$$

Using this result and applying (3.60) to $c(sI - \bar{A})^{-1}(b_i - b_j/\alpha_{ij})$, we obtain

$$c(sI - \bar{A})^{-1} \left(b_i - \frac{1}{\alpha_{ij}} b_j \right) = 0. \tag{3.77}$$

In view of (3.58) (with $\alpha_{ii} = 1$), we see that (3.77) is equivalent to (3.59): $c(sI - A - b_i k_{1i}^{*T})^{-1} b_j = \alpha_{ij} W_m(s)$. This result is true for all $j = 1, 2, \ldots, m,\ j \neq i$.

Necessity. Since $W_m(s)$ has relative degree n^*, it follows from (3.56) and (3.57) that $C(sI - \bar{A})^{-1} b_i$ and $C(sI - \bar{A})^{-1} b_j$ have the same relative degree n^*, where $\bar{A} = A + b_i k_{1i}^{*T}$. Using (3.60)–(3.62) for $b = b_i$ and $b = b_j$, we can derive that $c\bar{A}^k b_i = 0$, $c\bar{A}^k b_j = 0$, $k = 0, 1, \ldots, n^* - 2$, and $c\bar{A}^{n^*-1} b_i \neq 0$, $c\bar{A}^{n^*-1} b_j \neq 0$, which implies that $cA^k b_i = 0$, $cA^k b_j = 0$, $k = 0, 1, \ldots, n^* - 2$, and $cA^{n^*-1} b_i \neq 0$, $cA^{n^*-1} b_j \neq 0$, which means that $C(sI - A)^{-1} b_i$ and $C(sI - A)^{-1} b_j$ have relative degree n^*. ∇

The essence of Lemma 3.2.1 is that given any b_i and b_j such that $cA^k b_i = CA^k b_j = 0$, $k = 0, \ldots, n^* - 2$, $cA^{n^*-1} b_i = 1$ and $cA^{n^*-1} b_j \neq 0$, if $c(sI - A - b_i k_{1i}^{*T})^{-1} b_i = 1/P_m(s)$ for $P_m(s)$ in (3.65) and some $k_{1i}^* \in R^n$, then $c(sI - A - b_i k_{1i}^{*T})^{-1} b_j = -k_{3i}^* c(sI - A - b_i k_{1i}^{*T})^{-1} b_i$ for some $k_{3i}^* \in R$. On the other hand, if this equality holds, then (c, A, b_j) has relative degree n^*. For closed-loop stability, all zeros of $P_m(s)$ and (c, A, b_i) should be stable, that is, all eigenvalues of $A + b_i k_{1i}^{*T}$ are stable.

3.2.3 Adaptive Rejection of Unmatched Input Disturbance

In this section, we design an adaptive control scheme for the plant (3.47) of unknown parameters (c, A, b_i, b_j), with one control actuator $u_i(t) = v_i(t)$ and one disturbance actuator $u_j(t) = \sum_{l=1}^{q} \gamma_l g_l(t)$, where $g_l(t)$, $l = 1, 2, \ldots, q$, are known, bounded, and continuous signals, and γ_l, $l = 1, 2 \ldots, q$, are unknown constant parameters. The goal is to adaptively reject the effect of $u_j(t)$ by an adaptive feedback design $v_i(t)$, to ensure closed-loop stability and asymptotic tracking: $\lim_{t \to \infty} (y(t) - y_m(t)) = 0$ when (A, b_i) is controllable and all zeros of (c, A, b_i) are stable. Since b_i and b_j are not parallel, the disturbance $u_j(t)$ is not matched at the input $u_i(t)$.

Recall that, for the plant (3.47), from Lemma 3.2.1, if (c, A, b_i) and (c, A, b_j) have the same relative degree as that of $W_m(s)$ in (3.48), and the plant parameters (c, A, b_i, b_j) are known, and so are the disturbance parameters γ_l and signals $g_l(t)$, $l = 1, 2, \ldots, q$, then there exist $k_{1i}^* \in R^n$, $k_{2i}^* \in R$, and $f_i^*(t) = \sum_{l=1}^{q} \gamma_l^* g_l(t) \in R$, $\gamma_l^* = k_{3i}^* \gamma_l$, calculated from (3.51) and (3.54), such that the controller (3.49) leads to the desired plant-model output matching: $\lim_{t \to \infty} (y(t) - y_m(t)) = 0$ exponentially.

When the plant parameters (c, A, b_i, b_j) and disturbance parameters γ_l, $l = 1, 2, \ldots, q$, are unknown, to achieve the desired control objective, we use the following adaptive version of the controller (3.49):

$$u_i(t) = v_i(t) = k_{1i}^T(t)x(t) + k_{2i}(t)r(t) + f_i(t) \tag{3.78}$$

where

$$f_i(t) = \sum_{l=1}^{q} \hat{\gamma}_l(t)g_l(t) \tag{3.79}$$

and $k_{1i}(t), k_{2i}(t), \hat{\gamma}_l(t)$ are the estimates of $k_{1i}^*, k_{2i}^*, \gamma_l^*, l = 1, 2, \ldots, q$.

Introduce the parameter vectors θ^* and θ as

$$\theta^* = [k_{1i}^{*T}, k_{2i}^*, \gamma_1^*, \ldots, \gamma_q^*]^T, \ \theta = [k_{1i}^T, k_{2i}, \hat{\gamma}_1, \ldots, \hat{\gamma}_q]^T \tag{3.80}$$

and the regressor signal vector $\omega(t)$ as

$$\omega(t) = [x^T(t), r(t), g_1(t), \ldots, g_q(t)]^T. \tag{3.81}$$

We express the control signal $u_i(t)$ as

$$u_i(t) = u_i^*(t) + \tilde{\theta}^T(t)\omega(t), \ \tilde{\theta}(t) = \theta(t) - \theta^* \tag{3.82}$$

and the closed-loop system as

$$\dot{x}(t) = (A + b_i k_{1i}^{*T})x(t) + b_i k_{2i}^* r(t) + b_i \tilde{\theta}^T(t)\omega(t) + b_i f_i^*(t) + b_j u_j(t)$$
$$y(t) = cx(t). \tag{3.83}$$

Using (3.48), (3.51), and (3.54), we have the tracking error

$$e(t) = y(t) - y_m(t) = ce^{(A+b_i k_{1i}^{*T})t}x(0) + \rho^* W_m(s)[\tilde{\theta}^T\omega](t) \tag{3.84}$$

where $\rho^* = 1/k_{2i}^*$. Introducing the auxiliary signals

$$\zeta(t) = W_m(s)[\omega](t), \ \xi(t) = \theta^T(t)\zeta(t) - W_m[\theta^T\omega](t) \tag{3.85}$$

we define the estimation error

$$\epsilon(t) = e(t) + \rho(t)\xi(t) \tag{3.86}$$

where $\rho(t)$ is the estimate of ρ^*. From (3.84)–(3.86) it follows that

$$\epsilon(t) = \rho^*\tilde{\theta}^T(t)\zeta(t) + \tilde{\rho}(t)\xi(t) + \epsilon_p(t) \tag{3.87}$$

where $\tilde{\rho}(t) = \rho(t) - \rho^*$ and $\epsilon_p(t) = ce^{(A+b_i k_{1i}^{*T})t}x(0)$ is an exponentially decaying to zero term.

Based on (3.87), we choose the adaptive laws for θ and ρ as

$$\dot{\theta}(t) = -\frac{\text{sign}[\rho^*]\Gamma_\theta\zeta(t)\epsilon(t)}{1 + \zeta^T(t)\zeta(t) + \xi^2(t)}, \ \Gamma_\theta = \Gamma_\theta^T > 0 \tag{3.88}$$

$$\dot{\rho}(t) = -\frac{\gamma_\rho\xi(t)\epsilon(t)}{1 + \zeta^T(t)\zeta(t) + \xi^2(t)}, \ \gamma_\rho > 0. \tag{3.89}$$

This standard gradient adaptive scheme can be analyzed by using the positive definite function

$$V(\tilde{\theta}, \tilde{\rho}) = \frac{1}{2}|\rho^*|\tilde{\theta}^T \Gamma_\theta^{-1}\tilde{\theta} + \frac{1}{2}\gamma_\rho^{-1}\tilde{\rho}^2 \tag{3.90}$$

whose time derivative along the trajectories of (3.88) and (3.89) is

$$\dot{V} = -\frac{\epsilon^2(t)}{N^2(t)} + \frac{\epsilon(t)\epsilon_p(t)}{N^2(t)} \le -\frac{\epsilon^2(t)}{2N^2(t)} + \frac{\epsilon_p^2(t)}{2N^2(t)} \tag{3.91}$$

where $N(t) = \sqrt{1 + \zeta^T(t)\zeta(t) + \xi^2(t)}$. Since $\epsilon_p(t)$ decays to zero exponentially, it can be concluded from (3.91) that $V(\tilde{\theta}, \tilde{\rho}) \in L^\infty$ and $\epsilon(t)/N(t) \in L^2$. It in turn implies that $\theta(t), \rho(t) \in L^\infty$. With (3.87) and (3.88), we obtain that $\epsilon(t)/N(t) \in L^\infty$, $\dot{\theta} \in L^2 \cap L^\infty$, and $\dot{\rho} \in L^2 \cap L^\infty$. From these desired properties of the adaptive scheme (3.88) and (3.89), the closed-loop stability and asymptotic tracking: $\lim_{t\to\infty}(y(t) - y_m(t)) = 0$, can be proved by using a standard model reference adaptive control stability analysis [118].

Remark 3.2.1. The conditions of Lemma 3.2.1 are crucial for an adaptive disturbance rejection control scheme for the plant (3.47) with unknown parameters (c, A, b_i). If the plant parameters (c, A, b_i, b_j) and disturbance parameters γ_l and signals g_l, $l = 1, 2 \ldots, q$, are known, from (3.52) the choice of $f_i^*(s) = -k_{2i}^* P_m(s)c(sI - A - b_i k_{1i}^{*T})^{-1}b_j u_j(s)$ would lead to the desired output matching: $y(s) = y_m(s) + c(sI - A - b_i k_{1i}^{*T})^{-1}x(0)$, without needing the matching condition (3.54). However, the parametrization of this choice of $f_i^*(s)$ needs the knowledge of (c, A, b_i) when $n^* < n$ (that is, when (c, A, b_i) has finite zeros; in this case, they are in the denominator of $-k_{2i}^* P_m(s)c(sI - A - b_i k_{1i}^{*T})^{-1}b_j$, and are not convenient to be estimated), as well as that of the derivatives of $g_l(t)$ when the relative degree of $c(sI - A - b_i k_{1i}^{*T})^{-1}b_j$ is less than that of $c(sI - A - b_i k_{1i}^{*T})^{-1}b_i$. □

3.2.4 Application to Actuator Failure Compensation

We now consider the adaptive actuator failure compensation problem for the plant (3.47) with two actuators $u_i(t)$ and $u_j(t)$ and with up to one actuator failure as modeled in (3.55). We will develop an adaptive feedback control scheme to generate the applied input signals $v_i(t)$ and $v_j(t)$ such that closed-loop stability and asymptotic output tracking are ensured for all three possible situations that are unknown to the adaptive controller:

(a) $u_i(t) = v_i(t)$ and $u_j(t) = v_j(t)$;
(b) $u_i(t) = v_i(t)$ and $u_j(t)$ fails (that is, $k = j$ in (3.55)); and
(c) $u_j(t) = v_j(t)$ and $u_i(t)$ fails (that is, $k = i$ in (3.55)).

It is important to note that the adaptive compensator to be developed does not require the knowledge of the actuator failures: the failure values, failure time instants, and failure patterns (that is, which and how many actuators have failed) (also see the general case in Section 3.2.5).

To develop a solution to this adaptive control problem, we choose the equal-actuation scheme for generating control signals for the two actuators:

$$v_i(t) = v_j(t) = v_0(t) \tag{3.92}$$

and assume that (c, A, b_i), (c, A, b_j), and $(c, A, b_i + b_j)$ are all controllable and have the same relative degree n^* as that of $W_m(s)$ in (3.48), and their zeros are all stable, for a model reference control design.

Then we design the controller structure for $v_0(t)$ as

$$v_0(t) = k_{10}^T(t)x(t) + k_{20}(t)r(t) + f_0(t) \tag{3.93}$$

where

$$f_0(t) = \hat{u}_0(t) + \sum_{k=i}^{j} \sum_{l=1}^{q} \hat{d}_{kl}(t)f_{kl}(t) \tag{3.94}$$

$\hat{u}_0(t)$ is the estimate of

$$u_0^* = \begin{cases} 0 & \text{if } u_i(t) = v_i(t) \text{ and } u_j(t) = v_j(t) \\ k_{3i}^* \bar{u}_j & \text{if } u_i(t) = v_i(t) \text{ and } u_j(t) \text{ fails} \\ k_{3j}^* \bar{u}_i & \text{if } u_j(t) = v_j(t) \text{ and } u_i(t) \text{ fails} \end{cases} \tag{3.95}$$

with k_{3i}^* and k_{3j}^* defined from (3.54), $\hat{d}_{il}(t)$ is the estimate of

$$d_{il}^* = \begin{cases} k_{3j}^* \bar{d}_{il} & \text{if } u_j(t) = v_j(t) \text{ and } u_i(t) \text{ fails} \\ 0 & \text{otherwise} \end{cases} \tag{3.96}$$

and $\hat{d}_{jl}(t)$ is the estimate of

$$d_{jl}^* = \begin{cases} k_{3i}^* \bar{d}_{jl} & \text{if } u_i(t) = v_i(t) \text{ and } u_j(t) \text{ fails} \\ 0 & \text{otherwise} \end{cases} \tag{3.97}$$

for $\bar{u}_k$, $\bar{d}_{kl}$, $k = i, j$, $l = 1, 2, \ldots, q$, defined in the actuator failure model (3.55). The parameters $k_{10}(t)$ and $k_{20}(t)$ are the estimates of k_{10}^* and k_{20}^* that satisfy the matching equations

$$W_m(s) = \begin{cases} c(sI - A - (b_i + b_j)k_{10}^{*T})^{-1} \\ \quad \cdot (b_i + b_j)k_{20}^* & \begin{array}{l} \text{if } u_i(t) = v_i(t) \\ \text{and } u_j(t) = v_j(t) \end{array} \\ c(sI - A - b_ik_{10}^{*T})^{-1}b_ik_{20}^* & \begin{array}{l} \text{if } u_i(t) = v_i(t) \\ \text{and } u_j(t) \text{ fails} \end{array} \\ c(sI - A - b_jk_{10}^{*T})^{-1}b_jk_{20}^* & \begin{array}{l} \text{if } u_j(t) = v_j(t) \\ \text{and } u_i(t) \text{ fails.} \end{array} \end{cases} \tag{3.98}$$

Under the above assumption for (c, A, b_i, b_j), such matching parameters k_{10}^* and k_{20}^* exist, and the resulting zero-pole cancellation in (3.98) is stable.

It is clear that the matching parameters k_{10}^*, k_{20}^*, u_0^*, and d_{kl}^*, $k = i, j$, $l = 1, \ldots, q$, are piecewise constant, with one possible jump in their values when there is one possible actuator failure.

Introduce the parameter vectors θ^* and θ as

$$\theta^* = [k_{10}^{*T}, k_{20}^*, u_0^*, d_{i1}^*, \ldots, d_{iq}^*, d_{j1}^*, \ldots, d_{jq}^*]^T$$
$$\theta = [k_{10}^T, k_{20}, \hat{u}_0, \hat{d}_{i1}, \ldots, \hat{d}_{iq}, \hat{d}_{j1}, \ldots, \hat{d}_{jq}]^T \qquad (3.99)$$

and the regressor signal vector $\omega(t)$ as

$$\omega(t) = [x^T(t), r(t), 1, f_{i1}(t), \ldots, f_{iq}(t), f_{j1}(t), \ldots, f_{jq}(t)]^T. \qquad (3.100)$$

Different from that in (3.84), in this case, the tracking error equation is

$$e(t) = ce^{\bar{A}t}x(0) + W_m(s)[\rho^*\tilde{\theta}^T\omega](t) \qquad (3.101)$$

where $\bar{A} = A + \bar{b}k_{10}^{*T}$ with $\bar{b} = b_i + b_j$, b_i, or b_j for the three possible actuator failure cases in (3.98), and $\rho^* = 1/k_{20}^*$.

Introducing the auxiliary signals

$$\zeta(t) = W_m(s)[\omega](t), \quad \xi(t) = \theta^T(t)\zeta(t) - W_m[\theta^T\omega](t) \qquad (3.102)$$

we define the estimation error

$$\epsilon(t) = e(t) + \rho(t)\xi(t) \qquad (3.103)$$

where $\rho(t)$ is the estimate of ρ^*. From (3.101)–(3.103), it follows that

$$\epsilon(t) = \rho^*\tilde{\theta}^T(t)\zeta(t) + \tilde{\rho}(t)\xi(t) + \epsilon_a(t) \qquad (3.104)$$

where $\tilde{\rho}(t) = \rho(t) - \rho^*$, $\epsilon_a(t) = e^{\bar{A}t}x(0) + \epsilon_b(t)$, with

$$\epsilon_b(t) = \rho^*(\theta^{*T}(t)\zeta(t) - W_m(s)[\theta^{*T}\omega](t))$$
$$+ W_m(s)[\rho^*\tilde{\theta}^T\omega](t) - \rho^*W_m(s)[\tilde{\theta}^T\omega](t). \qquad (3.105)$$

Based on (3.105), we choose the adaptive laws for θ and ρ as

$$\dot{\theta}(t) = -\frac{\text{sign}[\rho^*]\Gamma_\theta\zeta(t)\epsilon(t)}{1 + \zeta^T(t)\zeta(t) + \xi^2(t)}, \quad \Gamma_\theta = \Gamma_\theta^T > 0 \qquad (3.106)$$

$$\dot{\rho}(t) = -\frac{\gamma_\rho\xi(t)\epsilon(t)}{1 + \zeta^T(t)\zeta(t) + \xi^2(t)}, \quad \gamma_\rho > 0. \qquad (3.107)$$

To implement this design, we need to assume that k_{20}^* in (3.98) has a known and constant sign for the three possible failure patterns.

Similarly, this standard gradient adaptive scheme can be analyzed by using the positive definite function

$$V(\tilde{\theta}, \tilde{\rho}) = \frac{1}{2}|\rho^*|\tilde{\theta}^T \Gamma_\theta^{-1}\tilde{\theta} + \frac{1}{2}\gamma_\rho^{-1}\tilde{\rho}^2 \tag{3.108}$$

whose time derivative along the trajectories of (3.106) and (3.107) is

$$\dot{V} = -\frac{\epsilon^2(t)}{N^2(t)} + \frac{\epsilon(t)\epsilon_p(t)}{N^2(t)} \leq -\frac{\epsilon^2(t)}{2N^2(t)} + \frac{\epsilon_p^2(t)}{2N^2(t)}, \ t \neq t_f \tag{3.109}$$

where $N(t) = \sqrt{1 + \zeta^T(t)\zeta(t) + \xi^2(t)}$, and t_f is the time instant when one of the two actuators possibly fails. At $t = t_f$, the matching parameters in θ^* defined in (3.99) change their values, causing a finite jump in θ^* and $V(\tilde{\theta}, \tilde{\rho})$ as well. It can be shown that $\epsilon_b(t)$ decays exponentially, and so does $\epsilon_a(t)$ as $e^{\bar{A}t}x(0)$ does. Then, it follows from (3.109) that $V(\tilde{\theta}, \tilde{\rho}) \in L^\infty$ (that is, $\theta(t), \rho(t) \in L^\infty$) and $\epsilon(t)/N(t) \in L^2$. From (3.105) and (3.106), we also have $\epsilon(t)/N(t) \in L^\infty$, $\dot{\theta} \in L^2 \cap L^\infty$, and $\dot{\rho} \in L^2 \cap L^\infty$. The closed-loop stability and asymptotic tracking: $\lim_{t\to\infty}(y(t) - y_m(t)) = 0$, can also be established.

3.2.5 Extension to the General Case

The above adaptive disturbance rejection and failure compensation results can be extended to the general case with a linear time-invariant plant

$$\dot{x}(t) = Ax(t) + Bu(t), \ y(t) = cx(t) \tag{3.110}$$

where $A \in R^{n\times n}$, $B = [b_1, \ldots, b_m] \in R^{n\times m}$, and $c \in R^{1\times n}$ are unknown constant parameter matrices, and with a disturbance (actuator failure) model

$$u_k(t) = \bar{u}_k + \sum_{l=1}^{q} \bar{d}_{kl} f_{kl}(t), \ t \geq t_k, \ k \in \{1, 2, \ldots, m\} \tag{3.111}$$

where $\bar{u}_k \in R$ is an unknown constants, $\bar{d}_{kl} \in R$ is an unknown constant, and $f_{kl}(t) \in R$ is known bounded signal, $k = 1, \ldots, m, l = 1, \ldots, q, q \geq 1$.

Adaptive Disturbance Rejection. The disturbance rejection problem is to design a feedback control law for one particular actuator $u_i(t)$ among the m actuators for the plant (3.110) with unknown parameters (A, B) to achieve closed-loop stability and asymptotic output tracking: $\lim_{t\to\infty}(y(t) - y_m(t)) = 0$, when all other inputs $u_j(t)$, $j = 1, 2, \ldots, i-1, i+1, \ldots, m$, are disturbance signals in the form (3.110) with unknown parameters $\bar{u}_j$ and $\bar{d}_{jl}$ and known signals $f_{jl}(t)$, $l = 1, \ldots, q$.

This problem can be solved under the conditions that (c, A, b_i) is controllable and has all its zeros stable, and (c, A, b_j), $j = 1, 2, \ldots, m$, have the same relative degree n^* as that of $W_m(s)$ in (3.48). Although the disturbance signals $u_j(t)$, $j = 1, 2, \ldots, i-1, i+1, \ldots, m$, are not matched to the control signal $u_i(t)$ as b_i and b_j may not be parallel to each other, Lemma 3.2.1 ensures that they are matched at output in the sense of (3.56) and (3.57) so that output disturbance rejection is possible.

For this control problem, the controller structure (3.78) can also be used but with the design signal $f_i(t)$ now parametrized in terms of all components of $u_j(t)$ of the form (3.111), that is, $f_i(t)$ is the estimate of

$$f_i^*(t) = \sum_{j \neq i} \alpha_{ij} u_j(t) \tag{3.112}$$

for α_{ij} in (3.77) and $u_j(t)$ in (3.111), $j = 1, 2, \ldots, i-1, i+1, \ldots, m$.

Then, an error equation of the form $y(t) - y_m(t) = \rho^* W_m(s)[\tilde{\theta}^T \omega](t)$ can be obtained for some parameter error vector $\tilde{\theta}(t) = \theta(t) - \theta^*$ and regressor vector $\omega(t)$. Based on such an error equation, an estimation error similar to that in (3.86) and an adaptive law similar to that in (3.88) and (3.89) can be chosen to update the parameter estimate $\theta(t)$, which contains $k_{1i}(t)$, $k_{2i}(t)$ and that of $f_i(t)$, for adaptive control to achieve the control objective.

Adaptive Actuator Failure Compensation. The actuator failure compensation problem is to design a set of m feedback control signals $v_i(t)$, $i = 1, 2, \ldots, m$, such that for all cases when there are up to $m-1$ actuator failures (that is, when $u_j(t) = \bar{u}_j + \sum_{l=1}^{q} \bar{d}_{jl} f_{jl}(t)$, $j \in \{j_1, \ldots, j_p\} \subset \{1, 2, \ldots, m\}$, $\forall p \in \{1, 2, \ldots, m-1\}$, and $u_j(t) = v_j(t)$, $j \notin \{j_1, \ldots, j_p\}$), as well as when there is no failure (that is, when $u_j(t) = v_j(t)$, $j = 1, 2, \ldots, m$), the closed-loop stability and asymptotic output tracking are ensured. This problem can be solved by an equal-actuation design $v_i(t) = v_0(t)$, $i = 1, 2 \ldots, m$, under the condition that $(c, A, \sum_{j \neq j_1, \ldots, j_p} b_j)$, $p \in \{0, 1, \ldots, m-1\}$, are controllable and have relative degree n^* and all their zeros are stable.

The adaptive control scheme consists of an equal-actuation control law

$$v_1(t) = v_2(t) = \cdots = v_m(t) = k_{11}^T(t)x(t) + k_{21}(t)r(t) + f_1(t) \tag{3.113}$$

which is similar to that in (3.15), where

$$f_1(t) = \hat{u}_1(t) + \sum_{k=1}^{m} \sum_{l=1}^{q} \hat{d}_{kl}(t) f_{kl}(t) \tag{3.114}$$

is the estimate of

$$f_1^*(t) = \bar{u}_1^* + \sum_{k=1}^{m} \sum_{l=1}^{q} \bar{d}_{kl}^* f_{kl}(t) \tag{3.115}$$

for the matching parameters $\bar{u}_1^*$ and $\bar{d}_{kl}^*$ defined to satisfy

$$\sum_{j \neq j_1,\ldots,j_p} b_j \left(\bar{u}_1^* + \sum_{k=1}^{m} \sum_{l=1}^{q} \bar{d}_{kl}^* f_{kl}(t) \right) + \sum_{j=j_1,\ldots,j_p} b_j u_j(t) = 0 \tag{3.116}$$

for $u_j(t)$ being the failed actuators of the form (3.111), $j = j_1, \ldots, j_p$.

The existence of such matching parameters is ensured by the design conditions and Lemma 3.2.1. These parameters are piecewise constant between two failure time instants and they can be adaptively estimated together with other controller parameters k_{11}^* and k_{21}^*, for desired adaptive control.

3.2.6 Boeing 747 Lateral Control Simulation II

To evaluate the adaptive actuator failure compensation control system performance, we consider a linearized Boeing 747 lateral motion dynamic model with actuator failures in the airplane's segmented rudder servomechanism. With two augmented actuation vectors $b_2 u_2$ and $b_3 u_3$ to simulate a three-piece rudder, the linearized Boeing 747 lateral dynamic equation [37] is

$$\dot{x}(t) = \begin{bmatrix} -0.0558 & -0.9968 & 0.0802 & 0.0415 \\ 0.598 & -0.115 & -0.0318 & 0 \\ -3.05 & 0.388 & -0.465 & 0 \\ 0 & 0.0805 & 1 & 0 \end{bmatrix} x(t) \tag{3.117}$$

$$+ \begin{bmatrix} 0.00729 & 0.01 & 0.005 \\ -0.475 & -0.5 & -0.3 \\ 0.153 & 0.2 & 0.1 \\ 0 & 0 & 0 \end{bmatrix} u(t) \tag{3.118}$$

$$y(t) = \begin{bmatrix} 0 & 1 & 0 & 0 \end{bmatrix} x(t). \tag{3.119}$$

Given $m = 3$, there are seven possible final failure patterns of up to two actuator failures: one for no failure, three for one failure, and three for two failures. For each two-failure case, there are three subpatterns: One actuator fails after another or they fail at the same time. A desirable adaptive controller should ensure closed-loop stability and output tracking for any of these 13 failure patterns. Such an adaptive controller can be derived based on the conditions that $(c, A, \sum_{j \neq j_1,\ldots,j_p} b_j)$, $p \in \{0, 1, 2\}$, are controllable and have relative degree n^* and all their zeros are stable.

Two failure patterns are compensated in this simulation study:

(i) $u_1(t)$ does not fail, while $u_2(t) = \bar{u}_2 + \bar{d}_{21} f_{21}(t)$ for $t \geq 50$ sec and $u_3(t) = \bar{u}_3 + \bar{d}_{31} f_{31}(t)$ for $t \geq 100$ sec; and

(ii) $u_1(t)$ does not fail, while $u_3(t)$ fails at $t = 50$ sec and $u_2(t)$ fails at $t = 100$ sec (with the above failure model).

The failure values are shown in Figure 3.3(b) and Figure 3.4(b).

For simulation, we used the reference system $W_m(s) = 1/(s + 0.5)$ with $r(t) = 0.01$, the design parameters $\Gamma_\theta = \text{diag}\{10, 10, 10, 10, 10, 1, 10^5, 10^5\}$, $\gamma_\rho = 10$, and initial conditions $\theta(0) = [0.8, 1.2, 0.1, -0.02, -2, 0, 0, 0]^T$, $\rho(0) = -0.5$, $x(0) = [0, -5 \times 10^{-3}, 0, 0]^T$.

Figure 3.5(a) shows the output tracking error $e(t) = y(t) - y_m(t)$ for failure pattern (i), and Figure 3.6(a) shows the error $e(t)$ for failure pattern (ii). Figure 3.5(b) and Figure 3.6(b) show the corresponding input signals (including $u_1(t) = v_0(t)$ and two failed input signals $u_2(t)$ and $u_3(t)$). In both cases, the tracking error converges to small values after a transient response caused by actuator failures, verifying the desired control system performance. The corresponding adaptive controller parameters are shown in Figures 3.7 and 3.8, indicating the adaptation action, especially at the actuator failure time instants.

As a comparison, for failure pattern (i), Figure 3.9 shows the performance of a linear feedback design based on the knowledge of (c, A, b_1, b_2, b_3) but without consideration of actuator failures, and Figure 3.10 shows the performance of an adaptive control design for unknown (c, A, b_1, b_2, b_3) without failure compensation, indicating the damaging effect of actuator failures.

To conclude this chapter, we note that output tracking control requires less restrictive design conditions for actuator failure compensation. Such design conditions are specified in terms of plant-model output matching based on the knowledge of plant parameters and actuator failure parameters. Compensation of constant failure values (the so-called lock-in-place failures) does not need relative degree matching between an active input and a failed input. Compensation of varying failure values does, however, need relative degree matching between an active input and a failed input.

In this chapter we present two solutions to the problem of adaptive state feedback output tracking compensation control of systems with unknown actuator failures of either constant values or varying values. For the varying failure case, a necessary and sufficient condition for the desired relative degree matching is given. Key issues such as controller parametrization, error model derivation, and stable adaptation are solved for this class of adaptive actuator failure compensation control problems, and desired system performance is verified by extensive simulation results.

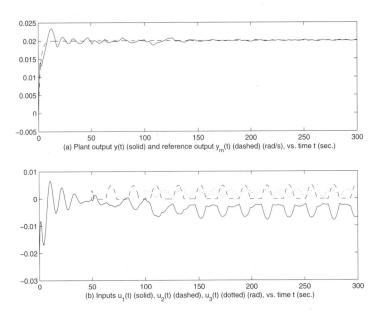

Fig. 3.5. System response for failure pattern (i).

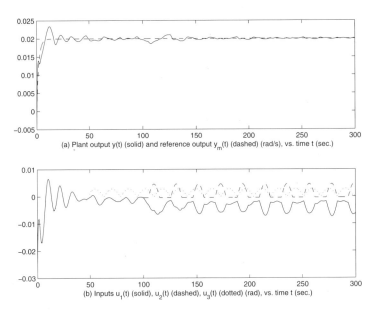

Fig. 3.6. System response for failure pattern (ii).

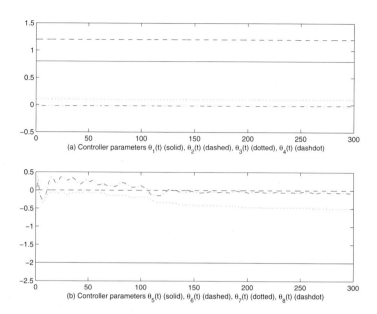

Fig. 3.7. Controller parameters for failure pattern (i).

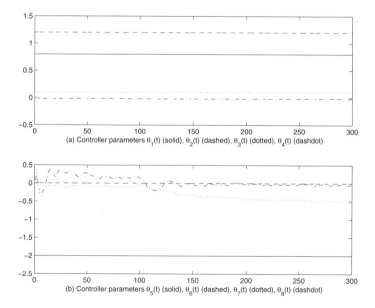

Fig. 3.8. Controller parameters for failure pattern (ii).

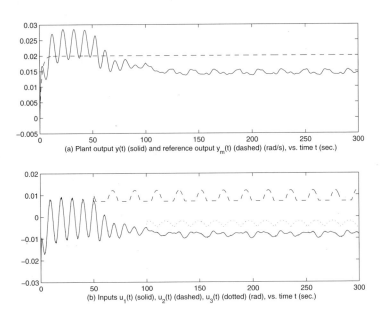

Fig. 3.9. System response with a linear feedback design.

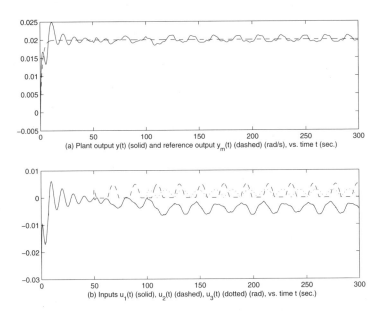

Fig. 3.10. System response with an adaptive design without failure compensation.

Chapter 4

Output Feedback Designs for Output Tracking

In Chapters 2 and 3, we developed adaptive *state feedback* control schemes for linear time-invariant plants with actuator failures characterized by some inputs being stuck at some values not influenced by control action, to achieve state tracking (Chapter 2) or output tracking (Chapter 3). The goal of this chapter is to develop an adaptive *output feedback* control scheme for systems with unknown parameters and unknown failures, to achieve output tracking. Some key issues such as controller structure, plant-model matching conditions, error model, adaptive law, stability analysis, and performance evaluation are addressed for output feedback based actuator failure compensation.

In Section 4.1, we formulate the output feedback and output tracking adaptive actuator failure compensation problems. First for the "lock-in-place" type of failures (of constant values), in Section 4.2, we propose a model reference controller structure capable of achieving plant-model output matching under a set of relative degree and minimum phase conditions on an input–output plant model in the presence of actuator failures. Next, in Section 4.3, we develop an adaptive law to update the parameters of the proposed actuator failure compensation controller, analyze the adaptive control system performance, and prove signal boundedness and asymptotic output tracking. In Section 4.4, we present simulation results from adaptive yaw rate control of a linearized Boeing 747 lateral dynamic model with actuator failures in a multisegment rudder servomechanism, to verify the desired performance of our adaptive actuator failure compensation scheme. Finally, in Section 4.5, we derive matching conditions and adaptive control designs for systems with actuator failures of varying values.

4.1 Problem Statement

Consider a linear time-invariant plant

$$\dot{x}(t) = Ax(t) + Bu(t), \ y(t) = Cx(t) \tag{4.1}$$

where $A \in R^{n \times n}$, $B = [b_1, \ldots, b_m] \in R^{n \times m}$, and $C \in R^{1 \times n}$ are unknown constant parameter matrices, $u(t) = [u_1, \ldots, u_m]^T(t) \in R^m$ is the input vector whose components may fail during system operation, and $y(t) \in R$ is the plant output. Unlike the cases studied in Chapters 2 and 3, for the control problem considered in this chapter, the state vector $x(t)$ is assumed not available for measurement, and only the plant output $y(t)$ can be measured.

In this study, we first consider the actuator failure model (1.2):

$$u_j(t) = \bar{u}_j, \ t \geq t_j, \ j \in \{1, 2, \ldots, m\} \tag{4.2}$$

with the constant value $\bar{u}_j$ and the failure time instant t_j unknown, and then deal with the more general failure model (1.6) in Section 4.5. We assume that the basic assumption (A1) stated in Section 1.3.3 is satisfied for the system (4.1), that is, for any up to $m - 1$ actuator failures, the remaining actuators can still achieve a desired control objective, when implemented with the knowledge of both plant parameters and failure parameters.

Recall from (1.3) that with actuator failures $u(t)$ can be expressed as

$$u(t) = v(t) + \sigma(\bar{u} - v(t)) \tag{4.3}$$

where $v(t) = [v_1, \ldots, v_m]^T(t)$ is an applied control input to be designed, and $\bar{u} = [\bar{u}_1, \bar{u}_2, \ldots, \bar{u}_m]^T$, $\sigma = \text{diag}\{\sigma_1, \sigma_2, \ldots, \sigma_m\}$ with $\sigma_i = 1$ if the ith actuator fails and $\sigma_i = 0$ otherwise.

The control objective is to design an output feedback control $v(t)$ for the plant (4.1) with unknown parameters and unknown actuator failures (4.2) or (1.6), such that despite the control error $u - v = \sigma(\bar{u} - v)$, all closed-loop signals are bounded, and the plant output $y(t)$ asymptotically tracks a given reference output $y_m(t)$ generated from the reference model

$$y_m(t) = W_m(s)[r](t), \ W_m(s) = \frac{1}{P_m(s)} \tag{4.4}$$

where $P_m(s)$ is a stable monic polynomial of degree n^*, and $r(t)$ is bounded. The key task is to design a controller structure capable of ensuring plant-model matching and suitable for desired adaptation under any failure pattern: $u_j(t) = \bar{u}_j$, $j = j_1, \ldots, j_p$, for any $p \in \{0, 1, \ldots, m - 1\}$.

Assume that the redundant actuators $u_i(t)$, $i = 1, 2, \ldots, m$, have proportional effects on the system dynamic behavior (the case when their effects are arbitrary is treated in Chapter 5 using multivariable adaptive control). We can choose an equal-actuation scheme for the actuators:

$$v_1(t) = \cdots = v_m(t) \triangleq v_0(t) \tag{4.5}$$

for a simplified control design. Based on this scheme, a solution to adaptive output feedback and output tracking actuator failure compensation can be developed, by an output feedback design for $v_0(t)$. This choice is realistic; for example, the multiple pieces of a segmented rudder or elevator on an aircraft are expected to move at an equal angle.

4.2 Plant-Model Output Matching

In this section, we develop a controller structure and derive conditions under which the desired plant-model output matching: $\lim_{t \to \infty} (y(t) - y_m(t)) = 0$ exponentially, can be achieved by the controller based on the knowledge of the plant and actuator failure parameters.

To derive a suitable controller parametrization using output feedback, we express the controlled plant in the input–output form

$$y(t) = \sum_{j=1}^{m} \frac{k_{pj} Z_j(s)}{P(s)} [u_j](t) \qquad (4.6)$$

where k_{pj} is a scalar, $Z_j(s)$ is a monic polynomial, and $P(s)$ is a monic polynomial of degree n, such that $C(sI - A)^{-1} b_j = k_{pj} Z_j(s)/P(s)$, $j = 1, 2, \ldots, m$. Conditions for plant-model output matching and closed-loop system stability depend on the choice of the control law for the plant (4.1). Based on the equal-actuation scheme (4.5), we generate the control signal $v_0(t)$ from

$$v_0(t) = v_0^*(t) = \theta_1^{*T} \omega_1(t) + \theta_2^{*T} \omega_2(t) + \theta_{20}^* y(t) + \theta_3^* r(t) + \theta_4^* \qquad (4.7)$$

where $\theta_1^* \in R^{n-1}$, $\theta_2^* \in R^{n-1}$, $\theta_{20}^* \in R$, and $\theta_3^* \in R$ are parameters to be defined for plant-model output matching, $\theta_4^* \in R$ is a constant to be chosen for compensation of the actuation error $u - v = \sigma(\bar{u} - v)$, and

$$\omega_1(t) = \frac{a(s)}{\Lambda(s)} [v_0](t), \quad \omega_2(t) = \frac{a(s)}{\Lambda(s)} [y](t) \qquad (4.8)$$

with $a(s) = [1, s, \ldots, s^{n-2}]^T$ and $\Lambda(s)$ being a monic stable polynomial of degree $n - 1$. We will give a model reference design [55] for (4.7).

Suppose that at a certain time instant t, there are p failed actuators, that is, $u_j(t) = \bar{u}_j$, $j = j_1, \ldots, j_p$, $1 \le p < m$. From (4.5) and (4.3), we can express the plant (4.1) or (4.6) as

$$y(t) = G(s)[v_0](t) + \bar{y}(t) \qquad (4.9)$$

where

$$G(s) = \frac{\sum_{j \neq j_1, \ldots, j_p} k_{pj} Z_j(s)}{P(s)} \triangleq \frac{k_p Z_a(s)}{P(s)} \tag{4.10}$$

for some scalar k_p and monic polynomial $Z_a(s)$, and

$$\bar{y}(t) = \sum_{j=j_1, \ldots, j_p} \frac{k_{pj} Z_j(s)}{P(s)} [\bar{u}_j](t). \tag{4.11}$$

For plant-model output matching and closed-loop system stability, we assume that for all the possible failure patterns,

(A4.1) $Z_a(s)$ has degree $n - n^*$, i.e., the systems $(C, A, \sum_{j \neq j_1, \ldots, j_p} b_j)$, $p \in \{0, 1, \ldots, m-1\}$, have the same relative degree n^*; and
(A4.2) $Z_a(s)$ is stable, i.e., the systems $(C, A, \sum_{j \neq j_1, \ldots, j_p} b_j)$, $p \in \{0, 1, \ldots, m-1\}$, are minimum phase.

To derive a concise expression of the closed-loop system, we define

$$F_1(s) = \theta_1^{*T} \frac{a(s)}{\Lambda(s)}, \quad F_2(s) = \theta_2^{*T} \frac{a(s)}{\Lambda(s)}. \tag{4.12}$$

Then the control signal $v_0(t) = v_0^*(t)$ from (4.7) is

$$\begin{aligned}
v_0(t) &= F_1(s)[v_0](t) + F_2(s)G(s)[v_0](t) + F_2(s)[\bar{y}](t) \\
&\quad + \theta_{20}^* G(s)[v_0](t) + \theta_{20}^* \bar{y}(t) + \theta_3^* r(t) + \theta_4^* \\
&= (F_1(s) + F_2(s)G(s) + \theta_{20}^* G(s)) [v_0](t) \\
&\quad + F_2(s)[\bar{y}](t) + \theta_{20}^* \bar{y}(t) + \theta_3^* r(t) + \theta_4^*
\end{aligned} \tag{4.13}$$

which can be further expressed as

$$\begin{aligned}
v_0(t) &= (1 - F_1(s) - F_2(s)G(s) - \theta_{20}^* G(s))^{-1} \\
&\quad \cdot [F_2(s)[\bar{y}] + \theta_{20}^* \bar{y} + \theta_3^* r + \theta_4^*](t).
\end{aligned} \tag{4.14}$$

From (4.9) and (4.14), the closed-loop system is

$$\begin{aligned}
y(t) &= G(s) (1 - F_1(s) - F_2(s)G(s) - \theta_{20}^* G(s))^{-1} \\
&\quad \cdot [F_2(s)[\bar{y}] + \theta_{20}^* \bar{y} + \theta_3^* r + \theta_4^*](t) + \bar{y}(t).
\end{aligned} \tag{4.15}$$

To proceed, we express

$$(1 - F_1(s) - F_2(s)G(s) - \theta_{20}^* G(s))^{-1} = \frac{\Lambda(s)P(s)}{P_0(s)} \tag{4.16}$$

where

$$\begin{aligned}
P_0(s) &= \Lambda(s)P(s) - \theta_1^{*T} a(s)P(s) \\
&\quad - \theta_2^{*T} a(s)k_p Z_a(s) - \Lambda(s)\theta_{20}^* k_p Z_a(s)
\end{aligned} \tag{4.17}$$

and determine the parameters θ_1^*, θ_2^*, θ_{20}^*, θ_3^* from

$$\theta_1^{*T} a(s) P(s) + (\theta_2^{*T} a(s) + \theta_{20}^* \Lambda(s)) k_p Z_a(s)$$
$$= \Lambda(s)(P(s) - k_p \theta_3^* Z_a(s) P_m(s)) \tag{4.18}$$

with $\theta_3^* = k_p^{-1}$, which leads to

$$(1 - F_1(s) - F_2(s)G(s) - \theta_{20}^* G(s))^{-1} = \frac{P(s)}{Z_a(s)P_m(s)}. \tag{4.19}$$

Substituting (4.10) and (4.19) into (4.15), we have

$$
\begin{aligned}
y(t) &= \frac{1}{\theta_3^* P_m(s)} [F_2(s)[\bar{y}] + \theta_{20}^* \bar{y} + \theta_3^* r + \theta_4^*](t) + \bar{y}(t) \\
&= \frac{1}{\theta_3^* P_m(s)} [F_2(s)[\bar{y}] + \theta_{20}^* \bar{y} + \theta_3^* r + \theta_4^* + \theta_3^* P_m(s)[\bar{y}]](t). \tag{4.20}
\end{aligned}
$$

Using (4.18), we have

$$
\begin{aligned}
F_2(s) + \theta_{20}^* + \theta_3^* P_m(s) &= \frac{\theta_2^{*T} a(s) + \theta_{20}^* \Lambda(s) + \theta_3^* P_m(s)\Lambda(s)}{\Lambda(s)} \\
&= \frac{P(s)}{k_p Z_a(s)} - \frac{\theta_1^{*T} a(s) P(s)}{\Lambda(s) k_p Z_a(s)}. \tag{4.21}
\end{aligned}
$$

Substituting (4.21) into (4.20), we have

$$
\begin{aligned}
y(t) &= \frac{1}{P_m(s)}[r](t) \\
&\quad + \frac{1}{\theta_3^* P_m(s)} \left[\frac{\Lambda(s) - \theta_1^{*T} a(s)}{\Lambda(s) k_p Z_a(s)} \sum_{j=j_1,\ldots,j_p} k_{pj} Z_j(s)[\bar{u}_j] + \theta_4^* \right](t). \tag{4.22}
\end{aligned}
$$

Since $P_m(s)$, $\Lambda(s)$, and $Z_a(s)$ are all stable, there exists a constant θ_4^* such that

$$f_p(t) \triangleq \frac{1}{\theta_3^* P_m(s)} \left[\frac{\Lambda(s) - \theta_1^{*T} a(s)}{\Lambda(s) k_p Z_a(s)} \sum_{j=j_1,\ldots,j_p} k_{pj} Z_j(s)[\bar{u}_j] + \theta_4^* \right](t) \tag{4.23}$$

converges to zero exponentially as $t \to \infty$, so that $\lim_{t\to\infty}(y(t) - y_m(t)) = 0$ exponentially. This defines the desired parameters θ_i^*, $i = 1, 2, 20, 3, 4$.

As in Chapter 2, we let (T_i, T_{i+1}), $i = 0, 1, \ldots, m_0$, with $T_0 = 0$, be the time intervals on which the actuator failure pattern is fixed, that is, actuators fail only at time T_i, $i = 1, \ldots, m_0$. Since there are m actuators and at least one of them does not fail, we have $m_0 < m$ and $T_{m_0+1} = \infty$. Then, at time

T_j, $j = 1, \ldots, m_0$, the unknown plant-model matching parameters θ_1^*, θ_2^*, θ_{20}^*, θ_3^*, and θ_4^*, similar to that in Chapter 2, change their values such that

$$\theta_1^* = \theta_{1(i)}^*, \ \theta_2^* = \theta_{2(i)}^*, \ \theta_{20}^* = \theta_{20(i)}^*, \ \theta_3^* = \theta_{3(i)}^*, \ \theta_4^* = \theta_{4(i)}^* \qquad (4.24)$$

for $t \in (T_i, T_{i+1})$, $i = 0, 1, \ldots, m_0$, that is, the plant-model matching parameters θ_1^*, θ_2^*, θ_{20}^*, θ_3^*, and θ_4^* are piecewise constant parameters, because the plant has different characterizations under different failure conditions so that the plant-model matching parameters are also different.

Remark 4.2.1. The analysis carried out in this section is for the closed-loop system (with a nominal controller) operating at a failure pattern that there are p failed actuators, that is, $u_j(t) = \bar{u}_j$, $j = j_1, \ldots, j_p$. When p or $j_1, \ldots, j_p$ change, the parameters of the equivalent plant transfer function $G(s)$ defined in (4.10) also change, and so do the matching parameters θ_1^*, θ_2^*, θ_{20}^*, θ_3^*, and θ_4^*. This parameter jumping also introduces additional transient response starting from each T_i defined above. Since there are only a finite number of actuator failures, the effect of such transient response is always bounded, and since the failure pattern will eventually be fixed (that is, no failure for t after T_{m_0}), such transient effect actually converges to zero exponentially with time. For stability and tracking performance analysis, for either a nominal or adaptive control design, this effect can be ignored. Therefore, the developed parametrization indeed is able to ensure plant-model output matching for any actuator failure pattern: $u_j(t) = \bar{u}_j$, $j = j_1, \ldots, j_p$, $1 \leq p < m$. More information about the effect of the jumping parameters on model reference adaptive control can be found in [132], where it is shown that instability can occur when the parameter jumping is persistent with sufficient high frequencies. It is, however, interesting to characterize such jumping parameter effect analytically, which needs more investigation and presentation. □

4.3 Adaptive Control

We now develop an adaptive control scheme for the plant (4.1) with unknown plant parameters and unknown actuator failures (4.2). In this case, the parameters θ_1^*, θ_2^*, θ_{20}^*, θ_3^*, and θ_4^* are unknown and need to be adaptively estimated. As an adaptive version of (4.7), we use the controller

$$v_0(t) = v_1(t) = v_2(t) = \cdots = v_m(t)$$
$$= \theta_1^T \omega_1(t) + \theta_2^T \omega_2(t) + \theta_{20} y(t) + \theta_3 r(t) + \theta_4 \qquad (4.25)$$

where $\theta_1(t) \in R^{n-1}$, $\theta_2(t) \in R^{n-1}$, $\theta_{20}(t) \in R$, $\theta_3(t) \in R$, and $\theta_4(t) \in R$ are the estimates of the unknown parameters θ_1^*, θ_2^*, θ_{20}^*, θ_3^*, and θ_4^*, respectively.

To derive an error equation, we define

$$\theta^* = [\theta_1^{*T}, \theta_2^{*T}, \theta_{20}^*, \theta_3^*, \theta_4^*]^T \in R^{2n+1} \tag{4.26}$$

$$\theta(t) = [\theta_1^T(t), \theta_2^T(t), \theta_{20}(t), \theta_3(t), \theta_4(t)]^T \in R^{2n+1} \tag{4.27}$$

$$\omega(t) = [\omega_1^T(t), \omega_2^T(t), y(t), r(t), 1]^T \in R^{2n+1} \tag{4.28}$$

$$\tilde{\theta}(t) = \theta(t) - \theta^*. \tag{4.29}$$

Ignoring the exponentially decaying effects such as $f_p(t)$, we have the output tracking error equation

$$e(t) = y(t) - y_m(t) = \frac{k_p}{P_m(s)}[\tilde{\theta}^T\omega](t). \tag{4.30}$$

To see this, operating both sides of (4.18) on $y(t)$, and using the system expression (4.9): $P(s)[y](t) = k_p Z_a(s)[v_0](t) + P(s)[\bar{y}](t)$, we have

$$\theta_1^{*T}a(s)k_p Z_a(s)[v_0](t) + \theta_1^{*T}a(s)P(s)[\bar{y}](t)$$
$$+(\theta_2^{*T}a(s) + \theta_{20}^*\Lambda(s))k_p Z_a(s)[y](t)$$
$$= \Lambda(s)k_p Z_a(s)[v_0](t) + \Lambda(s)P(s)[\bar{y}](t)$$
$$-\Lambda(s)k_p\theta_3^* Z_a(s)P_m(s)[y](t). \tag{4.31}$$

Because $\Lambda(s)$ and $Z_a(s)$ are stable, (4.31) can be expressed as

$$v_0(t) = \theta_1^{*T}\frac{a(s)}{\Lambda(s)}[v_0](t) + \theta_2^{*T}\frac{a(s)}{\Lambda(s)}[y](t) + \theta_{20}^*[y](t) + \theta_3^* P_m(s)[y](t) + \theta_4^*$$
$$-\left(\frac{(\Lambda(s) - \theta_1^{*T}a(s))P(s)}{\Lambda(s)k_p Z_a(s)}[\bar{y}](t) + \theta_4^*\right) + \epsilon_1(t) \tag{4.32}$$

for some initial condition-related exponentially decaying $\epsilon_1(t)$. Substituting (4.25) in (4.32) and ignoring exponentially decaying terms, we obtain

$$e(t) = \frac{1}{\theta_3^* P_m(s)}[\tilde{\theta}^T\omega](t) + f_p(t) \tag{4.33}$$

for $f_p(t)$ in (4.23) such that $\lim_{t\to\infty} f_p(t) = 0$ exponentially, and finally with $\theta_3^* = k_p^{-1}$, we obtain the error equation (4.30).

To develop an adaptive update law for $\theta(t)$, we introduce

$$\zeta(t) = W_m[\omega](t) \tag{4.34}$$

$$\xi(t) = \theta^T(t)\zeta(t) - W_m(s)[\theta^T\omega](t) \tag{4.35}$$

$$\epsilon(t) = e(t) + \rho(t)\xi(t) \tag{4.36}$$

where $\rho(t)$ is the estimate of $\rho^* = 1/\theta_3^*$. Then we choose the adaptive laws as

$$\dot{\theta}(t) = -\frac{\text{sign}[\theta_3^*]\Gamma\zeta(t)\epsilon(t)}{1+\zeta^T\zeta+\xi^2}, \quad \Gamma = \Gamma^T > 0 \tag{4.37}$$

$$\dot{\rho}(t) = -\frac{\gamma\xi(t)\epsilon(t)}{1+\zeta^T\zeta+\xi^2}, \quad \gamma > 0. \tag{4.38}$$

To implement (4.37), we need the following assumption:

(A4.3) For any $p \subset \{0,1,\ldots,m-1\}$, the high frequency gains $CA^{n^*-1}\sum_{j\neq j_1,\ldots,j_p} b_j$ have the same sign, which is known:

$$\text{sign}[\theta_3^*] = \text{sign}[k_p] = \text{sign}[CA^{n^*-1}\sum_{j\neq j_1,\ldots,j_p} b_j]. \tag{4.39}$$

For system stability and tracking performance analysis, we define the positive definite function

$$V(\tilde{\theta}, \tilde{\rho}) = \frac{1}{2}\left(|\rho^*|\tilde{\theta}^T\Gamma^{-1}\tilde{\theta} + \gamma^{-1}\tilde{\rho}^2\right), \quad \tilde{\rho} = \rho - \rho^*. \tag{4.40}$$

Then with (4.34)–(4.38), the time derivative of V for $t \in (T_i, T_{i+1})$, $i = 0,1,\ldots,m_0$, is

$$\dot{V} = -\frac{\rho^*\tilde{\theta}^T(t)\zeta(t)\epsilon(t) + \tilde{\rho}(t)\xi(t)\epsilon(t)}{1+\zeta^T\zeta+\xi^2} \tag{4.41}$$

because $\dot{\tilde{\theta}}(t) = \dot{\theta}(t)$ and $\dot{\tilde{\rho}}(t) = \dot{\rho}(t)$ for $t \in (T_i, T_{i+1})$, given that $\theta^*(t) = \theta^*_{(i)}$ and $\rho^*(t) = \rho^*_{(i)}$ are constant for $t \in (T_i, T_{i+1})$ (see (4.24); we should note that $V(\cdot)$ as a function of t is not continuous, because θ^* and ρ^* are piecewise constant parameters).

From (4.30) and (4.34)–(4.36), we have

$$\epsilon(t) = \rho^*\tilde{\theta}^T(t)\zeta(t) + \tilde{\rho}(t)\xi(t) + \epsilon_p(t) \tag{4.42}$$

where

$$\epsilon_p(t) = \rho^*(\theta^{*T}(t)\zeta(t) - W_m(s)[\theta^{*T}\omega](t)). \tag{4.43}$$

Denoting $w_m(t)$ as the impulse response function of $W_m(s)$, we express

$$\epsilon_p(t) = \rho^*(\theta^{*T}(t)\int_0^t w_m(t-\tau)\omega(\tau)d\tau$$
$$- \int_0^t w_m(t-\tau)\theta^{*T}(\tau)\omega(\tau)d\tau). \tag{4.44}$$

For $t \in (T_{m_0}, \infty)$ (in which $\theta^*(t) = \theta^*_{(m_0)}$ is constant), we have

$$\epsilon_p(t) = \rho^*\int_0^{T_{m_0}} w_m(t-\tau)(\theta^*_{(m_0)} - \theta^*(\tau))^T\omega(\tau)d\tau. \tag{4.45}$$

Since $W_m(s)$ is strictly proper and stable, $|w_m(t - \tau)| \leq \beta e^{-\alpha(t-\tau)}$, for some $\alpha > 0$, $\beta > 0$, so that

$$|\epsilon_p(t)| \leq \beta |\rho^*| e^{-\alpha t} \int_0^{T_{m0}} e^{\alpha\tau} |(\theta_{(m_0)}^* - \theta^*(\tau))^T \omega(\tau)| d\tau. \tag{4.46}$$

Since $\theta^*(t)$ is piecewise constant in $(0, T_{m_0})$, T_{m_0} is finite, and $\omega(t)$ is bounded in $(0, T_{m_0})$, there exists a finite $a_1 > 0$ such that

$$\int_0^{T_{m0}} e^{\alpha\tau} |(\theta_{(m_0)}^* - \theta^*(\tau))^T \omega(\tau)| d\tau \leq a_1 \tag{4.47}$$

which, together with (4.46), implies that $\lim_{t\to\infty} \epsilon_p(t) = 0$ exponentially. Substituting (4.42) into (4.41) and ignoring the effect of the exponentially decaying terms, we have

$$\dot{V} = -\frac{\epsilon^2(t)}{1 + \zeta^T(t)\zeta(t) + \xi^2(t)} \leq 0 \tag{4.48}$$

for $t \in (T_i, T_{i+1})$, $i = 0, 1, \ldots, m_0$. Since there are only a finite number of failures in the system, $V(T_{m_0})$ is finite, and, from

$$\dot{V} = -\frac{\epsilon^2(t)}{1 + \zeta^T(t)\zeta(t) + \xi^2(t)} \leq 0, \ t \in (T_{m_0}, \infty) \tag{4.49}$$

we have that $\theta(t)$, $\rho(t) \in L^\infty$, $\epsilon(t)/\sqrt{1 + \zeta^T(t)\zeta(t) + \xi^2(t)} \in L^2 \cap L^\infty$, $\dot{\theta}(t) \in L^2 \cap L^\infty$, and $\dot{\rho}(t) \in L^2 \cap L^\infty$.

Based on these desired properties, using the analysis method in [125], we have the following closed-loop system stability and asymptotic output tracking properties of this adaptive control system.

Theorem 4.3.1. *The adaptive controller (4.25) with adaptive law (4.37)–(4.38), applied to the plant (4.1) with actuator failures (4.2), guarantees that all closed-loop signals are bounded and the tracking error $e(t) = y(t) - y_m(t)$ goes to zero as t goes to infinity.*

In summary, the adaptive actuator failure compensation control scheme developed in this section consists of the controller structure (4.25), and the adaptive laws (4.37) and (4.38), and has the above desired properties. The adaptive design is based on the tracking error expression (4.30), which holds for any failure pattern $u_j(t) = \bar{u}_j$, $j = j_1, \ldots, j_p$, $1 \leq p < m$, with unknown $\bar{u}_j$, $j_1, \ldots, j_p$, and p. This is made possible by the equal-control design (4.25), that is, $v_1(t) = v_2(t) = \cdots = v_m(t)$, under the conditions (A4.1) and (A4.2) that also serve as design guidelines for constructing systems that can be compensated for unknown actuator failures. For adaptive actuator failure compensation, it is necessary and sufficient that there exists a controller that can achieve plant-model matching in the presence of actuator failures, with known plant and failure parameters.

4.4 Boeing 747 Lateral Control Simulation

As an example, we use the lateral dynamics model of a Boeing 747 airplane [37] as the controlled plant to which the adaptive actuator failure compensation control scheme is applied. In this study, output feedback is used, unlike the one in Chapter 3 where state feedback is used. Similar to that in Chapter 3, the Boeing 747 model [37] is modified with two augmented actuation vectors $b_2 u_2$ and $b_3 u_3$, for the study of actuator failure compensation.

Plant Model. The linearized model of the lateral dynamics of Boeing 747 with two augmented actuation vectors can be described as

$$\dot{x}(t) = Ax(t) + Bu(t), \ y(t) = x_2(t) = y_r(t)$$
$$x(t) = [\beta, y_r, p, \phi]^T, \ B = [b_1, b_2, b_3] \tag{4.50}$$

where β is the side-slip angle, y_r is the yaw rate, p is the roll rate, ϕ is the roll angle, y is the system output, which is the yaw rate in this case, and u is the control input vector, which contains three control signals: $u = [u_1, u_2, u_3]^T$ to represent three rudder servos: δ_{r1}, δ_{r2}, δ_{r3}, from a three-piece rudder for achieving compensation in the presence of actuator failures (the usual case [37] is $u_2 = u_3 = 0$, and $u_1 = \delta_r$, the rudder servo angle).

From the data provided in [37], in horizontal flight at 40,000 ft and nominal forward speed 774 ft/sec (Mach 0.8), the Boeing 747 lateral-perturbation dynamics matrices (also see Section 3.1.4) are

$$A = \begin{bmatrix} -0.0558 & -0.9968 & 0.0802 & 0.0415 \\ 0.598 & -0.115 & -0.0318 & 0 \\ -3.05 & 0.388 & -0.465 & 0 \\ 0 & 0.0805 & 1 & 0 \end{bmatrix} \tag{4.51}$$

$$b_1 = \begin{bmatrix} 0.00729 \\ -0.475 \\ 0.153 \\ 0 \end{bmatrix}, \ b_2 = \begin{bmatrix} 0.01 \\ -0.5 \\ 0.2 \\ 0 \end{bmatrix}, \ b_3 = \begin{bmatrix} 0.005 \\ -0.3 \\ 0.1 \\ 0 \end{bmatrix} \tag{4.52}$$

where b_2 and b_3 are the augmented actuation vectors actuator failure compensation. For this plant model, it can be verified that Assumptions (A4.1) and (A4.2) are satisfied and that $\text{sign}[\theta_3^*] = -1$.

Simulation Results. In the simulation, we considered the case wherein u_2 and u_3 fail during system operation. Two failure patterns were used in the simulation study (the same as those in Section 3.1). We first simulated the case wherein the failed actuators jump to new positions other than where they fail (Failure Pattern (I)): $u_2(t) = -0.02$ rad, $t \geq 50$ sec, and $u_3(t) = 0.03$ rad, $t \geq 100$ sec.

For simulation, $\Gamma = 20I$, $\gamma = 20$, $W_m(s) = 1/(s+3)$, $\Lambda(s) = (s+1)^3$, $y(0) = 0.02$, $y_m(0) = 0$, $\theta(0) = [0.7940, 2.4772, 2.2306, -1.6715, -4.66184, -3.5517, 3.7865, -0.7059, 0.0]^T$, $\rho(0) = 0$. The simulation results are shown in Figure 4.1 for $r(t) = 0.03$ and in Figure 4.2 for $r(t) = 0.03\sin(0.2t)$.

We then simulated the case wherein the failed actuators are stuck at the positions where they fail (Failure Pattern (II): $u_2(t) = u_2(50)$, $t \geq 50$ sec, and $u_3(t) = u_3(100)$, $t \geq 100$ sec. The simulation results are shown in Figures 4.3 and 4.4 for $r(t) = 0.03$ and $r(t) = 0.03\sin(0.2t)$, respectively.

All simulation results verified the desired system performance: At the time instant when one of the actuators fails, there is a transient system response, and as the time advances, the output tracking error $e(t) = y(t) - y_m(t)$, starting from a transient value, becomes smaller. The values of controller parameters $\theta_1, \theta_2, \theta_{20}, \theta_3$, and θ_4, which are not shown in the figures, also jump when one actuator failure occurs, and then converge to constant values.

4.5 Designs for Varying Failures

In this section, we derive solutions to the actuator failure compensation problem for the plant (4.1) in the presence of varying and parametrizable failure values, which are expressed as in (3.55), that is,

$$u_j(t) = \bar{u}_j + \sum_{l=1}^{q} \bar{d}_{jl} f_{jl}(t), \ t \geq t_j, \ j \in \{1, 2, \ldots, m\} \tag{4.53}$$

where $\bar{u}_j$ and $\bar{d}_{jl}$, $l = 1, 2, \ldots, q$, are unknown constant parameters, and $f_{jl}(t)$, $l = 1, 2, \ldots, q$, are known signals. The nominal controller structure is similar to that in (4.7) for the constant failure value case, that is,

$$v_0^*(t) = \theta_1^{*T}\omega_1(t) + \theta_2^{*T}\omega_2(t) + \theta_{20}^* y(t) + \theta_3^* r(t) + \theta_4^*(t) \tag{4.54}$$

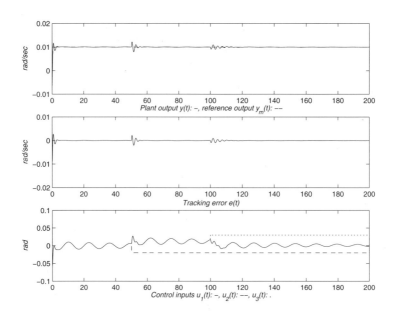

Fig. 4.1. System response for failure pattern (I) with $r(t) = 0.03$.

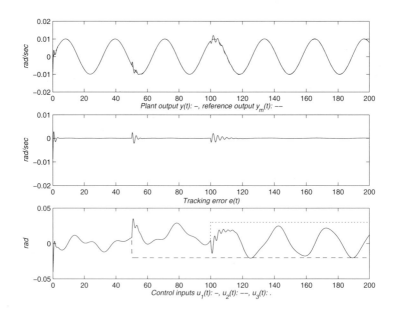

Fig. 4.2. System response for failure pattern (I) with $r(t) = 0.03 \sin(0.2t)$.

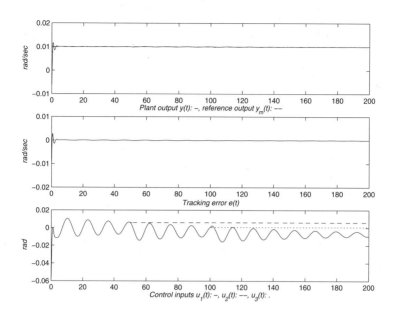

Fig. 4.3. System response for failure pattern (II) with $r(t) = 0.03$.

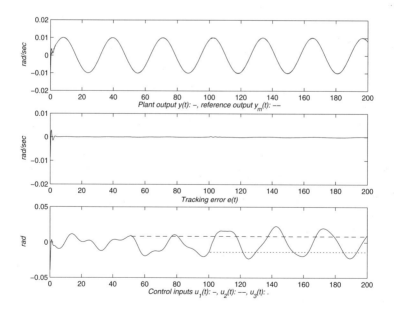

Fig. 4.4. System response for failure pattern (II) with $r(t) = 0.03 \sin(0.2t)$.

where $\theta_4^*(t)$ now is a time-varying signal for compensating the varying failure values. To handle actuator failures with varying and parametrizable failure values described by (4.53), the compensation term $\theta_4(t)$ in the corresponding adaptive version of the control law (4.54) needs to be designed from a dynamic compensator, based on a linear parametrization for the system parameters appearing bilinearly: one set of parameters from the nominal feedforward compensator $\theta_1^{*T}\omega_1(t)$ and the other set from the failure signals $\bar{u}_j(t)$.

4.5.1 Adaptive Disturbance Rejection

We start with the following adaptive output feedback disturbance rejection problem. Consider the input–output form of the plant (3.7), that is,

$$y(t) = \frac{k_{pi}Z_i(s)}{P(s)}[u_i](t) + \frac{k_{pj}Z_j(s)}{P(s)}[u_j](t) \tag{4.55}$$

where $u_i(t)$ is the control input and $u_j(t)$ is a disturbance of the form (4.53). The control objective is to find an output feedback control signal $u_i(t)$ such that the plant output $y(t)$ asymptotically tracking the reference output $y_m(t)$ from (4.4), in addition to closed-loop stability, under the assumptions that $Z_i(s)$ is stable, $P(s)$ has degree n, and $k_{pi}Z_i(s)/P(s)$ has relative degree n^*.

Nominal Control Design. For the nominal case when the plant parameters in $k_{pi}Z_i(s)/P(s)$ are known, we design the controller (4.54) with $v_0^*(t) = u_i(t)$, and

$$\omega_1(t) = \frac{a(s)}{\Lambda(s)}[u_i](t), \quad \omega_2(t) = \frac{a(s)}{\Lambda(s)}[y](t) \tag{4.56}$$

for $a(s) = [1, s, \ldots, s^{n-2}]^T$, $\Lambda(s)$ being a monic stable polynomial of degree $n - 1$, and $\theta_1^* \in R^{n-1}$, $\theta_2^* \in R^{n-1}$, $\theta_{20}^* \in R$, $\theta_3^* \in R$ being such that

$$\theta_1^{*T}a(s)P(s) + (\theta_2^{*T}a(s) + \theta_{20}^*\Lambda(s))k_{pi}Z_i(s)$$
$$= \Lambda(s)(P(s) - k_{pi}\theta_3^*Z_i(s)P_m(s)) \tag{4.57}$$

with $\theta_3^* = k_{pi}^{-1}$, and $\theta_4^* = \theta_4^*(t) \in R$ to be determined.

From (4.55) and (4.57), in the s-domain, we obtain

$$\theta_1^{*T}a(s)(k_{pi}Z_i(s)u_i(s) + k_{pj}Z_j(s)u_j(s))$$
$$+(\theta_2^{*T}a(s) + \theta_{20}^*\Lambda(s))k_{pi}Z_i(s)y(s)$$
$$= \Lambda(s)((k_{pi}Z_i(s)u_i(s) + k_{pj}Z_j(s)u_j(s)) - k_{pi}\theta_3^*Z_i(s)P_m(s)y(s)) \tag{4.58}$$

which leads to the parametrized plant

$$\theta_1^{*T} \frac{a(s)}{\Lambda(s)} [u_i](t) + \theta_2^{*T} \frac{a(s)}{\Lambda(s)} [y](t) + \theta_{20}^* y(t)$$

$$= u_i(t) - \theta_3^* P_m(s)[y](t) + \left(1 - \theta_1^{*T} \frac{a(s)}{\Lambda(s)}\right) \frac{k_{pj} Z_j(s)}{k_{pi} Z_i(s)} [u_j](t). \quad (4.59)$$

With the above nominal controller for $u_i(t)$, we have

$$\theta_3^* P_m(s)[y - y_m](t) = \theta_4^*(t) + \left(1 - \theta_1^{*T} \frac{a(s)}{\Lambda(s)}\right) \frac{k_{pj} Z_j(s)}{k_{pi} Z_i(s)} [u_j](t). \quad (4.60)$$

From (4.57), there is a monic polynomial $P_i(s)$ of degree $n^* - 1$ such that

$$\Lambda(s) - \theta_1^{*T} a(s) = Z_i(s) P_i(s) \quad (4.61)$$

from which we express

$$\left(1 - \theta_1^{*T} \frac{a(s)}{\Lambda(s)}\right) \frac{k_{pj} Z_j(s)}{k_{pi} Z_i(s)} [u_j](t) = \frac{k_{ij} Z_j(s) P_i(s)}{\Lambda(s)} [u_j](t), \ k_{ij} = \frac{k_{pj}}{k_{pi}}. \quad (4.62)$$

Then, the choice of the compensation signal

$$\theta_4^*(t) = -\frac{k_{ij} Z_j(s) P_i(s)}{\Lambda(s)} [u_j](t) \quad (4.63)$$

leads (4.60) to the desired error equation $\theta_3^* P_m(s)[y - y_m](t) = 0$, under the condition that the degree of $Z_j(s)$ is not greater than that of $Z_i(s)$.

Adaptive Control Design. For the case when the plant parameters in $k_{pi} Z_i(s)/P(s)$ are unknown, we use the adaptive controller

$$u_i(t) = \theta_1^T \omega_1(t) + \theta_2^T \omega_2(t) + \theta_{20} y(t) + \theta_3 r(t) + \theta_4(t) \quad (4.64)$$

where θ_1, θ_2, θ_{20}, θ_3, and $\theta_4(t)$ are the estimates of θ_1^*, θ_2^*, θ_{20}^*, θ_3^*, and $\theta_4^*(t)$.
For the disturbance $u_j(t)$ in (4.53), we express

$$u_j(t) = \theta_j^{*T} \omega_j(t) \quad (4.65)$$

where

$$\theta_j^* = [\bar{u}_j, \bar{d}_{j1}, \ldots, \bar{d}_{jq}]^T, \ \omega_j(t) = [1, f_{j1}(t), \ldots, f_{jq}(t)]^T \in R^{q+1}. \quad (4.66)$$

For $\theta_4^*(t)$ in (4.63), we express

$$\theta_4^*(t) = -\frac{k_{ij} Z_j(s) P_i(s)}{\Lambda(s)} [u_j](t) = \theta_{ij}^{*T} \frac{a(s)}{\Lambda(s)} [u_j](t) + \theta_{ij0}^* u_j(t) \quad (4.67)$$

for some parameters $\theta_{ij}^* \in R^{n-1}$ and $\theta_{ij0}^* \in R$ such that

$$\theta_{ij}^{*T} \frac{a(s)}{\Lambda(s)} + \theta_{ij0}^* = -\frac{k_{ij} Z_j(s) P_i(s)}{\Lambda(s)}. \tag{4.68}$$

We then introduce the new parametrization

$$\theta_4^*(t) = \theta_5^{*T} \omega_5(t) + \theta_{50}^{*T} \omega_j(t) \tag{4.69}$$

where

$$\begin{aligned}
\theta_5^* &= \theta_{ij}^* \otimes \theta_j^*, \ \theta_{50}^* = \theta_{ij0}^* \theta_j^* \\
\omega_5(t) &= \frac{A(s)}{\Lambda(s)} [\omega_j](t), \ A(s) = [I, sI, \ldots, s^{n-2} I]^T
\end{aligned} \tag{4.70}$$

with $\theta_{ij}^* \otimes \theta_j^* = [\theta_{ij1}^* \theta_j^{*T}, \ldots, \theta_{ij\,n-1}^* \theta_j^{*T}]^T \in R^{(n-1)(q+1)}$ being the Kronecker product of $\theta_{ij}^* = [\theta_{ij1}^*, \ldots, \theta_{ij\,n-1}^*]^T \in R^{n-1}$ and $\theta_j^* \in R^{q+1}$.

For the adaptive controller (4.64), we use

$$\theta_4(t) = \theta_5^T(t) \omega_5(t) + \theta_{50}^T(t) \omega_j(t) \tag{4.71}$$

where $\theta_5(t)$ and $\theta_{50}(t)$ are the estimates of θ_5^* and θ_{50}^*.

Using (4.59), (4.64), and (4.69), we obtain the error equation

$$\theta_3^* P_m(s)[y - y_m](t) = \tilde{\theta}^T(t) \omega(t) \tag{4.72}$$

where $\tilde{\theta}(t) = \theta(t) - \theta^*$,

$$\begin{aligned}
\theta(t) &= [\theta_1^T(t), \theta_2^T(t), \theta_{20}(t), \theta_3(t), \theta_5^T(t), \theta_{50}^T(t)]^T \\
\theta^* &= [\theta_1^{*T}, \theta_2^{*T}, \theta_{20}^*, \theta_3^*, \theta_5^{*T}, \theta_{50}^{*T}]^T \\
\omega(t) &= [\omega_1^T(t), \omega_2^T(t), y(t), r(t), \omega_5^T(t), \omega_j^T(t)]^T.
\end{aligned} \tag{4.73}$$

Based on the error equation (4.72) (which is similar to that in (4.30)), an estimation error, similar to that in (4.36), can be defined, which is suitable for the development of an adaptive law (similar to that in (4.37) and (4.38)) for updating the parameter estimate $\theta(t)$ so that the desired closed-loop stability and asymptotic output tracking are ensured.

4.5.2 Adaptive Failure Compensation

Now an adaptive control scheme is developed for the plant (4.1) with unknown plant parameters and varying actuator failures (4.53). Based on the equal-actuation scheme (4.5), from (4.10), (4.11), and (4.54) we have

$$\begin{aligned}
&\theta_3^* P_m(s)[y - y_m](t) \\
&= \theta_4^*(t) + \left(1 - \theta_1^{*T} \frac{a(s)}{\Lambda(s)}\right) \frac{\sum_{j=j_1,\ldots,j_p} k_{pj} Z_j(s)}{k_p Z_a(s)} [\bar{u}_j](t). \tag{4.74}
\end{aligned}$$

There exists a monic polynomial $P_a(s)$ of degree $n^* - 1$ such that

$$\Lambda(s) - \theta_1^{*T} a(s) = Z_a(s) P_a(s). \tag{4.75}$$

Then, we can choose the nominal compensation signal

$$\theta_4^*(t) = - \sum_{j=j_1,\dots,j_p} \frac{k_{pj} Z_j(s) P_a(s)}{k_p \Lambda(s)} [\bar{u}_j](t) \tag{4.76}$$

to ensure the desired error equation $\theta_3^* P_m(s)[y - y_m](t) = 0$, under Assumption (A4.1) that $Z_a(s)$ has the same degree $n - n^*$ as $Z_j(s)$, $j = j_1, j_2, \dots, j_p$, which is needed for solving the failure compensation problem.

For the failure signals $\bar{u}_j(t)$, $j = j_1, j_2, \dots, j_p$, in (4.76), with θ_j^* and $\omega_j(t)$ defined in (4.66), we parametrize $\theta_4^*(t)$ as

$$
\begin{aligned}
\theta_4^*(t) &= - \sum_{j=j_1,\dots,j_p} \frac{k_{pj} Z_j(s) P_a(s)}{k_p \Lambda(s)} [\bar{u}_j](t) \\
&= \sum_{j=1}^m \theta_{pj}^{*T} \frac{a(s)}{\Lambda(s)} [\bar{u}_j](t) + \sum_{j=1}^m \theta_{pj0}^* \bar{u}_j(t)
\end{aligned}
\tag{4.77}
$$

for some parameters $\theta_{pj}^* \in R^{n-1}$ and $\theta_{pj0}^* \in R$, where $\theta_{pj}^* = 0$, $\theta_{pj0}^* = 0$, for $j \neq j_1, j_2, \dots, j_p$. Using the Kronecker product $\otimes$ (see (4.70)), we define

$$\theta_{6j}^* = \theta_{pj}^* \otimes \theta_j^*, \; j = 1, 2, \dots, m \tag{4.78}$$

and introduce the new parametrization

$$\theta_4^*(t) = \theta_6^{*T} \omega_6(t) + \theta_{60}^{*T} \bar{\omega}(t) \tag{4.79}$$

where

$$
\begin{aligned}
\theta_6^* &= [\theta_{61}^{*T}, \dots, \theta_{6m}^{*T}]^T \in R^{m(n-1)(q+1)} \\
\theta_{60}^* &= [\theta_{p10}^* \theta_1^{*T}, \dots, \theta_{pm0}^* \theta_m^{*T}]^T \in R^{m(q+1)} \\
\omega_6(t) &= \frac{A(s)}{\Lambda(s)} [\bar{\omega}](t), \; A(s) = [I, sI, \dots, s^{n-2}I]^T \\
\bar{\omega}(t) &= [\omega_1^T, \omega_2^T, \dots, \omega_m^T]^T \in R^{m(q+1)}.
\end{aligned}
\tag{4.80}
$$

The equal-actuation based adaptive controller is then designed as

$$v_0(t) = \theta_1^T \omega_1(t) + \theta_2^T \omega_2(t) + \theta_{20} y(t) + \theta_3 r(t) + \theta_6^T \omega_6(t) + \theta_{60}^T \bar{\omega}(t) \tag{4.81}$$

where $\theta_1, \theta_2, \theta_{20}, \theta_3, \theta_6, \theta_{60}$ are estimates of $\theta_1^*, \theta_2^*, \theta_{20}^*, \theta_3^*, \theta_6^*, \theta_{60}^*$, and

$$\omega_1(t) = \frac{a(s)}{\Lambda(s)} [v_0](t), \; \omega_2(t) = \frac{a(s)}{\Lambda(s)} [y](t). \tag{4.82}$$

From (4.74), (4.76), and (4.81), we derive the tracking error equation as

$$\theta_3^* P_m(s)[y - y_m](t) = \tilde{\theta}^T(t)\omega(t) \tag{4.83}$$

where $\tilde{\theta}(t) = \theta(t) - \theta^*$,

$$\theta(t) = [\theta_1^T(t), \theta_2^T(t), \theta_{20}(t), \theta_3(t), \theta_6^T(t), \theta_{60}^T(t)]^T$$
$$\theta^* = [\theta_1^{*T}, \theta_2^{*T}, \theta_{20}^*, \theta_3^*, \theta_6^{*T}, \theta_{60}^{*T}]^T$$
$$\omega(t) = [\omega_1^T(t), \omega_2^T(t), y(t), r(t), \omega_6^T(t), \bar{\omega}^T(t)]^T. \tag{4.84}$$

Based on the error equation (4.84), we introduce

$$\zeta(t) = W_m[\omega](t) \tag{4.85}$$
$$\xi(t) = \theta^T(t)\zeta(t) - W_m(s)[\theta^T\omega](t) \tag{4.86}$$
$$\epsilon(t) = e(t) + \rho(t)\xi(t) \tag{4.87}$$

where $\rho(t)$ is the estimate of $\rho^* = 1/\theta_3^*$, and choose the adaptive laws

$$\dot{\theta}(t) = -\frac{\text{sign}[\theta_3^*]\Gamma\zeta(t)\epsilon(t)}{1 + \zeta^T\zeta + \xi^2}, \quad \Gamma = \Gamma^T > 0 \tag{4.88}$$
$$\dot{\rho}(t) = -\frac{\gamma\xi(t)\epsilon(t)}{1 + \zeta^T\zeta + \xi^2}, \quad \gamma > 0. \tag{4.89}$$

Similar to the analysis in Section 4.3, it can be proved that the adaptive controller (4.81) with adaptive laws (4.88) and (4.89), applied to the plant (4.1) with varying actuator failures (4.53), ensures closed-loop signal boundedness and asymptotic output tracking: $\lim_{t\to\infty}(y(t) - y_m(t)) = 0$.

In summary, in this chapter we present output feedback output tracking adaptive actuator failure compensation schemes for systems with actuator failures characterized by some unknown actuators being stuck at some unknown values at unknown time instants. We develop an effective controller structure for adaptive actuator failure compensation under a set of characterizable conditions. The controllers, when implemented with true parameters, ensure desired plant-model matching in the presence of actuator failures and, when implemented with adaptive parameter estimates, ensure asymptotic output tracking in the presence of unknown plant parameters and unknown actuator failure parameters. Simulation results verified the desired performance of adaptive actuator failure compensation. To handle actuator failures with varying and parametrizable failure values, a linear parametrization of the bilinear parameters (one set from the nominal feedforward compensator and the other set from the failure signals) is employed for adaptive failure compensation with desired stability and tracking performance.

Chapter 5

Designs for Multivariable Systems

In Chapters 3 and 4, we developed adaptive state feedback and output feedback actuator failure compensation control schemes for single-output linear time-invariant systems with possible failures of redundant actuators whose effect on the system dynamic behavior is similar (so that an equal-actuation (or proportional actuation, see (1.15)) scheme for individual actuators can be used: $v_1(t) = v_2(t) = \cdots = v_m(t)$). All those schemes guarantee the closed-loop signal boundedness as well as asymptotic output tracking in the presence of uncertain actuator failures.

In this chapter, we extend the adaptive actuator failure compensation approach to multi-input multi-output (MIMO) systems with M groups of inputs (actuators) and M outputs. For each input group, we assume that some actuators may fail during system operation, but that at least one actuator will remain fully functional and can be used to compensate for other actuators' failures. This problem is of both theoretical and practical significance. It is often the case that some actuators are used to perform the same or similar control tasks to provide actuator redundancy, and at the same time different types of actuators are used for different control actions.

A typical example is modern transport aircraft, which have two or more engines. In longitudinal motion control, the engines are used for forward speed control, and the elevator and stabilizer are used for pitch rate control (in normal flight or emergency situations). In some special designs, the elevator and stabilizer may consist of multiple independently operated segments in order to provide redundancy. We can consider the engines as one group of inputs and the elevator and stabilizer (possibly segmented) as another group. It should be noted that the plant is not fully decoupled so that engines have some effect on the pitching motion and the elevator and stabilizer also have some effect on forward speed. The longitudinal motion can be modeled as a plant with two groups of inputs and two outputs. Another example is an industrial boiler. For boiler control, fuel and air are adjusted for desired combustor temperature and air/fuel ratio. In some large modern boilers, multiple actuators may be used to feed fuel and air into the combustor. Such a system can also be considered as a plant with two groups of inputs and two outputs.

Actuator failure compensation control for such MIMO systems is based on multivariable adaptive control theory and our developed adaptive actuator failure compensation approach for single-output systems. Adaptive control of MIMO systems is an active research area in which the key issues are dynamic decoupling of system interactor matrices and static decoupling of system high frequency gain matrices, in addition to plant and controller parametrizations. These issues are also important in actuator failure compensation control designs for such MIMO systems with actuator failures.

In this chapter, we address such issues and develop two adaptive actuator failure compensation control designs for MIMO systems. In Section 5.1, we formulate the actuator failure compensation problem. In Section 5.2, we present the controller structure and design conditions for plant-model matching control when both plant parameters and actuator failure parameters are available. In Section 5.3, we develop two adaptive designs for asymptotic output tracking when plant and actuator failure parameters are unknown: a basic design and a design based on an SDU decomposition of the system high frequency gain matrix. In Section 5.4, we present simulation results to illustrate the design and performance of adaptive actuator failure compensation control for a MIMO aircraft system model.

5.1 Problem Statement

In this section, we formulate the control problem, including the controlled plant model, actuator failure model, and control objective.

Plant. We consider a linear time-invariant plant

$$y(t) = G_0(s)[u](t) \tag{5.1}$$

where $y(t) \in R^M$ is the plant output vector, $u(t) \in R^N$ is the plant input vector, and $G_0(s)$ is an $M \times N$ transfer matrix, with s being the Laplace transform variable or the time differentiation operator as the case may be. We assume that the N inputs can be separated into M groups. Each group contains n_i inputs, with $n_i > 1$, $i = 1, \ldots, M$, and $\sum_{i=1,\ldots,M} n_i = N$. Without loss of generality, we can express $G_0(s)$ and $u(t)$ in (5.1) as

$$\begin{aligned} G_0(s) &= [G_{11}(s), \ldots, G_{1n_1}(s), G_{21}(s), \ldots, G_{2n_2}(s), \ldots, \\ &\quad G_{M1}(s), \ldots, G_{Mn_M}(s)] \end{aligned} \tag{5.2}$$

$$u(t) = [u_{11}(t), \ldots, u_{1n_1}(t), u_{21}(t), \ldots, u_{2n_2}(t), \ldots,$$
$$u_{M1}(t), \ldots, u_{Mn_M}(t)]^T \tag{5.3}$$

where $G_{ij}(s)$, $i = 1, \ldots, M$, $j = 1, \ldots, n_i$, are $1 \times M$ vectors.

Failure Model. We consider the case that any actuator may fail during system operation, but at least one actuator in each group remains fully functional. For each group, the type of actuator failure is modeled as

$$u_{ij}(t) = \bar{u}_{ij}, \ t \geq t_{ij}, \ j \in \{1, 2, \ldots, n_i\}, \ i \in \{1, 2, \ldots, M\} \tag{5.4}$$

where the failed actuators, failure value $\bar{u}_{ij}$, and failure time instant t_{ij} are all unknown. The case when $M = 1$ is considered in Chapters 3 and 4.

Similar to the previous chapters, let $v_i(t) = [v_{i1}(t), \ldots, v_{in_i}(t)]^T$ be a designed control input vector for group i. Then, in the presence of actuator failures, the actual control input that group i produces is

$$u_i(t) = [u_{i1}(t), \ldots, u_{in_i}(t)]^T = v_i(t) + \sigma_i(\bar{u}_i - v_i(t)) \tag{5.5}$$

where $\bar{u}_i = [\bar{u}_{i1}, \bar{u}_{i2}, \ldots, \bar{u}_{in_i}]^T$, $\sigma_i = \text{diag}\{\sigma_{i1}, \sigma_{i2}, \ldots, \sigma_{in_i}\}$, and $\sigma_{ij} = 1$ if the jth actuator in group i fails, and $\sigma_{ij} = 0$ otherwise. The term $\sigma_i(\bar{u}_i - v_i(t))$ indicates the actuation error of the group i actuators.

The type of actuator failure (5.4) is the so-called lock-in-place failure, which is often encountered in control systems, such as the control surfaces of aircraft may be stuck at some places and cannot be influenced by the control inputs. The actuator failures not only decrease the system maneuverability but also introduce additional unwanted control effects into the system. More general types of actuators failures are those in (1.6) or (1.8), which have been treated in Chapters 2–4 and can be similarly treated for MIMO systems.

Control Objective. The control objective is to design an output feedback control $v(t) = [v_1^T(t), \ldots, v_M^T(t)]^T$ such that despite the unknown actuator failures (5.4), the plant output vector $y(t)$ asymptotically tracks a given reference output vector $y_m(t)$ generated from the reference model

$$y_m(t) = W_m(s)[r](t) \tag{5.6}$$

where $W_m(s)$ is an $M \times M$ stable rational transfer matrix, $r(t) \in R^M$ is a bounded reference input vector, and all closed-loop signals are bounded.

To achieve the control objective, we need to use the remaining actuators to achieve the desired plant-model matching (or tracking) as well as to cancel the unwanted control effect from the failed actuators. As stated previously, the basic assumption for the actuator failure compensation problem is that in the presence of actuator failures, the remaining actuators can still achieve the desired control objective when designed with the knowledge of both plant parameters and actuator failure parameters.

Following our adaptive actuator failure compensation approach developed in the previous chapters for single-output systems, for the case of MIMO systems, the first step is also to derive desired plant-model output matching conditions with *known* plant and actuator failure parameters. This step is presented in the next section. The second step is to develop adaptive laws for updating controller parameter estimates when both the plant parameters and failure parameters are *unknown*, which is done in Section 5.3.

5.2 Plant-Model Matching Control

For the actuator failure compensation problem considered in Section 5.1, as stated in the basic assumption, it is reasonable to assume that under certain conditions, we can design a nominal controller to achieve exact plant-model output matching when both plant parameters and actuator failure parameters are known. Such plant-model matching conditions (not the matching parameters) will be used as *a priori* knowledge for an adaptive design when the plant parameters and actuator failures are unknown.

In this section, we present the design and analysis of a plant-model output matching controller that also achieves closed-loop stability under some specified plant-model output matching conditions.

Actuation Scheme. Under the assumption that actuators within a group have a similar effect on the system dynamic behavior, for each group of actuators, we use an equal-actuation scheme

$$v_{i1}(t) = \cdots = v_{in_i}(t) \stackrel{\triangle}{=} v_{i0}(t), \ i = 1, \ldots, M \tag{5.7}$$

for output tracking control of MIMO systems with actuator failures. We define the design control input vector to be designed as follows:

$$v_0(t) \stackrel{\triangle}{=} [v_{10}(t), \ldots, v_{M0}(t)]^T. \tag{5.8}$$

This equal-actuation scheme is a natural selection inspired from practical experiences. For example, in the case for aircraft control, the two pieces of elevator (one on the left horizontal tail and one on the right) always move

at the same angle. If we further segment each into more subsegments (for redundancy), it is natural to make them move at an equal angle (but controlled
by different actuators). Such an equal-actuation scheme can be extended to
a proportional-actuation control scheme by appropriately scaling the control
signals to different actuators based on some physical consideration. The design and analysis of a proportional-actuation control design can be performed
in a way similar to this equal-actuation control design.

Plant Model. Starting from the case when there is no actuator failure, we
describe the controlled plant (5.1) as

$$y(t) = G(s)[v_0](t) \tag{5.9}$$

where

$$
\begin{aligned}
G(s) &= [G_1(s), \ldots, G_M(s)] \\
&= \left[\sum_{j=1,\ldots,n_1} G_{1j}(s), \ldots, \sum_{j=1,\ldots,n_M} G_{Mj}(s) \right]
\end{aligned} \tag{5.10}
$$

is an $M \times M$ rational transfer matrix.

In the presence of actuator failures, this expression needs to be modified.
Suppose that at time t, there are totally p failed actuators, that is,

$$u_{ij} = \bar{u}_{ij}, \ i = 1, \ldots, M, \ j = j_{i1}, \ldots, j_{ip_i} \tag{5.11}$$

where $0 \le p_i < n_i$, $\sum_{i=1,\ldots,M} p_i = p$. Then the plant (5.1) is described as

$$y(t) = G_a(s)[v_0](t) + \bar{y}(t) \tag{5.12}$$

where

$$
\begin{aligned}
G_a(s) &= [G_{a1}(s), \ldots, G_{aM}(s)] \\
&= \left[\sum_{j \ne j_{11}, \ldots, j_{1p_1}} G_{1j}(s), \ldots, \sum_{j \ne j_{M1}, \ldots, j_{Mp_M}} G_{Mj}(s) \right]
\end{aligned} \tag{5.13}
$$

is the $M \times M$ transfer matrix associated with the unfailed actuators, and

$$\bar{y}(t) = \sum_{i=1,\ldots,M} \sum_{j=j_{i1},\ldots,j_{ip_i}} G_{ij}(s)[\bar{u}_{ij}](t). \tag{5.14}$$

Clearly, the no-failure case (5.9) is a special case of (5.12) with $G_a(s) = G(s)$
and $\bar{y}(t) = 0$, so we can use (5.12) as a general expression in our analysis.

Assumptions. To design suitable control schemes for meeting the stated control objective, we assume that for each possible failure pattern, $G_a(s)$ is strictly proper and of full rank such that

(A5.1) an upper bound $\bar{\nu}_0$ on the observability indices of all possible $G_a(s)$ is known;

(A5.2) all zeros of $G_a(s)$ are stable;

(A5.3) the zero structure at infinity of $G_a(s)$ is known and does not change with actuator failures, that is, there is a known modified left interactor (MLI) matrix $\xi_m(s)$ for all failure patterns such that

$$\lim_{s\to\infty}\xi_m(s)G_a(s) = K_{pa} \qquad (5.15)$$

the high frequency gain matrix of $G_a(s)$ associated with each failure pattern, is finite and nonsingular;

(A5.4) the reference model transfer matrix is $W_m(s) = \xi_m^{-1}(s)$;

(A5.5) there exist two $M \times M$ polynomial matrices $P(s)$ and $P_l(s)$ for all failure patterns such that

$$G_a(s) = Z_a(s)P^{-1}(s) = P_l^{-1}(s)Z_{la}(s) \qquad (5.16)$$

where $Z_a(s)$ and $Z_{la}(s)$ are $M \times M$ polynomial matrices with $Z_a(s)$ and $P(s)$ right co-prime, $Z_{la}(s)$ and $P_l(s)$ left co-prime, and $P(s)$ column proper, and $P_l(s)$ row proper [65].

In these assumptions, $\xi_m(s)$, $\bar{\nu}_0$, $P(s)$, and $P_l(s)$ are fixed for all $t \in (0, \infty)$, which is reasonable given that the failures are from the actuators, but $G_a(s)$, K_{pa}, $Z_a(s)$, and $Z_{la}(s)$ change piecewise as failure patterns change.

These assumptions are based on those in [123] used for the design of MRAC schemes for MIMO systems without actuator failure, and they, together with an additional condition on the high frequency gain matrix K_{pa} to be stated later, are sufficient design conditions for plant-model output matching, which should be satisfied for each possible failure pattern.

Remark 5.2.1. Based on Assumption (A5.5), it is easy to see that $\bar{y}(t)$ in (5.14) can be described as

$$\bar{y}(t) = P_l^{-1}(s) \sum_{i=1,\dots,M} \sum_{j=j_{i1},\dots,j_{ip_i}} Z_{ij}(s)[\bar{u}_{ij}](t) \qquad (5.17)$$

for some M-dimensional polynomial vectors $Z_{ij}(s)$, $i = 1,\dots,M$, $j = j_{i1},\dots,j_{ip_i}$, which is important for plant-model matching analysis. $\qquad\square$

Plant-Model Matching Controller. Denote the nominal control input vector designed with the knowledge of the plant and failure parameters as

$$v_0(t) = v_0^*(t) \triangleq [v_{10}^*(t), \ldots, v_{M0}^*(t)]^T \qquad (5.18)$$

and choose the controller structure for $v_0^*(t)$ as

$$v_0(t) = v_0^*(t) = \Theta_1^{*T}\omega_1(t) + \Theta_2^{*T}\omega_2(t) + \Theta_{20}^*y(t) + \Theta_3^*r(t) + \Theta_4^* \qquad (5.19)$$

where $\omega_1(t) = F(s)[v_0](t)$, $\omega_2 = F(s)[y](t)$, $F(s) = A_0(s)/n(s)$, $A_0(s) = [I, sI, \ldots, s^{\bar{\nu}_0-2}I]^T$, and $n(s)$ is any monic stable polynomial of degree $\bar{\nu}_0 - 1$. $\Theta_1^* = [\Theta_{11}^*, \ldots, \Theta_{1(\bar{\nu}_0-1)}^*]^T$, $\Theta_2^* = [\Theta_{21}^*, \ldots, \Theta_{2(\bar{\nu}_0-1)}^*]^T$, Θ_{20}^*, Θ_3^*, $\Theta_{ij}^* \in R^{M \times M}$, $i = 1, 2$, $j = 1, \ldots, \bar{\nu}_0 - 1$, are for plant-model matching, and $\Theta_4^* \in R^M$ is used to cancel the effect of the failed actuators, to be defined next. It is clear that when there is no actuator failure, $\Theta_4^* = 0$ should be used.

Matching Parameters and Analysis. Now we show the existence of the parameters Θ_1^*, Θ_2^*, Θ_{20}^*, Θ_3^*, and Θ_4^* of the plant-model matching controller (5.19) based on Assumptions (A5.1)–(A5.5), and give the matching analysis, which is based on a fixed failure pattern (5.11) and applies to any possible failure patterns (including the no-failure case).

We rewrite the plant (5.12) as

$$y(t) = G_a(s)[v_0^*](t) + \bar{y}(t). \qquad (5.20)$$

Substituting (5.20) into (5.19), we get

$$\begin{aligned} v_0^*(t) = & \ \Theta_1^{*T}F(s)[v_0^*](t) + \Theta_2^{*T}F(s)G_a(s)[v_0^*](t) + \Theta_2^{*T}F(s)[\bar{y}](t) \\ & + \Theta_{20}^*G_a(s)[v_0](t) + \Theta_{20}^*\bar{y}(t) + \Theta_3^*r(t) + \Theta_4^* \end{aligned} \qquad (5.21)$$

which can be further expressed as

$$\begin{aligned} v_0^*(t) = & \left(I - \Theta_1^{*T}F(s) - \Theta_2^{*T}F(s)G_a(s) - \Theta_{20}^*G_a(s)\right)^{-1} \\ & \cdot \left(\Theta_2^{*T}F(s)[\bar{y}](t) + \Theta_{20}^*\bar{y}(t) + \Theta_3^*r(t) + \Theta_4^*\right). \end{aligned} \qquad (5.22)$$

Hence we have $y(t)$ as

$$\begin{aligned} y(t) = & \ G_a(s)\left(I - \Theta_1^{*T}F(s) - \Theta_2^{*T}F(s)G_a(s) - \Theta_{20}^*G_a(s)\right)^{-1} \\ & \cdot \left(\Theta_2^{*T}F(s)[\bar{y}](t) + \Theta_{20}^*\bar{y}(t) + \Theta_3^*r(t) + \Theta_4^*\right) + \bar{y}(t). \end{aligned} \qquad (5.23)$$

With the specification of $n(s), \xi_m(s), P(s)$, and $Z_a(s)$, it follows that there exist Θ_1^*, Θ_2^*, Θ_{20}^*, and $\Theta_3^* = K_{pa}^{-1}$ [32] such that

$$\Theta_1^{*T} A_0(s)P(s) + \Theta_2^{*T} A_0(s)Z_a(s) + \Theta_{20}^* n(s)Z_a(s)$$
$$= n(s)\left(P(s) - \Theta_3^* \xi_m(s)Z_a(s)\right) \tag{5.24}$$

from which we have the plant-model output matching equation as

$$I - \Theta_1^{*T} F(s) - \Theta_2^{*T} F(s)G_a(s) - \Theta_{20}^* G_a(s) = \Theta_3^* W_m^{-1}(s)G_a(s). \tag{5.25}$$

Substituting (5.25) into (5.23), we have

$$\begin{aligned}
y(t) &= G_a(s)(\Theta_3^* W_m^{-1}(s)G_a(s))^{-1} \\
&\quad \cdot \left(\Theta_2^{*T} F(s)[\bar{y}](t) + \Theta_{20}^* \bar{y}(t) + \Theta_3^* r(t) + \Theta_4^*\right) + \bar{y}(t) \\
&= W_m(s)[r](t) + W_m(s)K_{pa} \\
&\quad \cdot \left(\Theta_2^{*T} F(s)[\bar{y}](t) + \Theta_{20}^* \bar{y}(t) + \Theta_3^* \xi_m(s)[\bar{y}](t) + \Theta_4^*\right). \tag{5.26}
\end{aligned}$$

Using the identity

$$\begin{aligned}
&\Theta_2^{*T} F(s) + \Theta_3^* \xi_m(s) + \Theta_{20}^* \\
&= \frac{\Theta_2^{*T} A_0(s) + n(s)\Theta_3^* \xi_m(s) + \Theta_{20}^* n(s)}{n(s)} \\
&= \frac{(\Theta_2^{*T} A_0(s)Z_a(s) + n(s)(\Theta_3^* \xi_m(s) + \Theta_{20}^*)Z_a(s))Z_a^{-1}(s)}{n(s)} \\
&= \frac{(-\Theta_1^{*T} A_0(s)P(s) + n(s)P(s))Z_a^{-1}(s)}{n(s)} \tag{5.27}
\end{aligned}$$

and (5.17), we have

$$\begin{aligned}
f_p(t) &\triangleq W_m(s)K_{pa}[\Theta_2^{*T} F(s)[\bar{y}] + \Theta_{20}^* \bar{y} + \Theta_3^* \xi_m(s)[\bar{y}] + \Theta_4^*](t) \\
&= W_m(s)K_{pa}\left[\frac{(-\Theta_1^{*T} A_0(s)P(s) + n(s)P(s))Z_a^{-1}(s)}{n(s)}\right. \\
&\quad \left. \cdot P_l^{-1}(s) \sum_{i=1,\dots,M} \sum_{j=j_{i1},\dots,j_{ip_i}} Z_{ij}(s)[\bar{u}_{ij}] + \Theta_4^*\right](t) \\
&= W_m(s)K_{pa}\left[\frac{(-\Theta_1^{*T} A_0(s) + n(s)I)Z_{la}^{-1}(s)P_l(s)}{n(s)}\right. \\
&\quad \left. \cdot P_l^{-1}(s) \sum_{i=1,\dots,M} \sum_{j=j_{i1},\dots,j_{ip_i}} Z_{ij}(s)[\bar{u}_{ij}] + \Theta_4^*\right](t) \\
&= W_m(s)K_{pa}\left[\frac{n(s)I - \Theta_1^{*T} A_0(s)}{n(s)}\right. \\
&\quad \left. \cdot Z_{la}^{-1}(s) \sum_{i=1,\dots,M} \sum_{j=j_{i1},\dots,j_{ip_i}} Z_{ij}(s)[\bar{u}_{ij}] + O_4^*\right](t). \tag{5.28}
\end{aligned}$$

The stability of $n(s)$, $W_m(s)$, and $Z_{la}(s)$ guarantees that there exists a constant vector $\Theta_4^* \in R^M$ such that

$$\lim_{t \to \infty} f_p(t) = 0. \tag{5.29}$$

From (5.26), and ignoring the exponentially decaying term $f_p(t)$, we see

$$y(t) = W_m(s)[r](t) \tag{5.30}$$

so that plant-model output matching is achieved.

Remark 5.2.2. The analysis is based on a fixed failure pattern. When the actuator failure pattern changes, the parameters of the equivalent plant transfer matrix $G_a(s)$ and failure values change, and hence K_{pa}, $Z_a(s)$, and matching parameters Θ_1^*, Θ_2^*, Θ_{20}^*, Θ_3^*, and Θ_4^* also change, that is, they are all piecewise constant parameters. Such parameter jumps occur at the time instants when actuators fail and introduce some transient behavior into the closed-loop system. However, over any time interval when the failure pattern is fixed, such transient behavior is exponentially decaying, so that in the above analysis, the effect of such parameter jumps can be ignored.   □

5.3 Adaptive Control Designs

When plant parameters and actuator failures are unknown, the parameters Θ_1^*, Θ_2^*, Θ_{20}^*, Θ_3^*, and Θ_4^* are also unknown. The plant-model matching controller (5.19) cannot be used for failure compensation. We use its adaptive versions to achieve the control objective: stabilization and output tracking.

5.3.1 Basic Controller Structure

As an adaptive version of the nominal plant-model output matching controller (5.19), we use the controller structure

$$v_0(t) = \Theta_1^T(t)\omega_1(t) + \Theta_2^T(t)\omega_2(t) + \Theta_{20}(t)y(t) + \Theta_3(t)r(t) + \Theta_4(t) \tag{5.31}$$

where $\Theta_1(t)$, $\Theta_2(t)$, $\Theta_{20}(t)$, $\Theta_3(t)$, and $\Theta_4(t)$ are the estimates of the unknown parameters Θ_1^*, Θ_2^*, Θ_{20}^*, Θ_3^*, and Θ_4^*, respectively.

For a simple presentation of design and analysis, we define

$$\Theta^* = [\Theta_1^{*T}, \Theta_2^{*T}, \Theta_{20}^*, \Theta_3^*, \Theta_4^*]^T \tag{5.32}$$

$$\Theta(t) = [\Theta_1^T(t), \Theta_2^T(t), \Theta_{20}(t), \Theta_3(t), \Theta_4(t)]^T \tag{5.33}$$

$$\omega(t) = [\omega_1^T(t), \omega_2^T(t), y^T(t), r^T(t), 1]^T \tag{5.34}$$

$$\tilde{\Theta}(t) = \Theta(t) - \Theta^*. \tag{5.35}$$

Then the adaptive controller can be described in the compact form

$$v_0(t) = \Theta^T(t)\omega(t). \tag{5.36}$$

5.3.2 Error Equation

To develop adaptive laws updating the controller parameters, we first derive the tracking error equation. Operating both sides of (5.25) on $v_0(t)$, we have

$$v_0(t) - \Theta_1^{*T} F(s)[v_0](t) - \Theta_2^{*T} F(s)G_a(s)[v_0](t) - \Theta_{20}^* G_a(s)[v_0](t)$$
$$= \Theta_3^* W_m^{-1}(s)G_a(s)[v_0](t). \tag{5.37}$$

Substituting the plant description (5.20) into (5.37), we have

$$v_0(t) - \Theta_1^{*T} F(s)[v_0](t) - (\Theta_2^{*T} F(s) + \Theta_{20}^*)[y - \bar{y}](t)$$
$$= \Theta_3^* W_m^{-1}(s)[y - \bar{y}](t) \tag{5.38}$$

which can be rewritten as

$$v_0(t) = \Theta_1^{*T}\omega_1(t) + \Theta_2^{*T}\omega_2(t) + \Theta_{20}^* y(t) - \Theta_2^{*T} F(s)[\bar{y}](t) - \Theta_{20}^* \bar{y}(t)$$
$$+ \Theta_3^* W_m^{-1}(s)[y](t) - \Theta_3^* W_m^{-1}(s)[\bar{y}](t). \tag{5.39}$$

Substituting (5.31) into (5.39), we have

$$\Theta_1^T \omega_1(t) + \Theta_2^T \omega_2(t) + \Theta_{20} y(t) + \Theta_3 r(t) + \Theta_4$$
$$= \Theta_1^{*T}\omega_1(t) + \Theta_2^{*T}\omega_2(t) + \Theta_{20}^* y(t) - \Theta_2^{*T} F(s)[\bar{y}](t) - \Theta_{20}^*[\bar{y}](t)$$
$$+ \Theta_3^* W_m^{-1}(s)[y](t) - \Theta_3^* W_m^{-1}(s)[\bar{y}](t)$$
$$= \Theta_1^{*T}\omega_1(t) + \Theta_2^{*T}\omega_2(t) + \Theta_{20}^* y(t) + \Theta_3^* r(t) + \Theta_4^*$$
$$- \Theta_3^* r(t) - \Theta_4^* - \Theta_2^{*T} F(s)[\bar{y}](t) - \Theta_{20}^*[\bar{y}](t)$$
$$+ \Theta_3^* W_m^{-1}(s)[y](t) - \Theta_3^* W_m^{-1}(s)[\bar{y}](t) \tag{5.40}$$

which gives

$$\tilde{\Theta}^T \omega(t) = -\Theta_3^* r(t) - \Theta_4^* - \Theta_2^{*T} F(s)[\bar{y}](t) - \Theta_{20}^*[\bar{y}](t)$$
$$+ \Theta_3^* W_m^{-1}(s)[y](t) - \Theta_3^* W_m^{-1}(s)[\bar{y}](t). \tag{5.41}$$

Since $y_m(t) = W_m(s)[r](t)$ with $W_m(s) = \xi_m^{-1}(s)$, (5.41) can be rewritten as

$$\xi_m(s)[y - y_m](t)$$
$$= \Theta_3^{*-1}\left(\tilde{\Theta}^T \omega(t)\Theta_2^{*T} F(s)[\bar{y}](t) + \Theta_{20}^*[\bar{y}](t) + \Theta_3^* W_m^{-1}(s)[\bar{y}](t) + \Theta_4^*\right)$$
$$= \Theta_3^{*-1}\tilde{\Theta}^T \omega(t) + \xi_m(s)[f_p](t) \tag{5.42}$$

where $f_p(t)$ is an exponentially decaying term defined in (5.28).

Ignoring the exponentially decaying effect of $f_p(t)$, and defining the tracking error as $e(t) = y(t) - y_m(t)$, we have the tracking error equation

$$\xi_m(s)[e](t) = \xi_m(s)[y - y_m](t) = \Theta_3^{*-1}[\tilde{\Theta}^T \omega](t). \tag{5.43}$$

Letting d_m be the maximum degree of $\xi_m(s)$, $\Psi^* = \Theta_3^{*-1} = K_{pa}$, $f(s)$ be a stable polynomial of degree d_m, and $h(s) = 1/f(s)$, from (5.43) we have

$$\xi_m(s)h(s)[y - y_m](t) = \Psi^*(h(s)[v_0](t) - \Theta^{*T}h(s)[\omega](t)). \tag{5.44}$$

5.3.3 Design I: The Basic Scheme

We first present a basic adaptive design under the following assumption:

(A5.6) There exists a known $S_p \in R^{M \times M}$ such that for each possible failure pattern,

$$\Gamma_p = K_{pa}^T S_p^{-1} = \Gamma_p^T > 0. \tag{5.45}$$

In Assumption (A5.6), S_p is fixed for all $t \in (0, \infty)$, but Γ_p changes as K_{pa} changes due to the change of failure patterns. Assumption (A5.6) is a design condition for a stable adaptive law.

Adaptive Laws. We define the normalized estimation error as

$$\epsilon(t) = \frac{\xi_m(s)h(s)[y - y_m](t) + \Psi(t)\xi(t)}{m^2(t)} \tag{5.46}$$

where $\Psi(t)$ is the estimate of Ψ^*, and

$$\zeta(t) = h(s)[\omega](t) \tag{5.47}$$
$$\xi(t) = \Theta^T(t)\zeta(t) - h(s)[v_0](t) \tag{5.48}$$
$$m^2(t) = 1 + \zeta^T(t)\zeta(t) + \xi^T(t)\xi(t). \tag{5.49}$$

By ignoring some exponentially decaying terms, we can show that

$$\epsilon(t) = \frac{\Psi^*\tilde{\Theta}^T(t)\zeta(t) + \tilde{\Psi}(t)\xi(t)}{m^2(t)} \tag{5.50}$$

where $\tilde{\Psi} = \Psi - \Psi^*$.

The adaptive laws for the controller parameters can be chosen as

$$\dot{\Theta}^T(t) = -S_p\epsilon(t)\zeta^T(t) \tag{5.51}$$
$$\dot{\Psi}(t) = -\Gamma_\psi\epsilon(t)\xi^T(t) \tag{5.52}$$

where $\Gamma_\psi = \Gamma_\psi^T > 0$, and S_p satisfies Assumption (A5.6).

Stability Analysis. Suppose that during system operation, the actuator failures occur at time instants T_i, $i = 1, \ldots, m_0$, with $m_0 < N$ since at least M of the N actuators do not fail. Then for any time interval (T_i, T_{i+1}), $i = 0, 1, \ldots, m_0$, with $T_0 = 0$ and $T_{m_0+1} = \infty$, the actuator failure pattern is fixed. To analyze the stability and tracking performance of the adaptive control system, we define the positive definite function

$$V(\tilde{\Theta}, \tilde{\Psi}) = \frac{1}{2} \mathrm{tr}[\tilde{\Theta} \Gamma_p(t) \tilde{\Theta}^T] + \frac{1}{2} \mathrm{tr}[\tilde{\Psi}^T \Gamma_\psi^{-1} \tilde{\Psi}] \tag{5.53}$$

where $\mathrm{tr}[\cdot]$ denotes the trace of a matrix. We should note that $V(\tilde{\Theta}, \tilde{\Psi})$ as a function of t is not continuous, as both Θ^* and Γ_p are piecewise constant.

Then over any time interval $t \in (T_i, T_{i+1})$, $i = 0, 1, \ldots, m_0$, when the actuators do not fail, that is, the failure pattern is fixed, $\dot{\tilde{\Theta}} = \dot{\Theta}$, $\dot{\tilde{\Psi}} = \dot{\Psi}$, and $\dot{\Gamma}_p(t) = 0$, so that from (5.51) and (5.52), the time derivative of V is

$$
\begin{aligned}
\dot{V} &= \mathrm{tr}[\tilde{\Theta}(t) \Gamma_p(t) \dot{\tilde{\Theta}}^T(t)] + \frac{1}{2} \mathrm{tr}[\tilde{\Theta}(t) \frac{d}{dt}[\Gamma_p(t)] \tilde{\Theta}^T(t)] + \mathrm{tr}[\tilde{\Psi}^T(t) \Gamma_\psi^{-1} \dot{\tilde{\Psi}}(t)] \\
&= \mathrm{tr}[\tilde{\Theta}(t) \Gamma_p(t) \dot{\Theta}^T(t)] + \mathrm{tr}[\tilde{\Psi}^T(t) \Gamma_\psi^{-1} \dot{\Psi}(t)] \\
&= -\mathrm{tr}[\tilde{\Theta}(t) \Gamma_p(t) S_p \epsilon(t) \zeta^T(t)] - \mathrm{tr}[\tilde{\Psi}^T(t) \Gamma_\psi^{-1} \Gamma_\psi \epsilon(t) \xi^T(t)] \\
&= -\mathrm{tr}[\tilde{\Theta}(t) K_{pa}^T \epsilon(t) \zeta^T(t)] - \mathrm{tr}[\tilde{\Psi}^T(t) \epsilon(t) \xi^T(t)].
\end{aligned} \tag{5.54}
$$

Since for any $x, y \in R^n$, $\mathrm{tr}[yx^T] = y^T x$, we have

$$
\begin{aligned}
\dot{V} &= -(\tilde{\Theta}(t) K_{pa}^T \epsilon(t))^T \zeta(t) - (\tilde{\Psi}^T(t) \epsilon(t))^T \xi(t) \\
&= -\epsilon^T(t)(\Psi^* \tilde{\Theta}^T(t) \zeta(t) + \tilde{\Psi}(t) \xi(t))
\end{aligned} \tag{5.55}
$$

Finally, combining (5.55) with (5.50), we have

$$\dot{V} = -\epsilon^T(t) \epsilon(t) m^2(t) \leq 0, \ t \in (T_i, T_{i+1}), \ i = 0, 1, \ldots, m_0. \tag{5.56}$$

Since there are only a finite number of actuator failures and $V(\tilde{\Theta}, \tilde{\Psi})$ decreases between failures, $V(\tilde{\Theta}(T_{m_0}), \tilde{\Psi}(T_{m_0}), T_{m_0})$ is finite. Then, from

$$\dot{V} = -\epsilon^T(t) \epsilon(t) m^2(t) \leq 0, \ t \in (T_{m_0}, \infty) \tag{5.57}$$

we have $V(\tilde{\Theta}, \tilde{\Psi}) \in L^\infty$, which implies that $\Theta(t), \Psi(t) \in L^\infty$, and $\epsilon(t)m(t) \in L^2 \cap L^\infty$. From (5.51) and (5.52), in view of the fact that $\epsilon(t)m(t) \in L^2 \cap L^\infty$ and (5.49), we obtain $\dot{\Theta}(t) \in L^2 \cap L^\infty$ and $\dot{\Psi}(t) \in L^2 \cap L^\infty$. Based on these properties, the closed-loop stability and asymptotic tracking can be proved in a way similar to [123]. In summary, we have

Theorem 5.3.1. *The adaptive controller (5.31) with the adaptive laws (5.51) and (5.52), applied to the plant (5.1) with actuator failures (5.4), guarantees that all closed-loop signals are bounded and the tracking error* $e(t) = y(t) - y_m(t)$ *goes to zero as t goes to infinity.*

5.3.4 Design II: Based on SDU Factorization of K_{pa}

The adaptive control scheme in Section 5.3.2 needs Assumption (A5.6), which is a relatively restrictive condition. Assumption (A5.6) imposes a condition on the high frequency gain matrix, which leads to a relatively simple controller structure (5.31). The knowledge of the high frequency gain matrix plays a key role in multivariable MRAC designs. An MRAC scheme is proposed in [141], without the need for the knowledge of the high frequency gain matrix, using a family of estimators and a switching control algorithm. Recently, multivariable MRAC designs in [48], [52], [89] using high frequency gain matrix factorizations considerably reduce the requirement of *a priori* knowledge of the high frequency gain matrix, using a single adaptive controller. In this subsection, we present such an adaptive design based on the SDU factorization of the high frequency gain matrix K_{pa}, for actuator failure compensation.

SDU Factorization. The following result of an SDU factorization of square matrices is important for our adaptive controller design.

Lemma 5.3.1. [52] *Every $M \times M$ real matrix K_p with all leading principal minors Δ_1, Δ_2, ..., Δ_m nonzero can be factored as*

$$K_p = SD_sU \tag{5.58}$$

where S is symmetric positive definite, D_s is diagonal, and U is unit upper triangular, with $D_s = \Gamma \mathrm{sign}[D]$, where

$$D = \mathrm{diag}\,\{d_1, d_2, \ldots, d_M\} = \mathrm{diag}\left\{\Delta_1, \frac{\Delta_2}{\Delta_1}, \ldots, \frac{\Delta_M}{\Delta_{M-1}}\right\} \tag{5.59}$$

with $\Gamma = diag\{\gamma_1, \ldots, \gamma_M\}$ for arbitrary $\gamma_i > 0$, $i = 1, 2, \ldots, M$, and

$$\mathrm{sign}[D] = \mathrm{diag}\,\{\mathrm{sign}[d_1], \mathrm{sign}[d_2], \ldots, \mathrm{sign}[d_M]\}\,. \tag{5.60}$$

To develop an adaptive actuator failure compensation scheme, we need the following assumption for the high frequency gain matrix K_{pa}:

(A5.7) All leading principal minors of K_{pa} are nonzero, and their signs are known and do not change as failure patterns change.

Assumption (A5.7) and Lemma 5.3.1 guarantee that for any failure pattern, the high frequency gain matrix K_{pa} has the SDU factorization

$$K_{pa} = S_aD_aU_a, \quad D_a = \Gamma \mathrm{sign}[D] \tag{5.61}$$

where $S_a = S_a^T > 0$, U_a is unit upper triangular, $\Gamma = \text{diag}\{\gamma_1, \ldots, \gamma_M\} > 0$, $D = \text{diag}\{d_1, \ldots, d_M\}$, and $\text{sign}[D]$ does not change with actuator failure patterns. The matrix Γ is free and can be chosen as a design parameter for an adaptive law. The sign of D (i.e., the sign of D_a) is the only information we need from K_{pa} under Assumption (A5.7) for an adaptive design. By Assumption (A5.7) and Lemma 5.3.1, the sign of D is determined by the principal minors of K_{pa} and is known. Both S_a and U_a can be unknown and are allowed to change with failure patterns.

Reparametrized Error Equation. From the error equation (5.43), using (5.61), we see that for any fixed failure pattern,

$$\xi_m(s)[e](t) = K_{pa}\tilde{\Theta}^T \omega(t) = S_a D_a [U_a v_0 - U_a v_0^*](t). \tag{5.62}$$

Introducing the decomposition

$$U_a v_0 = v_0 - (I - U_a)v_0 \tag{5.63}$$

we can rewrite the error equation as

$$
\begin{aligned}
\xi_m(s)[e](t) &= S_a D_a [v_0 - U_a \Theta_1^{*T} \omega_1 - U_a \Theta_2^{*T} \omega_2 - U_a \Theta_{20}^* y \\
&\quad - U_a \Theta_3^* r - U_a \Theta_4^* - (I - U_a)v_0] \\
&= S_a D_a [v_0 - Q_1^{*T} \omega_1 - Q_2^{*T} \omega_2 - Q_{20}^* y - Q_3^* r - Q_4^* - Q_5^* v_0](t) \\
&= \bar{\Psi}^* [v_0 - Q^{*T} \bar{\omega}](t) \tag{5.64}
\end{aligned}
$$

where I is the $M \times M$ identity matrix, and

$$
\begin{aligned}
Q^* &= [Q_1^{*T}, Q_2^{*T}, Q_{20}^*, Q_3^*, Q_4^*, Q_5^*]^T \tag{5.65} \\
\bar{\omega}(t) &= [\omega_1^T, \omega_2^T, y^T, r^T, 1, v_0^T]^T(t) \tag{5.66}
\end{aligned}
$$

with $\bar{\Psi}^* = S_a D_a$, $Q_1^{*T} = U_a \Theta_1^{*T}$, $Q_2^{*T} = U_a \Theta_2^{*T}$, $Q_{20}^* = U_a \Theta_{20}^*$, $Q_3^* = U_a \Theta_3^*$, $Q_4^* = U_a \Theta_4^*$, and $Q_5^* = (I - U_a)$.

Since Q_5^* is strictly upper triangular, to remove the zero entries from its estimate Q_5 for control implementation, we introduce the new parameter vectors $\bar{\Theta}_i^*$, $i = 1, 2, \ldots, M$, via the identity

$$Q^{*T} \bar{\omega}(t) = [\bar{\Theta}_1^{*T} \bar{\Omega}_1, \bar{\Theta}_2^{*T} \bar{\Omega}_2, \ldots, \bar{\Theta}_M^{*T} \bar{\Omega}_M]^T(t) \tag{5.67}$$

where $\bar{\Theta}_i^{*T}$ is a row vector obtained by concatenating the ith row of the matrices $Q_1^{*T}, Q_2^{*T}, Q_{20}^*, Q_3^*$, and Q_4^* together with the nonzero entries (i.e., the entries above the diagonal) of the ith row of Q_5^*, and

$$\bar{\Omega}_1(t) = [\bar{\Omega}_M^T, v_{20}, v_{30}, \ldots, v_{(M-1)0}, v_{M0}]^T$$
$$\bar{\Omega}_2(t) = [\bar{\Omega}_M^T, v_{30}, \ldots, v_{(M-1)0}, v_{M0}]^T$$

$$\ldots$$

$$\bar{\Omega}_{M-1}(t) = [\bar{\Omega}_M^T, v_{M0}]^T$$
$$\bar{\Omega}_M(t) = [\omega_1^T, \omega_2^T, y^T, r^T, 1]^T. \tag{5.68}$$

Adaptive Controller. In view of the parametrization (5.67), we design the control input $v_0(t)$ from the adaptive controller:

$$v_0(t) = [\bar{\Theta}_1^T \bar{\Omega}_1, \bar{\Theta}_2^T \bar{\Omega}_2, \ldots, \bar{\Theta}_M^T \bar{\Omega}_M]^T(t) \tag{5.69}$$

where $\bar{\Theta}_i(t)$ is the estimate of $\bar{\Theta}_i^*$, $i = 1, 2, \ldots, M$.

Let $\bar{\Psi}(t)$ be the estimate of the unknown parameter $\bar{\Psi}^* = S_a D_a$. To develop the adaptive laws for updating the controller parameters $\bar{\Theta}_i(t)$, $i = 1, 2, \ldots, M$, we introduce the auxiliary signals

$$\xi(t) = [\bar{\Theta}_1^T \bar{\phi}_1, \ldots, \bar{\Theta}_M^T \bar{\phi}_M]^T(t) - h(s)[v_0](t) \tag{5.70}$$

$$\epsilon(t) = \frac{h(s)\xi_m(s)[e](t) + \bar{\Psi}(t)\xi(t)}{\eta_s^2(t)} \tag{5.71}$$

$$\bar{\phi}_i(t) = h(s)[\bar{\Omega}_i](t) \tag{5.72}$$

where $\eta_s(t)$ is the normalizing signal defined from

$$\eta_s^2(t) = 1 + \xi^T(t)\xi(t) + \sum_{i=1}^{M} \bar{\phi}_i^T(t)\bar{\phi}_i(t). \tag{5.73}$$

Then we choose the adaptive laws as

$$\dot{\bar{\Theta}}_i(t) = -\gamma_i \text{sign}[d_i]\epsilon(t)\bar{\phi}_i(t) \tag{5.74}$$

$$\dot{\bar{\Psi}}(t) = -\gamma\epsilon(t)\xi^T(t) \tag{5.75}$$

where $\gamma > 0$ and $\gamma_i > 0$, $i = 1, \ldots, M$, are design parameters.

Remark 5.3.1. The design parameters $\gamma > 0$ and $\gamma_i > 0$, $i = 1, \ldots, M$, which may influence the adaptive system response, can be made arbitrary. In particular, the parameters $\gamma_i > 0$, $i = 1, \ldots, M$, are nominally the parameters $\gamma_i > 0$, $i = 1, \ldots, M$, in the SDU factorization of K_{pa}: $K_{pa} = S_a D_a U_a$, $D_a = \Gamma\text{sign}[D]$, $\Gamma = \text{diag}\{\gamma_1, \ldots, \gamma_M\} > 0$ (see (5.61)). Such an SDU factorization of K_{pa} is not unique in terms of $\Gamma = \text{diag}\{\gamma_1, \ldots, \gamma_M\} > 0$ [52] and is used only for parametrization (see (5.64)). A different choice of $\gamma_i > 0$, $i = 1, \ldots, M$, would imply that a different parametrization is used for the adaptive control system, leading to a stable closed-loop system. $\square$

Stability Analysis. We choose the positive definite function

$$V(\bar{\Theta}, \bar{\Psi}) = \frac{1}{2}\left(\sum_{i=1}^{M} \tilde{\Theta}_i^T \tilde{\Theta}_i + \gamma^{-1}\mathrm{tr}[\tilde{\Psi} S_a^{-1} \tilde{\Psi}] \right) \tag{5.76}$$

where $\tilde{\Theta}_i = \bar{\Theta}_i - \bar{\Theta}_i^*$, $\tilde{\Psi} = \bar{\Psi} - \bar{\Psi}^*$. Then from (5.74) and (5.75), and using the fact that $\dot{\tilde{\Theta}}_i = \dot{\bar{\Theta}}_i$, $\dot{\tilde{\Psi}} = \dot{\bar{\Psi}}$, and $\dot{S}_a = 0$ over any time interval that the failure pattern is fixed, we have the time derivative of V as

$$\begin{aligned}
\dot{V} &= \sum_{i=1}^{M} \tilde{\Theta}_i^T(t)\dot{\bar{\Theta}}_i(t) + \gamma^{-1}\mathrm{tr}[\tilde{\Psi}(t)S_a^{-1}\dot{\tilde{\Psi}}(t)] \\
&= -\sum_{i=1}^{M} \tilde{\Theta}_i^T(t)(\gamma_i\mathrm{sign}[d_i]\epsilon_i(t)\bar{\phi}_i(t)) - \gamma^{-1}\mathrm{tr}[\tilde{\Psi}(t)S_a^{-1}\gamma\epsilon(t)\xi^T(t)] \\
&= -\epsilon^T(t)D_a[\tilde{\Theta}_1^T\bar{\phi}_1, \tilde{\Theta}_2^T\bar{\phi}_2, \dots, \tilde{\Theta}_M^T\bar{\phi}_M]^T(t) - \epsilon^T(t)S_a^{-1}\tilde{\Psi}(t)\xi(t) \\
&= -\epsilon^T(t)S_a^{-1}\left(\Psi^*[\tilde{\Theta}_1^T\bar{\phi}_1, \tilde{\Theta}_2^T\bar{\phi}_2, \dots, \tilde{\Theta}_M^T\bar{\phi}_M]^T(t) + \tilde{\Psi}(t)\xi(t) \right). \tag{5.77}
\end{aligned}$$

From (5.70)–(5.72), (5.64), and (5.66), we have

$$\begin{aligned}
\epsilon(t) &= \frac{1}{\eta_s^2(t)}(\bar{\Psi}^*(h(s)[v_0])(t) - [\tilde{\Theta}_1^T\bar{\phi}_1, \tilde{\Theta}_2^T\bar{\phi}_2, \dots, \tilde{\Theta}_M^T\bar{\phi}_M]^T) \\
&\quad + \bar{\Psi}(t)([\bar{\Theta}_1^T\bar{\phi}_1, \dots, \bar{\Theta}_M^T\bar{\phi}_M]^T - h(s)[v_0](t))) \\
&= \frac{1}{\eta_s^2(t)}\left(\Psi^*[\tilde{\Theta}_1^T\bar{\phi}_1, \tilde{\Theta}_2^T\bar{\phi}_2, \dots, \tilde{\Theta}_M^T\bar{\phi}_M]^T + \tilde{\Psi}(t)\xi(t) \right). \tag{5.78}
\end{aligned}$$

Substituting (5.78) into (5.77), we have

$$\dot{V} = -\epsilon^T(t)S_a^{-1}\epsilon(t)\eta_s^2(t) \le 0, \ t \in (T_i, T_{i+1}), \ i = 0, 1, \dots, m_0 \tag{5.79}$$

which implies that $V(\bar{\Theta}, \bar{\Psi}) \in L^\infty$, $\bar{\Theta}_i(t)$, $i = 1, \dots, M$, $\bar{\Psi}(t) \in L^\infty$, and $\epsilon(t)\eta_s(t) \in L^2 \cap L^\infty$. Based on these properties, the closed-loop stability and asymptotic tracking can be proved. In summary, we have the following result:

Theorem 5.3.2. *The adaptive controller (5.69) with the adaptive laws (5.74) and (5.75), applied to the plant (5.1) with actuator failures (5.4), guarantees that all closed-loop signals are bounded and the tracking error $e(t) = y(t) - y_m(t)$ goes to zero as t goes to infinity.*

5.4 Boeing 737 Lateral Control Simulation

To verify the performance of the developed multivariable adaptive actuator failure compensation schemes, we now present a simulation study of the lateral motion control of a Boeing 737 aircraft.

Aircraft Model. A linearized lateral motion model of the Boeing 737 is described in state-space form as

$$\dot{x} = Ax + Bu, \ x = [v_b, p_b, r_b, \phi, \psi]^T, \ u = [d_r, d_a]^T. \tag{5.80}$$

The five state variables are lateral velocity v_b, roll rate p_b, yaw rate r_b, roll angle ϕ, and yaw angle ψ. We choose roll angle ϕ and yaw angle ψ as plant outputs. The control inputs are the rudder position d_r and aileron position d_a. In the case of landing, the matrices A and B are

$$A = \begin{bmatrix} -0.13858 & 14.326 & -219.04 & 32.167 & 0 \\ -0.02073 & -2.1692 & 0.91315 & 0.000256 & 0 \\ 0.00289 & -0.16444 & -0.15768 & -0.00489 & 0 \\ 0 & 1 & 0.00618 & 0 & 0 \\ 0 & 0 & 1 & 0 & 0 \end{bmatrix}$$

$$B = [b_1, b_2], \ b_1 = \begin{bmatrix} 0.15935 \\ 0.01264 \\ -0.12879 \\ 0 \\ 0 \end{bmatrix}, \ b_2 = \begin{bmatrix} 0.00211 \\ 0.21326 \\ 0.00171 \\ 0 \\ 0 \end{bmatrix}. \tag{5.81}$$

Simulation Data for Adaptive Design I. For the adaptive design in Section 5.3.2, we suppose that both rudder and aileron are double-redundant, so we have four actuators such that $u = [d_{r1}, d_{r2}, d_{a1}, d_{a2}]^T$ and $B = [b_{11}, b_{12}, b_{21}, b_{22}]$. Hence the matrix B used in the simulation study is

$$b_{11} = b_{12} = \begin{bmatrix} 0.15935 \\ 0.01264 \\ -0.12879 \\ 0 \\ 0 \end{bmatrix}, \ b_{21} = b_{22} = \begin{bmatrix} 0.00211 \\ 0.21326 \\ 0.00171 \\ 0 \\ 0 \end{bmatrix}. \tag{5.82}$$

By considering d_{r1} and d_{r2} as one group of actuators and d_{a1} and d_{a2} as another group, we have a plant with two group of inputs and two outputs. It can be verified that the aircraft model satisfies all the Assumptions (A5.1)–(A5.6) if at least one actuator in each group remains active.

Simulation Data for Adaptive Design II. For the adaptive design based on SDU factorization of the high frequency gain matrix in Section 5.3.3, we still consider the case wherein both rudder and aileron are augmented with a redundant actuator. The matrix B used in the simulation study is

$$b_{11} = \begin{bmatrix} 0.15935 \\ 0.01264 \\ -0.12879 \\ 0 \\ 0 \end{bmatrix}, b_{12} = \begin{bmatrix} 0.16 \\ 0.012 \\ -0.13 \\ 0 \\ 0 \end{bmatrix}$$

$$b_{21} = \begin{bmatrix} 0.00211 \\ 0.021326 \\ 0.00171 \\ 0 \\ 0 \end{bmatrix}, b_{22} = \begin{bmatrix} 0.002 \\ 0.02 \\ 0.0015 \\ 0 \\ 0 \end{bmatrix}. \tag{5.83}$$

Unlike the simulation study for adaptive design I, here b_{11} is not parallel to b_{12}, and b_{21} is not parallel to b_{22}, which means that the design conditions are relaxed. It can be verified that Assumptions (A5.1)–(A5.5) and (A5.7) are satisfied if at least one actuator in each group remains active.

Reference Model and Failure Pattern. From the simulation data for both adaptive designs, it can be verified that for any possible failure pattern, all entries in the transfer matrix are of the same relative degree 2 and the MLI matrix $\xi_m(s)$ can take the form

$$\xi_m(s) = \text{diag}\{\xi_{m1}(s), \xi_{m2}(s)\} \tag{5.84}$$

and both $\xi_{m1}(s)$ and $\xi_{m2}(s)$ are monic stable polynomials of degree 2, resulting in a nonsingular high frequency gain matrix K_{pa}. For example, for the no-failure case, the high frequency gain matrix is

$$K_{pa} = \begin{bmatrix} 0.0253 & 0.0426 \\ -0.0258 & 0.0034 \end{bmatrix}. \tag{5.85}$$

In the simulation study, we choose

$$\xi_m(s) = \text{diag}\{(s+1)^2, (s+1)^2\} \tag{5.86}$$

for both adaptive designs. Hence the reference model is

$$W_m(s) = \xi_m^{-1}(s) = \text{diag}\left\{\frac{1}{(s+1)^2}, \frac{1}{(s+1)^2}\right\} \tag{5.87}$$

which is fully decoupled, while the controlled plant is cross-coupled for every possible failure pattern. The reference input is selected as $r(t) = [0.1, 0.15]^T$ (radians), which gives modest reference outputs.

For both adaptive designs, we consider the failure pattern:

$d_{r2} = 4$ (deg) for $t \geq 50$ (sec) and
$d_{a2} = 12$ (deg) for $t \geq 100$ (sec).

Simulation Results. The simulation results are shown in Figures 5.1 and 5.2. Despite the transient behavior at the beginning of simulation (because the initial controller parameter estimates are not close to the matching controller parameters) and at the time instants of actuator failures (because of the jumps of matching controller parameters), we see that asymptotic output tracking is achieved for both adaptive designs with unknown rudder and aileron failures and unknown plant parameters.

5.5 Concluding Remarks

In this chapter, we study the actuator failure compensation problem for multi-input multi-output linear time-invariant systems with possible failures of redundant actuators that belong to different groups having different physical features. We propose a controller structure that effectively achieves plant-model matching when implemented with a known plant and known actuator failure parameters. Conditions are derived for plant-model output matching. The existence of such a nominal matching controller is sufficient for adaptive control. Under a unified framework, two adaptive control schemes are developed for the case where both the plant and actuator failure parameters are unknown. Key issues such as controller parametrization, error model, adaptive laws, and stability are resolved for this multivariable case. The developed adaptive control schemes are capable of achieving asymptotic output tracking and closed-loop signal boundedness despite actuator failure and plant parameter uncertainties, which is verified by simulation results.

Although the adaptive designs are derived for actuator failures characterized by unknown inputs being stuck at some unknown fixed values at unknown time instants, similar designs can be developed for varying actuator failures, based on the results of Chapters 2–4.

Multivariable adaptive control is an important area of research [32], [43], [52], [89], [95], [99], [118], [123], [141]. Other design techniques of multivariable adaptive control can also be used for actuator failure compensation; for example, the LDU and LDS factorizations of a high frequency gain matrix (see [52], [89]) can be used to relax conditions on a high frequency gain matrix for adaptive control, in a way similar to that with an SDU factorization.

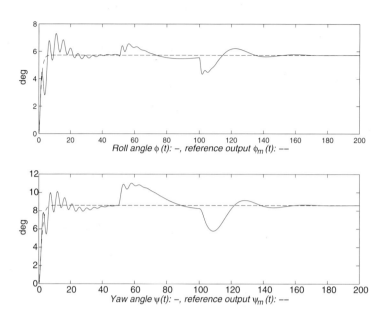

Fig. 5.1. System response with Adaptive Design I

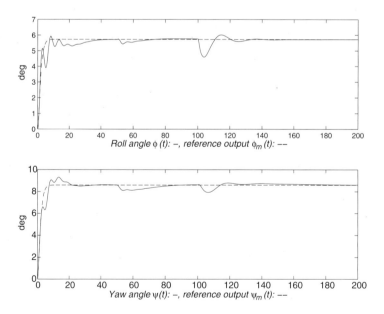

Fig. 5.2. System response with Adaptive Design II

Chapter 6

Pole Placement Designs

In the previous chapters model reference control based adaptive actuator failure compensation schemes were developed for linear time-invariant (LTI) plants with unknown actuator failures. Model reference adaptive control (MRAC) requires that the controlled plant is minimum phase. However, many physical plants are nonminimum phase, and the adaptive control schemes based on MRAC cannot be applied to such systems. For example, the linearized aircraft model may contain both minimum phase and nonminimum phase control channels. For a Boeing 737-100 aircraft, in the case of cruise flight, the control channels from aileron, spoiler, and rudder to roll rate are all nonminimum phase. In the lateral motion model of the DC-8 aircraft, the control channel from aileron to side-slip angle is also nonminimum phase. To compensate for actuator failures in nonminimum phase plants, effective control schemes such as pole placement are needed. In this chapter, we present pole placement based actuator failure compensation control designs for systems with unknown actuator failures, which are applicable to both minimum phase and nonminimum phase LTI plants. In Section 6.1, we formulate the control problem. In Section 6.2, we present a nominal pole placement control scheme for actuator failure compensation. In Section 6.3, we develop an adaptive control scheme for the case when both plant parameters and failure parameters are unknown. In Section 6.4, we present simulation results to verify the desired adaptive system performance.

6.1 Problem Statement

We consider a linear time-invariant plant

$$\dot{x}(t) = Ax(t) + Bu(t), \; y(t) = Cx(t) \tag{6.1}$$

where $A \in R^{n \times n}$, $B = [b_1, \ldots, b_m] \in R^{n \times m}$, $C \in R^{1 \times n}$ are unknown constant matrices, $u(t) = [u_1, \ldots, u_m]^T \in R^m$ is the input vector whose components may fail during system operation, and $y(t) \in R$ is the plant output.

To demonstrate a pole placement based failure compensation design, in this chapter, we consider the actuator failure model (1.2):

$$u_j(t) = \bar{u}_j, \ t \geq t_j, \ j \in \{1, 2, \ldots, m\} \tag{6.2}$$

where the failed actuators, constant value $\bar{u}_j$, and failure time instant t_j are all unknown. In this case, the basic actuator failure compensation assumption is that the system (6.1) is so constructed that for any up to $m - 1$ actuator failures, the remaining actuators can still achieve a desired control objective stated below. The objective of adaptive failure compensation is to adjust the remaining controls to achieve the desired system performance when the plant parameters as well as failure pattern and parameters are unknown.

The control objective is to design a feedback control $v(t)$ for the plant (6.1) with actuator failures (6.2), such that despite the control error, all closed-loop signals are bounded and the plant output $y(t)$ asymptotically tracks a given reference output $y_m(t)$ satisfying

$$Q_m(s)[y_m](t) = 0 \tag{6.3}$$

where

$$Q_m(s) = s^{n_q} + q_{n_q-1}s^{n_q-1} + \cdots + q_1 s + q_0 \tag{6.4}$$

is a monic polynomial of degree n_q with no zeros in $Re[s] > 0$ and no repeated zeros on the $j\omega$-axis so that $y_m(t)$ is bounded. For example, $Q_m(s) = s$ for $y_m(t) = 1$, and $Q_m(s) = s^2 + \omega_0^2$ for $y_m(t) = \sin(\omega_0 t)$.

Unlike the case considered in the previous chapters, where the controlled system is minimum phase, in this chapter the controlled plant can be either minimum phase or nonminimum phase, for which a pole placement control design is needed for both output matching and adaptive tracking. Moreover, the reference output $y_m(t)$ now is a prespecified signal satisfying (6.3).

6.2 Nominal Matching Controller Design

In this section, we develop a nominal pole placement controller that achieves both asymptotic output tracking and closed-loop signal boundedness in the presence of actuator failures, based on the knowledge of plant parameters and actuator failures. We resolve some new issues related to this controller.

Plant Model. To derive a suitable controller parametrization using output feedback, we express the controlled plant in the input–output form

$$P(s)[y](t) = \sum_{j=1}^{m} Z_j(s)[u_j](t) \tag{6.5}$$

where

$$
\begin{aligned}
P(s) &= s^n + p_{n-1}s^{n-1} + \cdots + p_1 s + p_0 &\tag{6.6}\\
Z_j(s) &= z_{j(n-1)}s^{n-1} + \cdots + z_{j1}s + z_{j0}, \; j = 1, \ldots, m &\tag{6.7}
\end{aligned}
$$

are polynomials in s (with s being the time differentiation operator: $s[x](t) = \dot{x}(t)$, or the Laplace transform variable) such that

$$C(sI - A)^{-1}b_j = \frac{Z_j(s)}{P(s)}, \; j = 1, 2, \ldots, m. \tag{6.8}$$

Since A, B, and C are all constant parameters, we have that p_i, z_{ji}, $i = 0, 1, \ldots, n - 1$, $j = 1, 2, \ldots, m$, are also constant parameters.

Actuation Scheme. The design of an actuator failure compensation scheme depends on the selection of a control actuation scheme. In this case, we choose the equal-actuation scheme for a group of physically similar actuators:

$$v_1(t) = v_2(t) = \cdots = v_m(t) \stackrel{\triangle}{=} v_0(t). \tag{6.9}$$

Suppose that at time t, there are p failed actuators, that is,

$$u_j = \bar{u}_j, \; j = j_1, j_2, \ldots, j_p, \; p \in \{0, 1, 2, \ldots, m - 1\}. \tag{6.10}$$

Then, we can describe the plant model (6.5) as

$$P(s)[y](t) = \sum_{j \neq j_1, \ldots, j_p} Z_j(s)[v_0](t) + \sum_{j=j_1, \ldots, j_p} Z_j(s)[\bar{u}_j](t). \tag{6.11}$$

For each failure pattern, defining

$$Z_a(s) = \sum_{j \neq j_1, \ldots, j_p} Z_j(s) = z_{n-1}s^{n-1} + \cdots + z_1 s + z_0 \tag{6.12}$$

and noting that $\bar{u}_j$, $j = j_1, \ldots, j_p$, is constant for each pattern, we have

$$P(s)[y](t) = Z_a(s)[v_0](t) + \theta_f^* \tag{6.13}$$

where

$$\theta_f^* = \sum_{j=j_1, \ldots, j_p} z_{j0}\bar{u}_j. \tag{6.14}$$

Controller Structure. Inspired by the structure of a standard pole placement controller [55] for single-input single-output plants without actuator failures, we choose the controller structure as

$$v_0(t) = (\Lambda_1(s) - C(s)Q_m(s)) \frac{1}{\Lambda_1(s)} [v_0](t) + D(s) \frac{1}{\Lambda_1(s)} [y_m - y](t) + k^* \quad (6.15)$$

where $k^* \in R$ is an additional term to compensate actuator failures,

$$C(s) = s^{n-1} + c_{n-2}s^{n-2} + \cdots + c_1 s + c_0 \quad\quad (6.16)$$
$$D(s) = d_{n+n_q-1}s^{n+n_q-1} + \cdots + d_1 s + d_0 \quad\quad (6.17)$$

are polynomials whose parameters are to be determined from a design equation, and $\Lambda_1(s)$ is a monic stable polynomial

$$\Lambda_1(s) = s^{n+n_q-1} + \bar{\lambda}_{n+n_q-2}s^{n+n_q-2} + \cdots + \bar{\lambda}_1 s + \bar{\lambda}_0. \quad (6.18)$$

We denote the parameter vectors of $C(s)$ and $D(s)$ by

$$\psi_c = [c_0, c_1, \ldots, c_{n-2}]^T \in R^{n-1} \quad\quad (6.19)$$
$$\psi_d = [d_0, d_1, \ldots, d_{n+n_q-1}]^T \in R^{n+n_q}. \quad\quad (6.20)$$

Then ψ_c and ψ_d can be obtained by solving the Diophantine equation

$$C(s)Q_m(s)P(s) + D(s)Z_a(s) = A^*(s) \quad\quad (6.21)$$

where $A^*(s)$ is a given stable polynomial of degree $2n + n_q - 1$ that characterizes the desired closed-loop system performance based on the knowledge of the prespecified closed-loop poles.

For the Diophantine equation (6.21) to be solvable for pole placement control, the following common assumption is needed:

(A6.1) For all failure patterns, $Q_m(s)P(s)$ and $Z_a(s)$ are co-prime.

Assumption (A6.1) guarantees the existence and uniqueness of $C(s)$ and $D(s)$ that satisfy the Diophantine equation (6.21).

The controller structure (6.15) is equivalent to

$$v_0(t) = \psi_1^T \frac{a(s)}{\Lambda_1(s)} [v_0](t) + \psi_2^T \frac{a(s)}{\Lambda_1(s)} [y - y_m](t) + \psi_3(y - y_m)(t) + k^* \quad (6.22)$$

where

$$
\psi_1 =
\begin{bmatrix}
\bar{\lambda}_0 \\
\bar{\lambda}_1 \\
\cdot \\
\cdot \\
\cdot \\
\cdot \\
\bar{\lambda}_{n+n_q-3} \\
\bar{\lambda}_{n+n_q-2}
\end{bmatrix}
-
\begin{bmatrix}
\psi_c & 0 & \cdots & 0 & 0 \\
1 & \psi_c & \cdots & & 0 \\
0 & 1 & \cdots & 0 & \cdot \\
\cdot & 0 & \cdots & 0 & 0 \\
0 & \cdot & \cdots & 0 & 0 \\
0 & 0 & \cdots & \psi_c & 0 \\
0 & 0 & \cdots & 1 & \psi_c
\end{bmatrix}
\begin{bmatrix}
q_0 \\
q_1 \\
\cdot \\
\cdot \\
\cdot \\
\cdot \\
q_{n_q-1} \\
1
\end{bmatrix}
\tag{6.23}
$$

$$
\psi_2 = d_{n+n_q-1}
\begin{bmatrix}
\bar{\lambda}_0 \\
\bar{\lambda}_1 \\
\cdot \\
\cdot \\
\cdot \\
\bar{\lambda}_{n+n_q-3} \\
\bar{\lambda}_{n+n_q-2}
\end{bmatrix}
-
\begin{bmatrix}
d_0 \\
d_1 \\
\cdot \\
\cdot \\
\cdot \\
d_{n+n_q-3} \\
d_{n+n_q-2}
\end{bmatrix}
\tag{6.24}
$$

$$
\psi_3 = -d_{n+n_q-1}. \tag{6.25}
$$

Asymptotic Tracking. We first show that the controller structure (6.15) can achieve asymptotic output tracking. From (6.15), we have

$$
C(s)Q_m(s)[v_0](t) = D(s)[y_m - y](t) + \Lambda_1(s)[k^*]. \tag{6.26}
$$

For a fixed failure pattern, using (6.21), we have

$$
\begin{aligned}
& A^*(s)[y - y_m](t) \\
= \; & A^*(s)[y](t) - A^*[y_m](t) \\
= \; & C(s)Q_m(s)P(s)[y](t) + Z_a(s)D(s)[y](t) - A^*[y_m](t). \tag{6.27}
\end{aligned}
$$

Substituting the plant description (6.13) into (6.27), we have

$$
\begin{aligned}
& A^*(s)[y - y_m](t) \\
= \; & C(s)Q_m(s)Z_a(s)[v_0](t) + C(s)Q_m(s)[\theta_f^*](t) \\
& + Z_a(s)D(s)[y](t) - A^*[y_m](t) \tag{6.28}
\end{aligned}
$$

which, combined with (6.26) and (6.21), gives

$$
\begin{aligned}
& A^*(s)[y - y_m](t) \\
= \; & Z_a(s)D(s)[y_m - y](t) + Z_a(s)\Lambda_1(s)[k^*] \\
& + C(s)Q_m(s)[\theta_f^*] + Z_a(s)D(s)[y](t) - A^*[y_m](t) \\
= \; & Z_a(s)D(s)[y_m](t) + Z_a(s)\Lambda_1(s)[k^*] + C(s)Q_m(s)[\theta_f^*] - A^*[y_m](t) \\
= \; & -C(s)Q_m(s)P(s)[y_m](t) + Z_a(s)\Lambda_1(s)[k^*] + C(s)Q_m(s)[\theta_f^*] \\
= \; & Z_a(s)\Lambda_1(s)[k^*] + C(s)Q_m(s)[\theta_f^*]. \tag{6.29}
\end{aligned}
$$

Since k^* and θ_f^* are constant, from (6.4), (6.12), (6.16), and (6.18), we have

$$A^*(s)[y - y_m](t) = z_0\bar{\lambda}_0 k^* + c_0 q_0 \theta_f^*. \tag{6.30}$$

For $\lim_{t\to\infty}(y(t) - y_m(t)) = 0$ exponentially, we need to ensure the existence of a constant k^* such that $z_0\bar{\lambda}_0 k^* + c_0 q_0 \theta_f^* = 0$, where $z_0 = \sum_{j\neq j_1, j_2, \ldots, j_p} z_{j0}$ in (6.12) and $\theta_f^* = \sum_{j=j_1, \ldots, j_p} z_{j0}\bar{u}_j$ in (6.14), for the following three cases.

Case (i): If $q_0 = 0$, that is, $Q_m(s)$ has one zero at $s = 0$, then we have

$$A^*(s)[y - y_m](t) = z_0\bar{\lambda}_0 k^* \tag{6.31}$$

which indicates that for asymptotic output tracking, we should set $k^* = 0$. The effect of the failure values $\bar{u}_j$, $j = j_1, \ldots, j_p$, is nullified by $Q_m(s)$. For this case, the term k^* in the controller (6.15) or (6.22) is not needed, and the controller structure is simplified as

$$v_0(t) = \psi_1^T \frac{a(s)}{\Lambda_1(s)}[v_0](t) + \psi_2^T \frac{a(s)}{\Lambda_1(s)}[y - y_m](t) + \psi_3(y - y_m)(t). \tag{6.32}$$

Case (ii): If $q_0 \neq 0$ and $z_0 \neq 0$, then from (6.30) there exists a bounded

$$k^* = -\frac{c_0 q_0 \theta_f^*}{z_0\bar{\lambda}_0} \tag{6.33}$$

such that

$$A^*(s)[y - y_m](t) = 0 \tag{6.34}$$

which implies that $\lim_{t\to\infty}(y(t) - y_m(t)) = 0$ exponentially.

Case (iii): If $q_0 \neq 0$ and $z_0 = 0$, then for the existence of k^* to make $z_0\bar{\lambda}_0 k^* + c_0 q_0 \theta_f^* = 0$, we need $\theta_f^* = \sum_{j=j_1, \ldots, j_p} z_{j0}\bar{u}_j = 0$ (see (6.14)).

Since up to $m - 1$ of the m actuators may fail at arbitrary values, to ensure that $\theta_f^* = 0$, we need $z_{j0} = 0$, for all $j = 1, \ldots, m$. For this case, we can also set $k^* = 0$ and use the controller structure (6.32).

On the other hand, if one of z_{j0}, $j = 1, \ldots, m$, is nonzero, then θ_f^* is nonzero (for arbitrary $\bar{u}_j$, $j = 1_1, j_2, \ldots, j_p$), so that z_0 should be nonzero (see Case (ii)). Hence, to guarantee the existence of a constant k^* to make $z_0\bar{\lambda}_0 k^* + c_0 q_0 \theta_f^* = 0$, the following assumption is needed:

(A6.2) If $q_0 \neq 0$, then the terms z_{j0} in (6.7), $j = 1, \ldots, m$, should either be all zero (so that $\theta_f^* = \sum_{j=j_1, \ldots, j_p} z_{j0}\bar{u}_j = 0$) or be such that $z_0 = \sum_{j\neq j_1, j_2, \ldots, j_p} z_{j0} \neq 0$, for all $p \in \{0, 1, 2, \ldots, m - 1\}$ (so that (6.33) can be used to define a desired k^*).

Signal Boundedness. Operating on $y(t)$ by both sides of (6.21), we have

$$C(s)Q_m(s)P(s)[y](t) + D(s)Z_a(s)[y](t) = A^*(s)[y](t). \tag{6.35}$$

From (6.13), (6.26), and (6.35), we have

$$
\begin{aligned}
& A^*(s)[y](t) \\
&= C(s)Q_m(s)P(s)[y](t) + Z_a(s)[D(s)[y_m] \\
& \quad -C(s)Q_m(s)[v_0] + \Lambda_1(s)[k^*]](t) \\
&= Z_a(s)D(s)[y_m](t) + C(s)Q_m(s)[\theta_f^*] + Z_a(s)\Lambda_1(s)[k^*]. \tag{6.36}
\end{aligned}
$$

Since $A^*(s)$ is stable, $y_m(t)$, θ_f^*, and k^* are all bounded signals, we have $y(t) \in L^\infty$. Similarly, operating on $v_0(t)$ by both sides of (6.21), we have

$$C(s)Q_m(s)P(s)[v_0](t) + D(s)Z_a(s)[v_0](t) = A^*(s)[v_0](t). \tag{6.37}$$

Substituting the plant description (6.13) into (6.37), we have

$$
\begin{aligned}
A^*(s)[v_0](t) &= C(s)Q_m(s)P(s)[v_0](t) + D(s)[P(s)[y] - \theta_f^*](t) \\
&= P(s)C(s)Q_m(s)[v_0](t) + D(s)[P(s)[y] - \theta_f^*](t). \tag{6.38}
\end{aligned}
$$

Substituting (6.26) into (6.38), we have

$$
\begin{aligned}
A^*(s)[v_0](t) &= P(s)D(s)[y_m - y](t) + P(s)\Lambda_1(s)[k^*] + D(s)[P(s)[y] - \theta_f^*](t) \\
&= P(s)D(s)[y_m](t) + P(s)\Lambda_1(s)[k^*] - D(s)[\theta_f^*]. \tag{6.39}
\end{aligned}
$$

Since $A^*(s)$ is stable, $y_m(t)$ is bounded, and k^* and θ_f^* are finite constants, we have $v_0(t) \in L^\infty$. Hence, all closed-loop signals are bounded.

Remark 6.2.1. The analysis carried out above is based on the steady-state analysis over a fixed failure pattern and the effect of initial conditions is ignored. When failure patterns change, $Z_a(s)$ and failure values change, which results in the change of nominal controller parameters ψ_c, ψ_d, and k^*, that is, such parameters are all piecewise constant and change their values at the time instants that actuator failures occur. The parameter jumps also introduce some additional transient behavior into the closed-loop system. Such transient behavior caused by initial conditions and parameter jumps is exponentially decaying, so it does not destroy the properties of asymptotic output tracking and closed-loop signal boundedness established above. □

Thus far, we have developed a nominal controller that achieves the desired pole placement, asymptotic output tracking, and closed-loop signal boundedness when both plant parameters and actuator failure values are available. The design conditions on the controlled plant are given by Assumptions

(A6.1) and (A6.2), which do not require the plant to be minimum phase, and hence it is applicable to possibly nonminimum phase plants. The controller does not need the plant relative degree knowledge so that the plant with different actuator failures can have different relative degrees.

6.3 Adaptive Control Scheme

When the plant or actuator failure parameters are unknown, the controller (6.15) or (6.22) cannot be implemented because its parameters depend on the plant and failure parameters. In this section, we develop an adaptive pole placement control design for which we first estimate the unknown parameters, and then use the parameter estimates to obtain the controller parameters.

Plant Parametric Model. Denote the true plant parameters by

$$\theta_p^* = [p_0, p_1, \ldots, p_{n-1}]^T, \ \theta_z^* = [z_0, z_1, \ldots, z_{n-1}]^T. \tag{6.40}$$

Operating on both sides of the plant (6.13) by the stable filter $\frac{1}{\Lambda(s)}$, where

$$\Lambda(s) = s^n + \lambda_{n-1} s^{n-1} + \cdots + \lambda_1(s) + \lambda_0 \tag{6.41}$$

is a monic stable polynomial, and using

$$\theta^* = [\theta_z^{*T}, (\theta_\lambda - \theta_p^*)^T, \theta_f^*]^T, \ \theta_\lambda = [\lambda_0, \lambda_1, \ldots, \lambda_{n-1}]^T \tag{6.42}$$

$$\phi(t) = \left[\frac{1}{\Lambda(s)}[v_0](t), \frac{s}{\Lambda(s)}[v_0](t), \cdots, \frac{s^{n-1}}{\Lambda(s)}[v_0](t), \right.$$

$$\left. \frac{1}{\Lambda(s)}[y](t), \frac{s}{\Lambda(s)}[y](t), \cdots, \frac{s^{n-1}}{\Lambda(s)}[y](t), \frac{1}{\Lambda(s)}[1](t) \right]^T \tag{6.43}$$

where $\frac{1}{\Lambda(s)}[1](t)$ is the step response of $\frac{1}{\Lambda(s)}$, we parametrize (6.13) as

$$y(t) = \frac{Z_a(s)}{\Lambda(s)}[v_0](t) + \frac{\Lambda(s) - P(s)}{\Lambda(s)}[y](t) + \frac{\theta_f^*}{\Lambda(s)}[1](t) = \theta^{*T}\phi(t). \tag{6.44}$$

Parameter Estimation. Let θ_p, θ_z, and θ_f be the estimates of θ_p^*, θ_z^*, and θ_f^*, and $\theta = [\theta_z^T, (\theta_\lambda - \theta_p)^T, \theta_f]^T$, and define the estimation error

$$\epsilon(t) = \theta^T(t)\phi(t) - y(t) \tag{6.45}$$

which can be expressed as

$$\epsilon(t) = \tilde{\theta}^T(t)\phi(t) \tag{6.46}$$

where $\tilde{\theta}^T(t) = \theta(t) - \theta^*$. Using the normalized gradient algorithm [55], we have the adaptive law updating the parameter estimates as

$$\dot{\theta}(t) = -\frac{\Gamma \phi(t)\epsilon(t)}{m^2(t)} \tag{6.47}$$

where $\Gamma = \Gamma^T > 0$, and $m(t)$ is the normalizing signal

$$m(t) = \sqrt{1 + \alpha \phi^T(t)\phi(t)}, \ \alpha > 0. \tag{6.48}$$

This adaptive parameter estimation algorithm has the desired properties.

Lemma 6.3.1. *The adaptive algorithm (6.47) guarantees that*

(i) $\theta(t)$, $\dot{\theta}(t)$, $\epsilon(t)/m(t)$ are bounded, and

(ii) $\epsilon(t)/m(t)$ and $\dot{\theta}(t)$ belong to L^2.

This lemma can be proved using the positive definite function $V(\tilde{\theta}) = \tilde{\theta}^T \Gamma^{-1} \tilde{\theta}/2$, whose time derivative of $V = V(\tilde{\theta}(t))$, along (6.47), is

$$\dot{V} = -\frac{\epsilon^2(t)}{m^2(t)} \tag{6.49}$$

over any time interval where the actuator failure pattern does not change.

With the parameter estimate $\theta(t)$, we can obtain the estimated parameters $\theta_z(t)$, $\theta_p(t)$, and $\theta_f(t)$, since θ_λ is a design parameter vector and is known.

Design Procedure. Once the estimates of the plant parameters and actuator failure values are obtained, we can use them to design an adaptive version of the actuator failure compensation controller (6.15).

The adaptive controller is implemented in the following steps:

(i) Use the parameter estimation algorithm (6.47) to get θ;

(ii) get the estimates of plant parameters and actuator failure values as

$$\theta_z = [\hat{z}_0, \hat{z}_1, \ldots, \hat{z}_{n-1}]^T, \ \theta_p = [\hat{p}_0, \hat{p}_1, \ldots, \hat{p}_{n-1}]^T \tag{6.50}$$

where $\theta = [\theta_z^T, (\theta_\lambda - \theta_p)^T, \theta_f]^T$ and θ_λ is a design parameter vector defined in (6.42) such that the polynomial $\Lambda(s)$ in (6.41) is stable;

(iii) let the estimates of $P(s)$ and $Z_a(s)$ be

$$\widehat{P}(s) = s^n + \hat{p}_{n-1}s^{n-1} + \cdots + \hat{p}_1 s + \hat{p}_0$$
$$\widehat{Z}_a(s) = \hat{z}_{n-1}s^{n-1} + \cdots + \hat{z}_1 s + \hat{z}_0 \tag{6.51}$$

and solve the Diophantine equation

$$\hat{C}(s)Q_m(s)\hat{P}(s) + \hat{D}(s)\hat{Z}_a(s) = A^*(s) \tag{6.52}$$

to get $\hat{\psi}_c$ and $\hat{\psi}_d$, the parameter vectors of $\hat{C}(s)$ and $\bar{D}(s)$, which are the estimates of $C(s)$ and $D(s)$ defined in (6.19)–(6.21);

(iv) calculate k from

$$k = -\frac{c_0 q_0 \theta_f}{\hat{z}_0 \bar{\lambda}_0}; \tag{6.53}$$

(v) design the control input $v_0(t)$ from

$$v_0(t) = \frac{\Lambda_1(s) - C(s)Q_m(s)}{\Lambda_1(s)}[v_0](t) + \frac{D(s)}{\Lambda_1(s)}[y_m - y](t) + k. \tag{6.54}$$

In (6.52) and (6.54), $Q_m(s)$ is a monic polynomial of degree n_q defined in (6.4) such that $Q_m(s)[y_m](t) = 0$, $A^*(s)$ is a stable polynomial of degree $2n + n_q - 1$ characterizing the desired closed-loop system performance, and $\Lambda_1(s)$ is a monic stable polynomial of degree $n + n_q - 1$ defined in (6.18).

As stated in Section 6.3, for two cases: (i) $q_0 = 0$, and (ii) $z_0 = 0$ (see Assumption (A6.2); we need to know whether $z_0 = 0$ as a design condition), we know that $k^* = 0$, we can safely set $k = 0$ and the calculation in (6.53) is not needed. For the case that $q_0 \neq 0$ and $z_0 \neq 0$, a parameter projection may be needed for calculating k in (6.53).

Remark 6.3.1. To implement an adaptive controller, it is required that the Diophantine equation (6.52) with parameter estimates has a uniformly bounded solution. This is a singularity-free condition for adaptive pole placement control problems [55]. Here, we need to assume that Assumption (A6.1) is also satisfied with on-line parameter estimates so that (6.52) can be solved. Although there are many schemes in the literature for ensuring a singularity-free control scheme, its solution remains to be derived for this adaptive actuator failure compensation control problem.   □

Under the singularity-free condition, the desired properties of closed-loop stability and asymptotic output tracking can be proved.

6.4 DC-8 Lateral Control Simulation

As an illustrative example, we use the lateral dynamics model of a DC-8 aircraft [111] as the controlled plant to which the adaptive actuator failure compensation control scheme is applied. In this simulation study, we use the aileron as the control input, and the side-slip angle as the plant output. The original DC-8 lateral dynamics model [111] is modified with one augmented actuation vector (aileron) for the study of actuator failure compensation.

Linearized DC-8 Lateral Dynamic Model. The linearized lateral dynamics of DC-8 with one augmented piece of aileron can be described as

$$\dot{x}(t) = Ax(t) + Bu(t), \; y(t) = Cx(t)$$
$$x(t) = [\beta, p, \phi, r]^T, \; B = [b_1, b_2] \tag{6.55}$$

where β is the side-slip angle, p is the roll rate, ϕ is the roll angle, r is the yaw rate, y is the system output, which is the side-slip angle β in this case, and u is the control input vector that contains two control signals: $u = [u_1, u_2]^T$, to represent two aileron servos' angles, for achieving compensation in the presence of actuator failures. In the model, β, ϕ, u_1, and u_2 are in degrees and p and r are in deg/sec.

From the data provided in [111] (p. 252), in a cruise-flight at altitude 33,000 ft, Mach number 0.84, nominal forward speed 825 ft/sec, the DC-8 lateral-perturbation dynamics matrices are

$$A = \begin{bmatrix} -0.0869 & 0.0 & 0.0390 & -1.0 \\ -4.424 & -1.184 & 0.0 & 0.335 \\ 1.0 & 1.0 & 0.0 & 0.0 \\ 2.148 & -0.021 & 0.0 & -0.228 \end{bmatrix} \tag{6.56}$$

$$b_1 = \begin{bmatrix} 0.0 \\ 2.120 \\ 0.0 \\ 0.065 \end{bmatrix}, \; b_2 = \begin{bmatrix} 0.0 \\ 2.0 \\ 0.0 \\ 0.06 \end{bmatrix} \tag{6.57}$$

$$C = \begin{bmatrix} 1 & 0 & 0 & 0 \end{bmatrix} \tag{6.58}$$

where b_2 is the augmented actuation vector for studying actuator failure compensation (note that the normal case [111] without actuator failure is $u_2 = 0$, and $u_1 =$ the single aileron servo angle).

The control objective is to design an output feedback control vector $v = [v_1, v_2]^T$ applied to the plant (6.55) with actuator failures (6.1), to ensure that the plant output $y(t)$ (side-slip angle β) tracks a reference signal $y_m(t)$, in the presence of actuator failures.

For the plant (6.55)–(6.58), in the case when there is no actuator failure, the transfer function from the designed control input (aileron angle) to the plant output (side-slip angle) is

$$G(s) = C(sI - A)^{-1}(b_1 + b_2)$$
$$= \frac{-0.125s^2 + 0.0922s + 0.0383}{s^4 + 1.4989s^3 + 2.5477s^2 + 2.8327s + 0.0113} \tag{6.59}$$

which has a zero at $s = 1.0777$, that is, the plant is nonminimum phase. It can be verified that when one actuator (u_1 or u_2) fails, the plant is also nonminimum phase. We will use the adaptive pole placement control scheme developed in this chapter to compensate for the aileron segment failure.

Simulation Condition. We performed simulation by considering that u_2 fails during system operation with the following failure pattern:

$$u_2(t) = -2.0 \text{ deg, for } t \geq 100 \text{ sec.}$$

We performed simulation for two cases: (i) to make $y(t)$ track $y_m(t) = 1$ deg, and (ii) to make $y(t)$ track $y_m(t) = 0.5 \sin(0.1t)$ deg.

For case (i), $Q_m(s) = s$ and $k^* = 0$, we used the controller structure (6.32) and chose polynomials $\Lambda_1(s)$ in (6.15) and $\Lambda(s)$ in (6.41) as

$$\Lambda_1(s) = \Lambda(s) = (s+0.3)^4. \tag{6.60}$$

For case (ii), $Q_m(s) = s^2 + 0.01$, we used the controller structure (6.15) and chose $\Lambda_1(s)$ and $\Lambda(s)$ as

$$\Lambda(s) = (s+0.3)^4, \ \Lambda_1(s) = (s+0.3)\Lambda(s). \tag{6.61}$$

The closed-loop characteristic polynomial $A^*(s)$ was chosen as

$$A^*(s) = \Lambda_1(s)P_c(s) \tag{6.62}$$
$$P_c(s) = (s^2 + 0.6s + 2.3196)(s+1.258)(s+0.4) \tag{6.63}$$

which gives much better closed-loop system characteristics.

For both cases, we selected $\alpha = 1$, $y(0) = y_m(0) = 0$, and $\theta(0) = 0.9\theta^*$, where θ^* is the true parameter vector as in (6.42). We set the adaptive gain matrix $\Gamma = 0.2I$ for case (i), and $\Gamma = 0.5I$ for case (ii).

Simulation Results. The simulation results are shown in Figures 6.1 and 6.2. The results are as expected, that is, at the time instants of actuator failures, there are some transients in both the plant output and control input, and as time progresses (when the failure pattern does not change), the tracking error converges (eventually to zero). There are also some transients at the beginning of the system response due to the initial parameter errors. It is shown that such transients also converge to zero when the failure pattern does not change.

For comparison, the system response with a fixed parameter controller is shown in Figure 6.3. The fixed controller is the nominal controller whose parameters are designed to track $y_m(t) = 0.5\sin(0.1t)$ when there is no actuator

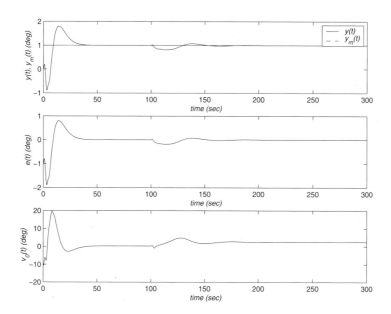

Fig. 6.1. System response for $y_m(t) = 1$ deg.

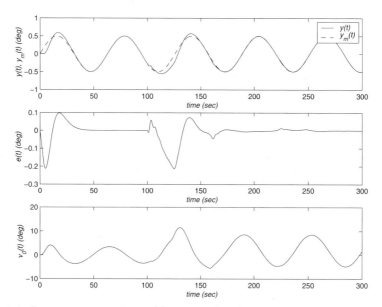

Fig. 6.2. System response for $y_m(t) = 0.5\sin(0.1t)$ deg.

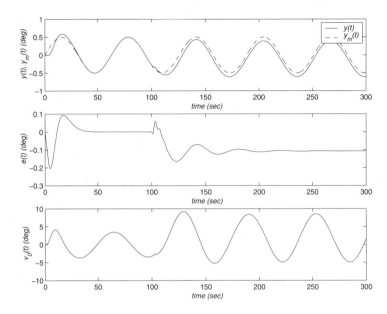

Fig. 6.3. System response for a nominal controller.

failure. From the simulation results, we can see that the nominal controller works well initially when there is no actuator failure. However there is a tracking error that does not converge to zero after the failure of u_2.

In summary, the adaptive output feedback output tracking problem for possibly nonminimum phase systems with actuator failures is formulated in this chapter. A nominal pole placement controller, based on relaxed design conditions as compared with those for model reference adaptive control, is developed to achieve the desired nominal system performance: asymptotic output tracking and closed-loop signal boundedness, when implemented with known plant and actuator failure parameters. For the case when the plant and actuator failure parameters are unknown, an adaptive pole placement control scheme is derived for adaptive stabilization and asymptotic output tracking. However, a singularity problem in solving the Diophantine equation for controller parameters remains to be further addressed to take into account the actuator failure uncertainty. Nevertheless, the simulation results demonstrated the desired adaptive control system performance.

Finally, we note that pole placement based compensation designs may also be developed to handle actuator failures with varying values.

Chapter 7

Designs for Linearized Aircraft Models

Actuator failures encountered in aircraft flight control systems may cause severe performance deterioration and even catastrophic instability. The nature of such failures, in terms of which actuators have failed, failure parameters (e.g., position of a stuck control surface), and the time of failures are usually not known. Adaptive failure compensation designs are desirable for handling such system uncertainties, as shown by recent research results (see Section 1.1.1). The goal of this chapter is to apply several adaptive actuator failure compensation control schemes to a linearized aircraft model to verify the desired system performance under different actuator failure situations.

7.1 Linearized Aircraft Models

An aircraft in flight has six degrees-of-freedom, and there are many forces and moments acting on it. In general, the dynamics of aircraft is nonlinear and time-varying. However, for control of such complex systems, usually one can trim the aircraft dynamics around steady-state flight conditions and extract linear state-space and transfer function descriptions of aircraft models [113]. In this chapter, we consider such linearized state-space models of a Boeing 737-100 transport aircraft for adaptive actuator failure compensation.

The state-space aircraft models may be described in different sets of coordinates, such as body axes, stability axes, and wind axes, with different state variables. We are interested in the models described in body axes, which are briefly presented in this section and whose longitudinal part is used for our study in the subsequent sections.

The Body-Axis Models. Aircraft models in the body axes [93] are of the form $\dot{x} = Ax + Bu$, where $A \in R^{9 \times 9}$, $B \in R^{9 \times 7}$ are constant matrices for a specific flight condition (different for different flight conditions), and $x = [U_b, W_b, Q_b, \theta, V_b, P_b, R_b, \phi, \psi]^T$, $u = [thrust, dels, delr, dele, dela, dspl, dspr]^T$.

These state variables are U_b: forward velocity (x-axis), V_b: lateral velocity (y-axis), W_b: vertical velocity (z-axis), P_b: roll rate (x-axis), Q_b: pitch rate (y-axis), R_b: yaw rate (z-axis), ϕ: Euler roll angle (x-axis), θ: Euler pitch angle (y-axis), ψ: Euler yaw angle (z-axis).

The input variables are dels: stablizer position, delr: rudder position, dele: elevator position, dela: aileron position, dspl: left spoiler position, dspr: right spoiler position, thrust: engine thrust (left and right).

Longitudinal Dynamics. The aircraft dynamics can be uncoupled into longitudinal dynamics and lateral-directional dynamics. The longitudinal dynamics consists of forward (axial) (U_b), vertical (W_b), and pitching (θ, Q_b) motion, and the lateral dynamics consists of lateral (V_b), rolling (ϕ, P_b), and yawing (ψ, R_b). In this chapter, we consider the control problems for the longitudinal dynamics in the presence of actuator failures.

The linearized Boeing 737 longitudinal dynamics model has the form

$$\dot{x}(t) = Ax(t) + Bu(t) \tag{7.1}$$

where $x(t) = [U_b, W_b, \theta, Q_b]^T$, $u(t) = [thrust, dels, dele, ssp]^T$. This subsystem consists of forward (axial) (U_b), vertical (W_b), and pitching (θ, Q_b) motion. The four control inputs are $thrust$, engine thrust, $dels$, the position of the stabilizer, $dele$, the position of the elevator, and ssp, the position of spoilers.

In longitudinal motion control, the elevator is the primary actuator. The failure of the elevator can result in catastrophic accidents. One important type of elevator failure is that the elevator may get stuck at an unknown position at an unknown time instant. In order to compensate for actuator failures, it is necessary to have some redundancy in actuators. Redundancy can be obtained by having different types of actuators so that the system remains controllable despite some actuators' failing. Examples of such a redundancy is the use of the stabilizer and spoilers (to a limited extent) to compensate for a failed elevator and the use of differential engine thrust to compensate for a failed rudder. Redundancy can also be obtained by dividing a control surface into a number of segments, each of which can be controlled separately. For example, we use the concept that the elevator is segmented into multiple segments, and we design control inputs for each segment. This is equivalent to having multiple actuators of the same type. While it may be difficult to retrofit existing aircraft in this manner, our objective is to investigate possible safety-enhancing features in aircraft designs. In this chapter, we assume that the elevator can consist of multiple similar segments. We also assume that the spoiler and stabilizer can be used as actuators, although in normal flight control, they are customarily used only to trim the equilibrium.

As illustrated in this chapter, with different knowledge of the controlled plant parameters, different patterns of actuator redundancy, and different control objectives (state tracking or output tracking), different adaptive control schemes will be used for actuator failure compensation.

When the plant parameters are known and the elevator is divided into parallel segments, a simple adaptive state feedback control scheme is used to accommodate the unknown failure of elevator segments to achieve asymptotic state tracking (see Section 7.2).

When the plant dynamics is unknown, a state feedback design (see Chapter 2) is used to achieve asymptotic state tracking in the presence of elevator failure. Two cases of actuator redundancy are considered. One is that there is only one elevator, which may fail during system operation, and other actuators (engine, spoiler, and stabilizer) are used to compensate for the elevator failure. The second case is that there are two segments of elevator, one of them may fail, and the other will compensate for the failure (but it is unknown which one failed) (see Section 7.3).

When output tracking is the control objective, both adaptive schemes developed in Chapter 3 (using state feedback) and Chapter 4 (using output feedback) are used to accommodate the elevator failure. The plant output is the pitch angle. We also consider two cases for elevator failure compensation: using the stabilizer to compensate for the possible failure of elevator when the elevator in not segmented; and using the remaining functional elevator segment to compensate for the failed elevator segment when the elevator is divided into two segments (see Section 7.4).

For comparison, for each design in Sections 7.2 to 7.4, we also present the simulation results for the corresponding nominal controller, which achieves plant-model matching in the no-failure case, but not for suitable for actuator failure compensation. The response of a such nominal controller can be viewed to examine the typical effect of actuator failures on system performance.

7.2 Design for Uncertain Actuator Failures

We first demonstrate a simple state feedback adaptive actuator failure compensation control design for the linearized B737 model with known plant parameters but unknown actuator failure parameters.

7.2.1 Problem Statement

For compensation of possible elevator failure, we segment the elevator into m parallel segments so that the longitudinal motion subsystem (7.1) becomes

$$\dot{x}(t) = Ax(t) + B_1 u_1(t) + B_2 u_2(t) \tag{7.2}$$

where

$$B_1 = [b_{11}, \ldots, b_{1m}],\ u_1 = [u_{11}, \ldots, u_{1m}]^T \tag{7.3}$$
$$B_2 = [b_{21}, b_{22}, b_{23}]\ u_2 = [thrust, dels, ssp]^T. \tag{7.4}$$

Such a segmentation makes it possible to use the functional elevator segments to compensate for the failure of other elevator segments. For the plant (7.2), A, B_2, C are known constant parameter matrices, the state vector $x(t)$ is available for measurement, $y(t)$ is the plant output, $u_1(t)$ is the input vector (elevator segments) whose components may fail during system operation as described by the following failure model:

$$u_{1j}(t) = \bar{u}_{1j},\ t \geq t_j,\ j \in \{1, 2, \ldots, m\} \tag{7.5}$$

with the constant value $\bar{u}_{1j}$ and the failure time instant t_j unknown, and u_2 are other inputs, which do not fail by assumption. Similar to (1.3), the actuator failures in (7.4) can be expressed as

$$u_1(t) = v_1(t) + \sigma(\bar{u}_1 - v_1(t)) \tag{7.6}$$

where $v_1(t)$ is the applied control input for u_1, and

$$\bar{u}_1 = [\bar{u}_{11}, \bar{u}_{12}, \ldots, \bar{u}_{1m}]^T,\ \sigma = \mathrm{diag}\{\sigma_1, \sigma_2, \ldots, \sigma_m\} \tag{7.7}$$
$$\sigma_i = \begin{cases} 1 & \text{if the } i\text{th actuator fails, i.e., } u_{1i} = \bar{u}_{1i} \\ 0 & \text{otherwise.} \end{cases} \tag{7.8}$$

Conditions needed for adaptive compensation of actuator failures are

(A7.1) there is a known $b \in R^n$ such that $b_{1i} = \alpha_i b, i = 1, \ldots, m$, for some $\alpha_i \in R$ and all α_i have the same sign, and
(A7.2) the pair $(A, [b|B_2])$ is controllable.

The reference state vector $x_m(t)$ is generated from the reference system

$$\dot{x}_m(t) = A_M x_m(t) + b k_{12} r_1(t) + B_2 k_{22} r_2(t) \tag{7.9}$$

where

$$A_M = A + b k_{11}^T + B_2 k_{21} \tag{7.10}$$

$r_1(t) \in R$ and $r_2(t) \in R^3$ are reference inputs for the two groups of actuators, respectively, $k_{11} \in R^4$, $k_{12} \in R$, $k_{21} \in R^{3 \times 4}$, and $k_{22} \in R^{3 \times 3}$ are constant gains (such that the eigenvalues of A_M are stable and desired).

When there are p failed elevator segments in $u_1(t)$ at time t, that is, $u_{1j}(t) = \bar{u}_{1j}, j = j_1, \ldots, j_p$, $1 \leq p < m$, with the knowledge of plant and failure parameters, the nominal failure compensation control signals are

$$v_{1i}(t) = v_{10}(t) = \alpha^*(k_{11}^T x(t) + k_{12} r_1(t)) + \beta^*, \ i = 1, \ldots, m$$

$$\alpha^* = \frac{1}{\sum_{j \neq j_1, \ldots, j_p} \alpha_j}, \ \beta^* = -\frac{\sum_{j=j_1, \ldots, j_p} \alpha_j \bar{u}_{1j}}{\sum_{j \neq j_1, \ldots, j_p} \alpha_j} \tag{7.11}$$

for actuator group $u_1(t)$, and

$$u_2(t) = k_{21} x(t) + k_{22} r_2(t) \tag{7.12}$$

for actuator group $u_2(t)$, which can achieve plant-model matching.

7.2.2 Adaptive Compensation Scheme

When actuator failure parameters are unknown, the parameters α^* and β^* are also unknown. In this case, we use the adaptive controller

$$v_{1i}(t) = v_{10}(t) = \alpha(k_{11}^T x(t) + k_{12} r_1(t)) + \beta, \ i = 1, \ldots, m \tag{7.13}$$

to compensate for possible actuator failures in u_1, where α and β are estimates of α^* and β^*, and use the controller (7.12) for the input u_2.

The adaptive laws for updating α and β in (7.13) is

$$\dot{\alpha}(t) = -\text{sign}[\alpha^*]\gamma_1(k_{11}^T x + k_{12} r_1)e_s^T(t)Pb, \ \gamma_1 > 0 \tag{7.14}$$

$$\dot{\beta}(t) = -\text{sign}[\alpha^*]\gamma_2 e_s^T(t)Pb, \ \gamma_2 > 0 \tag{7.15}$$

where $e_s(t) = x(t) - x_m(t)$ is the state tracking error, and $P = P^T > 0$ is the solution of $PA_M + A_M^T P = -Q$, for some constant $Q = Q^T > 0$.

This adaptive control scheme guarantees the asymptotic state tracking and closed-loop signal boundedness. Its implementation needs the knowledge of the parameters k_{11}, k_{12}, k_{12}, and k_{22} to the closed-loop system with $\alpha = \alpha^*$ and $\beta = \beta^*$ to match (7.8) (that is, that of the plant parameters A, b, B_2), and the sign of α^* (see (7.11) and Assumption (A7.1)). This is a simple adaptive compensation design, as the only uncertainty is from actuator failures. Additional plant uncertainties is treated in Sections 7.3 and 7.4.

7.2.3 Longitudinal Control Simulation I

To study failure compensation of actuators that are elevator segments in longitudinal control, we simply set $u_2(t) = 0$ in (7.2) and $r_2(t) = 0$ in (7.9) so that the linearized longitudinal motion model (7.2) becomes

$$\dot{x}(t) = Ax(t) + B_1 u_1(t). \tag{7.16}$$

In the case of landing, the dynamics matrices for model (7.16) are

$$A = \begin{bmatrix} -0.026373 & 0.12687 & -12.926 & -32.169 \\ -0.25009 & -0.80174 & 220.55 & -0.16307 \\ 0.000171 & -0.00754 & -0.5510 & -0.000334 \\ 0 & 0 & 1.0000 & 0 \end{bmatrix} \tag{7.17}$$

$$b = \begin{bmatrix} 0.010887 \\ -0.18577 \\ -0.022966 \\ 0 \end{bmatrix}, \quad C = \begin{bmatrix} 0 & 0 & 0 & 1 \end{bmatrix}. \tag{7.18}$$

It is to check that the (A, b) is controllable.

For simulation study, we split the elevator into four pieces so that

$$b_1 = 0.2b, \ b_2 = 0.3b, \ b_3 = 0.3b, \ b_4 = 0.2b. \tag{7.19}$$

For longitudinal control, we choose the pitch angle as the plant output. The gain k_{11} is determined from an LQR design, which leads to

$$A_M = \begin{bmatrix} -0.02625 & 0.12353 & -10.073 & -21.2826 \\ -0.25219 & -0.74473 & 171.8676 & -185.9225 \\ 0.000088 & -0.00049 & -6.5694 & -22.9650 \\ 0 & 0 & 1.0000 & 0 \end{bmatrix}. \tag{7.20}$$

With this choice the step response from elevator to pitch angle is good: rising time less than 1 sec, and there is very small overshoot. The parameter $k_{12} = -10$ is selected, which ensures a positive step response from r_1.

We use the following failure pattern:
$u_{14} = -4.0$ (deg), $t \geq 20$ (sec), and $u_{13} = 2.0$ (deg), $t \geq 40$ (sec).

The initial conditions used in simulation are

$$x(0) = 0, \ x_m(0) = 0, \ \alpha(0) = 1.0, \ \beta(0) = 0.0. \tag{7.21}$$

The reference input is $r_1(t) = 4$. The adaptation gains are $\gamma_1 = \gamma_2 = 20$ and $Q = 20I_4$. The simulation results are shown in Figure 7.1 for the tracking error $e_s(t) = x(t) - x_m(t)$, control input $v_1(t)$ from (7.13). For comparison, we present the simulation results for the fixed parameter plant-model match-ing controller (7.11). The results are shown in Figure 7.2. We see that the adaptive failure compensation control schemes can effectively eliminate the effect of actuator failures, but the fixed plant-model matching controller can-not. The plant-model matching controller only works well when there is no failure, that is the condition under which the controller is designed.

This study is for the case when the elevator can be divided into parallel segments (see Assumption (A7.1)) and the plant parameters are known. In the following sections, we study some actuator failure compensation schemes for systems with unknown parameters and without Assumption (A7.1).

7.3 State Tracking Design

In this section, we study an adaptive state feedback design for state tracking with uncertain system dynamics and actuator failures.

7.3.1 Problem Statement

In this section, we study the control problem: how to make the plant state vector $x(t)$ to asymptotically track a given reference state vector $x_m(t)$ in the presence of elevator failure when the elevator cannot be divided into parallel segments. The failure is characterized as the elevator (or one segment of it) fails at some unknown fixed value at some unknown time instant.

We consider two types of actuator redundancy. One is that there is only one elevator (i.e., we do not segment the elevator), and we use other actuators (engine, stabilizer, and spoiler) to compensate for the possible elevator failure. In this case, the linearized longitudinal model is

$$\dot{x}(t) = Ax(t) + Bu(t), B = [b_1, b_2, b_3, b_4] \tag{7.22}$$

where $x(t) = [U_b, W_b, \theta, Q_b]^T$, $u(t) = [thrust, dels, dele, ssp]^T$, the same as (7.1). The actuator failure uncertainties are the failure time and value.

Another case is that we divide the elevator into two segments, and we use $dele_1$ and $dele_2$ to present the positions of the two segments. To compensate for the possible failure of one elevator segment, we will use other actuators (stabilizer and spoiler) and the remaining functional elevator segment (engine will not be used here). Then the linearized longitudinal model is

$$\dot{x}(t) = Ax(t) + Bu(t), B = [b_1, b_2, b_3, b_4] \tag{7.23}$$

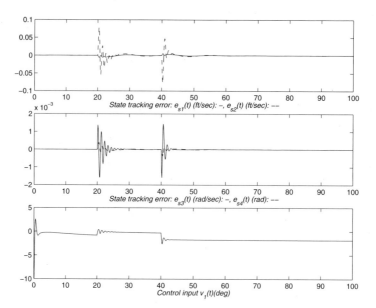

Fig. 7.1. System response with adaptive failure compensation for (7.16).

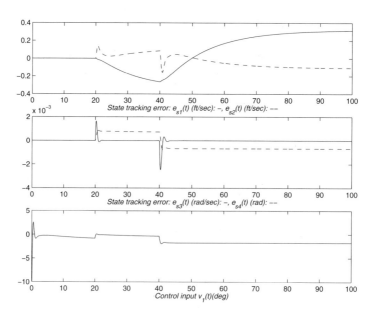

Fig. 7.2. System response with nominal controller for (7.16).

where $x(t)$ is the same as that in (7.22), $u(t) = [dels, dele_1, dele_2, ssp]^T$. Unlike the segmentation in Section 7.2, b_2 and b_3, the columns of B corresponding to the control inputs $dele_1$ and $dele_2$ are not required to be parallel, a more realistic segmentation. For this case, the actuator failure uncertainties are the failed elevator segment, failure time, and failure value.

7.3.2 Adaptive Control Scheme

Here we give a brief description of the adaptive control scheme in Chapter 2. This scheme is designed for the linear time-invariant plant

$$\dot{x}(t) = Ax(t) + Bu(t) \tag{7.24}$$

where $A \in R^{n \times n}, B = [b_1, \ldots, b_m] \in R^{n \times m}$, with actuator failure

$$u_j(t) = \bar{u}_j, \ t \geq t_j, \ j \in \{1, 2, \ldots, m\} \tag{7.25}$$

where the constant value $\bar{u}_j$ and the failure time instant t_j are unknown. From (1.3), the system input $u(t)$ can be expressed as

$$u(t) = v(t) + \sigma(\bar{u} - v(t)) \tag{7.26}$$

where $v(t)$ is the applied control input, and σ represents a failure pattern (see (1.3)) that is uncertain in adaptive control.

The reference state vector $x_m(t)$ is generated from the reference model

$$\dot{x}_m(t) = A_M x_m(t) + B_M r(t) \tag{7.27}$$

where $A_M \in R^{n \times n}$ is stable, $B_M \in R^{n \times l}$, and $r(t) \in R^l$ is bounded.

To compensate for up to $m - q$ $(q \geq 1)$ actuator failures, the following plant-model matching conditions (see Chapter 2) are needed: For every $B_a \in R^{n \times q}$ consisting of q columns of B,

$$\text{rank}(B_a) = \text{rank}([B_a|A_M - A]) \tag{7.28}$$
$$\text{rank}(B_a) = \text{rank}([B_a|B_M]) \tag{7.29}$$
$$\text{rank}(B_a) = \text{rank}(B). \tag{7.30}$$

For plant and failure parameters unknown, the controller structure is

$$v(t) = K_1^T(t)x(t) + K_2^T(t)r(t) + k_3(t) \tag{7.31}$$

where K_1, K_2, and k_3 are updated from the adaptive laws

$$\dot{k}_{1j}(t) = -\Gamma_{1j}x(t)e^T(t)Pb_j, \ \Gamma_{1j} = \Gamma_{1j}^T > 0 \qquad (7.32)$$

$$\dot{k}_{2j}(t) = -\Gamma_{2j}r(t)e^T(t)Pb_j, \ \Gamma_{2j} = \Gamma_{2j}^T > 0 \qquad (7.33)$$

$$\dot{k}_{3j}(t) = -\gamma_{3j}e^T(t)Pb_j, \ \gamma_{3j} > 0. \qquad (7.34)$$

for $j = 1, \ldots, m$, $e(t) = x(t) - x_m(t)$ is the state tracking error, and $P \in R^{n \times n}$, $P = P^T > 0$ such that $PA_M + A_M^T P = -Q < 0$. This adaptive control scheme guarantees that all closed-loop signals are bounded and $\lim_{t \to \infty} e(t) = 0$.

7.3.3 Longitudinal Control Simulation II

We use the linearized Boeing 737 longitudinal motion model (7.22) and (7.23) to verify the performance of the above adaptive control scheme.

Plant and Reference Models. In the case of landing, the dynamics matrix A and actuation matrix B for longitudinal motion model (7.22) are

$$A = \begin{bmatrix} -0.026373 & 0.12687 & -12.926 & -32.169 \\ -0.25009 & -0.80174 & 220.55 & -0.16307 \\ 0.000171 & -0.00754 & -0.5510 & -0.000334 \\ 0 & 0 & 1.0000 & 0 \end{bmatrix} \qquad (7.35)$$

$$B = \begin{bmatrix} 0.40220 & 0.022768 & 0.010887 & -0.015744 \\ -0.00049881 & -0.38849 & -0.18577 & 0.13972 \\ 0.0063014 & -0.048047 & -0.022966 & 0.0016106 \\ 0 & 0 & 0 & 0 \end{bmatrix}. \qquad (7.36)$$

For the plant model (7.23), the dynamics matrix A is the same as (7.35). We segment the elevator into two separate pieces such that

$$B = \begin{bmatrix} 0.022768 & 0.0065322 & 0.0043548 & -0.015744 \\ -0.38849 & -0.1021735 & -0.0835965 & 0.13972 \\ -0.048047 & -0.0149279 & -0.0080381 & 0.0016106 \\ 0 & 0 & 0 & 0 \end{bmatrix}. \qquad (7.37)$$

The elevator is segmented so that the two columns of B resulting from the segment are not parallel but the sum of their control effects is the same as that of the original elevator.

The reference dynamics matrix A_M is

$$A_M = \begin{bmatrix} -0.02625 & 0.12353 & -10.073 & -21.2826 \\ -0.25219 & -0.74473 & 171.8676 & -185.9225 \\ 0.000088 & -0.00049 & -6.5694 & -22.9650 \\ 0 & 0 & 1.0000 & 0 \end{bmatrix}. \qquad (7.38)$$

The reference dynamics matrix A_M is from an LQR design (the same as (7.20)) such that the step response of pitch angle is good: rising time less than 1 sec with very small overshoot. The reference actuation matrix B_M is the same as b_3 of the plant actuation matrix B in (7.36), which corresponds to the unsegmented elevator. By this choice of B_M, we have $\Gamma_{1j} \in R^{4 \times 4}$, $\Gamma_{2j}, \gamma_{3j} \in R$, $k_{1j} \in R^4$, $k_{2j}, k_{3j} \in R$, $j = 1, 2, 3, 4$.

Failure Patterns and Plant-Model Matching Conditions. For plant model (7.22) with parameters (7.35) and (7.36), it can be verified that when the elevator fails, the plant-model matching conditions (7.28)–(7.30) still hold, that is, we can apply the adaptive control scheme to compensate for the elevator failure to achieve asymptotic state tracking. The failure pattern is

(I) $dele = -2$ (deg), for $t \geq 30$ (sec).

For plant model (7.23) with parameters (7.35) and (7.37), it can also be verified that when one segment of the elevator fails, the plant-model matching conditions (7.28)–(7.30) still hold. We do simulation by assuming that one segment of the elevator ($dele_2$) fails. The failure pattern is

(II) $dele_2 = -5$ (deg), for $t \geq 30$ (sec).

From the matching conditions (7.28)–(7.30), and rank(B) $= 3$ for both cases, we see that at least three redundant actuators are needed for elevator failure compensation. To implement the adaptive laws, the matrix B is assumed to be known for both cases.

Simulation Results. The reference input is selected as $r(t) = -20$ such that the reference state vector $x_m(t)$ is neither too big nor too small.

For plant model (7.22) and failure pattern (I), we select $Q = 10I_4$, and use $\Gamma_{11} = 0.0005I_4$, $\Gamma_{12} = 0.005I_4$, $\Gamma_{13} = 0.05I_4$, $\Gamma_{14} = 0.005I_4$, $\Gamma_{21} = \Gamma_{22} = \Gamma_{23} = 0.005$, $\Gamma_{24} = 0.0005$, $\gamma_{31} = 0.1$, and $\gamma_{32} = \gamma_{33} = \gamma_{34} = 0.05$.

For plant model (7.23) and failure pattern (II), we also choose $Q = 10I_4$, and use $\Gamma_{11} = \Gamma_{14} = 0.05I_4$, $\Gamma_{12} = \Gamma_{13} = 0.3I_4$, $\Gamma_{21} = \Gamma_{22} = \Gamma_{23} = \Gamma_{24} = 0.05$, and $\gamma_{31} = \gamma_{32} = \gamma_{33} = \gamma_{34} = 0.1$.

With the initial values $x(0) = x_m(0) = 0$, the simulation results are shown in Figures 7.3–7.5 for plant model (7.22) with failure pattern (I), and in Figures 7.7–7.9 for plant model (7.23) with failure pattern (II).

The simulation results show that at the time instant of elevator failure, there is a transient behavior in the state tracking error. As time goes on, the tracking error goes to zero asymptotically. At the time instant when the stabilizer fails, the controller parameters (e.g., k_3 in Figures 7.5 and 7.9) also have a transient behavior, and then go close to some constant values.

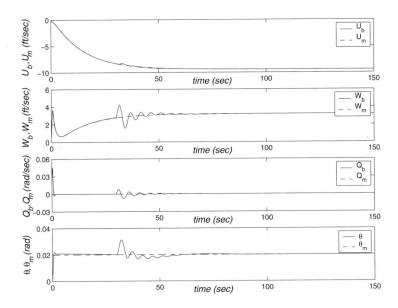

Fig. 7.3. System response (state feedback state tracking design for (7.22)).

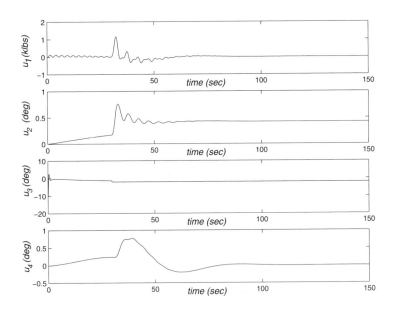

Fig. 7.4. Actuating inputs (state feedback state tracking design for (7.22)).

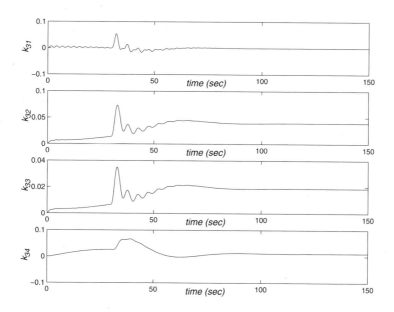

Fig. 7.5. Parameter k_3 (state feedback state tracking design for (7.22)).

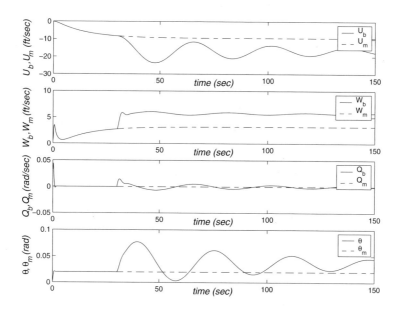

Fig. 7.6. System response (nominal controller for (7.22)).

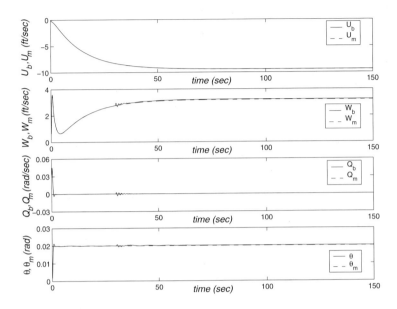

Fig. 7.7. System response (state feedback state tracking design for (7.23)).

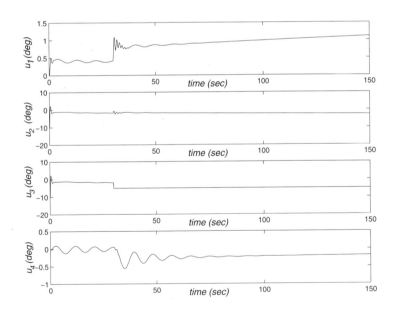

Fig. 7.8. Actuating inputs (state feedback state tracking design for (7.23)).

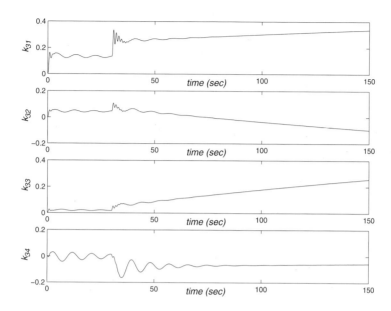

Fig. 7.9. Parameter k_3 (state feedback state tracking design for (7.23)).

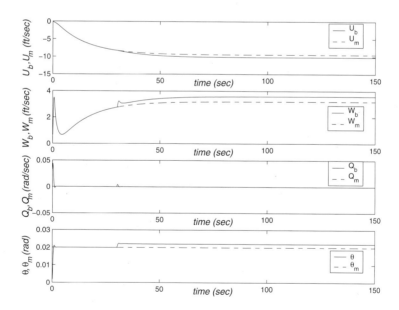

Fig. 7.10. System response (nominal controller for (7.23)).

These values are not necessarily the true matching parameters. All signals in the adaptive control system are bounded, and stability and convergence are ensured. The simulation results verified that when the elevator fails, other actuators can be used to compensate for the failure and achieve asymptotic state tracking by using the adaptive control scheme.

For comparison, we also did simulations using the fixed plant-model matching controllers (for the no-failure case) for both models. The results are shown in Figures 7.6 and 7.10. The parameters of the nominal controllers can achieve plant-model matching when there is no actuator failure. From the simulation results, we see that the fixed controller works well when there is no actuator failure, but they cannot achieve asymptotic state tracking in the presence of actuator failures.

7.4 Output Tracking Designs

In this section, we consider the output tracking problem for the linearized aircraft longitudinal model without the knowledge of both plant parameters (in particular, the B matrix) and actuator failure parameters, under some relaxed design conditions (as compared with (7.28)–(7.30)).

7.4.1 Problem Statement

For output tracking, we select θ (pitch angle) as the plant output $y(t)$, and consider two actuators: stabilizer and elevator as control inputs since the engine and spoiler have little effect on pitch angle.

The control problem is how to make the plant output $y(t)$ to asymptotically track a given reference output $y_m(t)$ in the presence of elevator failure without the knowledge of both plant and actuator failure parameters.

As in Section 7.3, we also consider two types of actuator redundancy. When the elevator is not segmented, the linearized longitudinal model is

$$\dot{x}(t) = Ax(t) + Bu(t), \ y(t) = Cx(t), \ B = [b_1, b_2] \qquad (7.39)$$

with $x = [U_b, W_b, Q_b, \theta]^T$ the state vector, $u = [dels, dele]^T$ the control inputs, $C = [0 \ 0 \ 0 \ 1]$, and $y = \theta$ the plant output.

If the aircraft has two segments of elevator operated individually, namely $dele_1$ and $dele_2$, we will consider how to use the remaining functional elevator segment to compensate for the possible failure of one elevator segment, and the stabilizer is not used as actuator in this case. Therefore, the linearized aircraft longitudinal model can be described as

$$\dot{x}(t) = Ax(t) + Bu(t), \ y(t) = Cx(t), \ B = [b_1, b_2] \qquad (7.40)$$

where x, C, y are the same as those in (7.39), and the control input vector is $u = [dele_1, dele_2]^T$. As in Section 7.3, we do not need b_1 and b_2, the two columns of B resulting from the segmentation of elevator, to be parallel.

For a linear time-invariant plant

$$\dot{x}(t) = Ax(t) + Bu(t), \ y(t) = Cx(t) \qquad (7.41)$$

where $A \in R^{n \times n}$, $B = [b_1, \ldots, b_m] \in R^{n \times m}$, $C \in R^{1 \times n}$, $y \in R$ the plant output, with actuator failure (7.25), two adaptive actuator failure compensation schemes for output tracking, using state feedback and output feedback, respectively, are developed in Chapter 3 and Chapter 4. Both schemes ensure that the plant output $y(t)$ asymptotically tracks a given reference output $y_m(t)$ and the boundedness of closed-loop signals. The reference output $y_m(t)$ is generated from the reference model

$$y_m(t) = W_m(s)[r](t), \ W_m(s) = \frac{1}{P_m(s)} \qquad (7.42)$$

where $P_m(s)$ is a stable monic polynomial of degree n^*, and $r(t)$ is bounded.

In this section, we will apply the adaptive schemes to the linearized aircraft longitudinal models (7.39) and (7.40), and design controllers to compensate for the unknown elevator failure and achieve asymptotic output tracking when plant parameters are unknown. For plant model (7.39), we consider the case that the elevator fails while the stabilizer remains functioning. Here the actuator failure uncertainties are the unknown elevator failure time and failure value. For plant model (7.40), we consider the case that one segment of elevator fails and another segment of elevator remains functioning. For this case, the actuator failure uncertainties are the failed elevator segment, failure time, and failure value. The elevator failures considered here can be described by the actuator failure model (7.25).

7.4.2 Adaptive Control Schemes

There are two adaptive schemes applicable to the plant (7.39) or (7.40): state feedback and output feedback designs, for output tracking.

State Feedback Design. The adaptive control scheme of Chapter 3 is

$$v_0(t) = v_1(t) = v_2(t) = \cdots = v_m(t) = k_{11}^T(t)x(t) + k_{21}(t)r(t) + k_{31}(t) \quad (7.43)$$

where $k_{11}(t)$, $k_{21}(t)$, and $k_{31}(t)$ are updated by the following adaptive laws.

Introducing the auxiliary signals

$$w(t) = [x^T(t), r(t), 1]^T \tag{7.44}$$

$$\theta(t) = [k_{11}^T, k_{21}, k_{31}]^T \tag{7.45}$$

$$\zeta(t) = W_m[w](t) \tag{7.46}$$

$$\xi(t) = \theta^T(t)\zeta(t) - W_m(s)[\theta^T w](t) \tag{7.47}$$

$$\epsilon(t) = e(t) + \rho(t)\xi(t) \tag{7.48}$$

where $e(t) = y(t) - y_m(t)$ is the output tracking error, $\rho(t)$ is the estimate of $\rho^* = 1/k_{21}^*$, the adaptive laws updating $\theta(t)$ and $\rho(t)$ are

$$\dot{\theta}(t) = -\frac{\text{sign}[k_{21}^*]\Gamma\zeta(t)\epsilon(t)}{1 + \zeta^T\zeta + \xi^2}, \quad \Gamma = \Gamma^T > 0 \tag{7.49}$$

$$\dot{\rho}(t) = -\frac{\gamma\xi(t)\epsilon(t)}{1 + \zeta^T\zeta + \xi^2}, \quad \gamma > 0. \tag{7.50}$$

Output Feedback Design. When state variables are not available and only plant output can be measured, the adaptive scheme using output feedback in Chapter 4 can be used. With the filters

$$w_1(t) = \frac{a(s)}{\Lambda(s)}[v_0](t), \quad w_2(t) = \frac{a(s)}{\Lambda(s)}[y](t) \tag{7.51}$$

with $a(s) = [1, s, \cdots, s^{n-2}]^T$, and $\Lambda(s)$ being a monic stable polynomial of degree $n - 1$, the adaptive compensation controller is

$$v_0(t) = v_1(t) = v_2(t) = \cdots = v_m(t)$$
$$= \theta_1^T w_1(t) + \theta_2^T w_2(t) + \theta_{20}y(t) + \theta_3 r(t) + \theta_4 \tag{7.52}$$

where $\theta_1(t)$, $\theta_2(t)$, $\theta_{20}(t)$, $\theta_3(t)$, and $\theta_4(t)$ are parameter estimates.

To form an adaptive scheme, we define

$$\theta(t) = [\theta_1^T(t), \theta_2^T(t), \theta_{20}(t), \theta_3(t), \theta_4(t)]^T \tag{7.53}$$

$$w(t) = [w_1^T(t), w_2^T(t), y(t), r(t), 1]^T \tag{7.54}$$

$$\zeta(t) = W_m[w](t) \tag{7.55}$$

$$\xi(t) = \theta^T(t)\zeta(t) - W_m(s)[\theta^T w](t) \tag{7.56}$$

$$\epsilon(t) = y(t) - y_m(t) + \rho(t)\xi(t) \tag{7.57}$$

and choose the adaptive laws

$$\dot{\theta}(t) = -\frac{\text{sign}[\theta_3^*]\Gamma\zeta(t)\epsilon(t)}{1 + \zeta^T\zeta + \xi^2}, \quad \Gamma = \Gamma^T > 0 \tag{7.58}$$

$$\dot{\rho}(t) = -\frac{\gamma\xi(t)\epsilon(t)}{1 + \zeta^T\zeta + \xi^2}, \quad \gamma > 0. \tag{7.59}$$

7.4.3 Longitudinal Control Simulation III

Plant Model. In the case of landing, the parameters of (7.39) are

$$
A = \begin{bmatrix}
-0.026373 & 0.12687 & -12.926 & -32.169 \\
-0.25009 & -0.80174 & 220.55 & -0.16307 \\
0.000171 & -0.00754 & -0.5510 & -0.000334 \\
0 & 0 & 1.0000 & 0
\end{bmatrix}
$$

$$
B = [b_1, b_2] = \begin{bmatrix}
0.022768 & 0.010887 \\
-0.38849 & -0.18577 \\
-0.048047 & -0.022966 \\
0 & 0
\end{bmatrix}, \; C = [\, 0 \;\; 0 \;\; 0 \;\; 1 \,]. \tag{7.60}
$$

We segment the elevator into two pieces so that the matrix B is

$$
B = [b_2, b_3] = \begin{bmatrix}
0.0065322 & 0.0043548 \\
-0.1021735 & -0.0835965 \\
-0.0149279 & -0.0080381 \\
0 & 0
\end{bmatrix} \tag{7.61}
$$

with $u = [dele_1, dele_2]^T$. The elevator is segmented as in Section 7.2 so that the two columns of B resulting from the segment are not parallel, but the sum of their control effects on pitch rate is the same as that of the original elevator. Other parameter matrices A and C are the same as in (7.60). For simulation study, the reference model is with

$$
W_m(s) = \frac{1}{(s+1)(s+2)}, \; r(t) = 0.1. \tag{7.62}
$$

Failure Patterns and Plant-Model Matching Conditions. In this study, we first consider the plant model (7.39) with parameters (7.60). The transfer function from $dels$ (stabilizer) to θ (pitch angle) is

$$
G_1(s) = \frac{-0.0480s^2 - 0.0369s - 0.024}{s^4 + 1.3791s^3 + 2.1744s^2 + 0.989s + 0.0651} \tag{7.63}
$$

and from $dele$ (elevator) to θ is

$$
G_2(s) = \frac{-0.0230s^2 - 0.0176s - 0.012}{s^4 + 1.3791s^3 + 2.1744s^2 + 0.989s + 0.0651}. \tag{7.64}
$$

Since the plant-model matching conditions are satisfied (see Chapters 3 and 4), we can apply the above two adaptive control schemes to this longitudinal model for asymptotic output tracking in the presence of unknown actuator failures. We consider the case when the elevator fails and the stabilizer remains fully functional. The actuator failure pattern is

(I) $dele = -2$ (deg) for $t \geq 30$ (sec).

For the plant model (7.40), which has two elevator segments, we consider the case when one segment of elevator $dele_2$ fails during system operation. The actuator failure pattern is

(II) $dele_2 = -5$ (deg) for $t \geq 30$ (sec).

In all the simulations, the actuator failure, including the failed actuator, failure time, and failure value, is unknown to the adaptive controller.

Simulation Results. For all simulations, we set the initial values of $y_m(t)$ and $y(t)$ to be zero, and the initial values of controller parameters to be 90% of the true matching parameters for the no-failure case.

For state feedback design, the adaptation gains are $\Gamma = 10I$ and $\gamma = 10$. The simulation results are shown in Figure 7.11 for plant model (7.39) with failure pattern (I), with plant output $y(t)$, reference output $y_m(t)$, tracking error $e(t)$, and designed control output $v(t)$, and in Figure 7.13 for plant model (7.40) with actuator failure (II).

For output tracking design, $\Gamma = 100I$ and $\gamma = 100$. The simulation results are shown in Figure 7.15 for plant model (7.39) with failure pattern (I), and in Figure 7.17 for plant model (7.40) with actuator failure (II).

For both designs, the simulation results are as expected. At the time instant of elevator failure, there is a transient behavior in tracking error and controller parameters. As time goes on, the tracking error goes to zero. The simulation results verified that when the elevator (or one elevator segment) fails, the stabilizer (or the functional elevator segment) can be used for failure compensation, and the asymptotic output tracking and the boundedness of the closed-loop signals can be achieved by the adaptive control schemes.

For comparison, the system responses for the corresponding fixed parameter controllers (plant-model matching controllers for the no-failure case) are shown in Figures 7.12, 7.14, 7.16, and 7.18. By comparing the results of adaptive controllers and fixed parameter controllers, we can see that while adaptive controllers can effectively compensate for the elevator failure and achieve asymptotic tracking, the fixed parameter controllers cannot.

7.5 Concluding Remarks

In this chapter, we study four adaptive actuator failure compensation control schemes for a linearized longitudinal model of a transport aircraft. It is shown

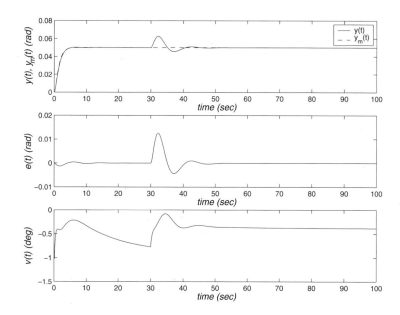

Fig. 7.11. System response (state feedback output tracking design for (7.39)).

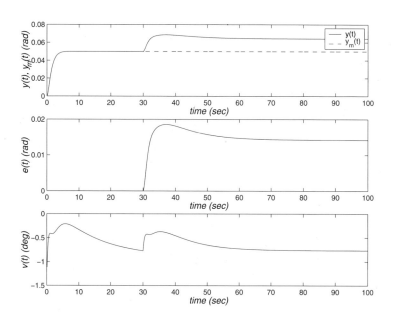

Fig. 7.12. System response (nominal controller for (7.39)).

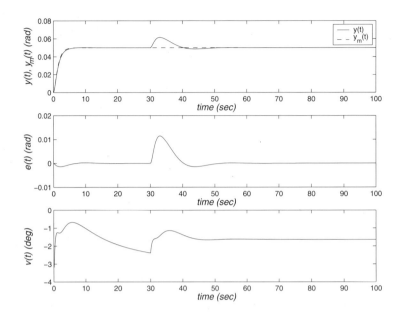

Fig. 7.13. System response (state feedback output tracking design for (7.40)).

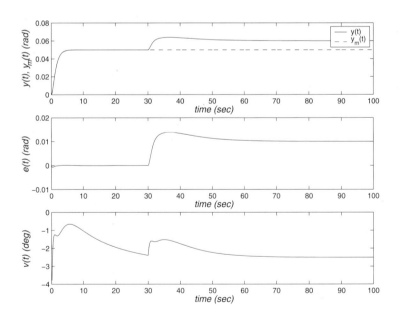

Fig. 7.14. System response (nominal controller for (7.40)).

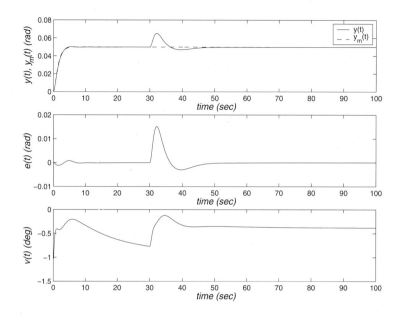

Fig. 7.15. System response (output feedback output tracking design for (7.39)).

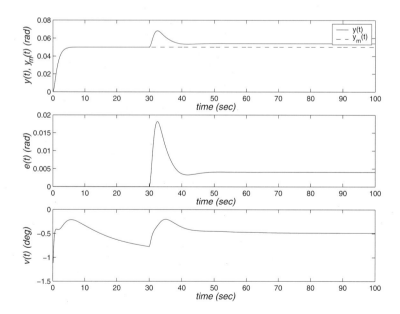

Fig. 7.16. System response (nominal controller for (7.39)).

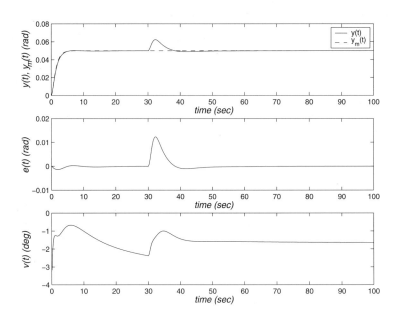

Fig. 7.17. System response (output feedback output tracking design for (7.40)).

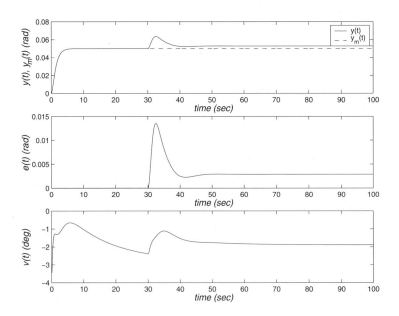

Fig. 7.18. System response (nominal controller for (7.40)).

that with different knowledge of plant parameters, different patterns of actuator redundancy, and different control objectives, different adaptive actuator control schemes can be used to compensate for unknown elevator failures. When the plant parameters are known, and the elevator can be divided into parallel segments, adaptive failure compensation can be designed for asymptotic state or output tracking (see Section 7.2). With only the knowledge of the actuation matrix B, asymptotic state tracking can also be achieved by an advanced adaptive compensation scheme (see Section 7.3), while for output tracking even less system knowledge is needed for implementing adaptive failure compensation schemes (see Section 7.4).

In addition to what has been shown in this chapter, we also did simulations to test the robustness of the developed adaptive actuator failure compensation schemes with respect to uncertain actuator dynamics. In the simulation study, the actuator dynamics of the elevator, stabilizer, and spoiler are modeled as a first-order system and are included in the controlled plant models. Such actuator dynamics are supposed to be unknown and are not considered in the adaptive controller designs. Three adaptive control schemes are tested: state feedback for state tracking with unknown plant dynamics and unknown failures; state feedback for output tracking with unknown plant parameters and unknown failures; and output feedback for output tracking with unknown plant parameters and unknown failures. The simulation results show that the closed-loop system performance is deteriorated by the unknown actuator dynamics, but the closed-loop system stability and asymptotic tracking can still be achieved, which means that the developed control schemes have some robustness with respect to uncertain actuator dynamics. It is also shown by the simulation results that the faster the actuator dynamics are, the better the closed-loop system performance is. As an illustration, we redo the simulation in Section 7.2.3 for the plant (7.28) with actuator dynamics

$$u(t) = \frac{1}{t_s s + 1}[v](t) \tag{7.65}$$

where $v(t)$ is the designed control input, and $u(t)$ is the actuation input. First, we consider the actuator dynamics as $t_s = 0.05$ for the elevator, $t_s = 0.5$ for the stabilizer and spoiler. The system response is shown in Figure 7.19 for the plant model (7.22) with failure pattern (I). Then, we consider a fast actuator dynamics as $t_s = 0.02$ for the elevator, $t_s = 0.2$ for the stabilizer and spoiler. The simulation results are shown in Figure 7.20.

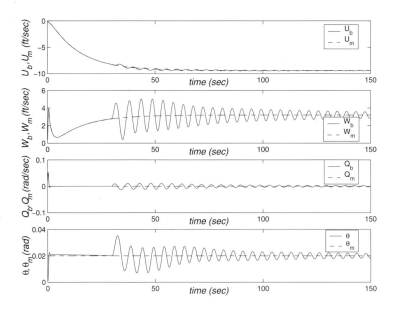

Fig. 7.19. System response with slow actuator dynamics.

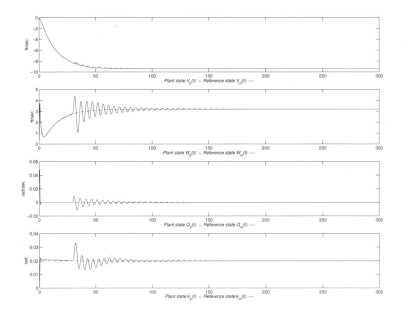

Fig. 7.20. System response with fast actuator dynamics.

Chapter 8

Robust Designs for Discrete-Time Systems

In the previous chapters, we presented various adaptive actuator failure compensation schemes for *continuous-time* linear time-invariant systems with uncertain actuator failures. In this chapter, we will develop adaptive actuator failure compensation schemes for *discrete-time* systems, which are useful for real-life implementation of actuator failure control compensation using digital computers. In addition, we will address the *robustness* issue for adaptive actuator failure compensation in the presence of system modeling errors, whose results are also important and analogous for the continuous-time designs.

It is the goal of this chapter to present a robust model reference adaptive control scheme for discrete-time systems with unknown parameters and unknown failures, in the presence of unmodeled dynamics and output disturbances. The related key issues such as plant and controller parametrization, plant-model matching, design of adaptive laws, stability analysis, and robustness are solved using a model reference adaptive control method. In Section 8.1, we present the adaptive actuator failure compensation problem in the discrete-time framework. In Section 8.2, we derive plant-model matching conditions for the plant without unmodeled dynamics and output disturbances, in the presence of actuator failures. In Section 8.3, we develop an adaptive actuator compensation control scheme and analyze the adaptive control system performance for the nominal plant with unknown parameters and unknown actuator failures. In Section 8.4, we propose a robust adaptive control law to update the parameters of the proposed actuator compensation controller in the presence of unmodeled dynamics and output disturbances. In Section 8.5, we conduct a case study for adaptive compensation of rudder servomechanism failures of a Boeing 747 lateral dynamics model [37] derived in discrete time, verifying the desired discrete-time adaptive system performance in the presence of uncertain actuator failures.

To ensure the robustness of closed-loop stability in the presence of unmodeled dynamics and disturbances, adaptive laws for updating the controller parameters are modified based on robust adaptive control theory. The presented discrete-time results are also analogous in continuous time.

8.1 Problem Statement

In this chapter, we consider a discrete-time multi-input single-output lin-ear time-invariant plant with external output disturbance and additive and multiplicative unmodeled dynamics, in addition to actuator failures. The con-trolled plant is described in the input–output form

$$y(k) = G(z)[u](k) + d(k) \qquad (8.1)$$

where $y(k) \in R$ is the measured plant output, $u(k) \in R^m$ is the plant input vector whose components may fail during system operation, $d(k) \in R$ is a bounded external output disturbance, and $G(z)$ is an $m \times 1$ transfer matrix. The symbol z is used to denote, as the case may be, the z-transform variable z or the time advance operator: $z[x](k) = x(k+1)$, for $k = 0, 1, 2, \ldots$.

The plant transfer matrix $G(z)$ is

$$G(z) = G_0(z)(I + \mu \Delta_m(z)) + \mu \Delta_a(z), \ \mu \geq 0 \qquad (8.2)$$

where

$$G_0(z) = [G_{01}(z), \ldots, G_{0j}(z)], \ G_{0j}(z) = k_{pj} \frac{z_j(z)}{P(z)}, \ j = 1, 2, \ldots, m \qquad (8.3)$$

$$\Delta_m(z) = \operatorname{diag}\{\Delta_{m1}(z), \ldots, \Delta_{mm}(z)\}, \ \Delta_a(z) = [\Delta_{a1}(z), \ldots, \Delta_{am}(z)].(8.4)$$

In (8.2), $G_0(z)$ is the nominal plant transfer matrix, and $\mu \Delta_m(z)$ and $\mu \Delta_a(z)$ are multiplicative and additive unmodeled dynamics. We assume that $G_{0j}(z)$, $j = 1, 2, \ldots, m$, are all strictly proper and rational.

To demonstrate actuator failure compensation, we only consider the "lock-in-place" type of failures (see (1.2)), which are modeled in discrete-time as

$$u_j(k) = \bar{u}_j, \ k \geq k_j, \ j \in \{1, 2, \ldots, m\} \qquad (8.5)$$

where the constant value $\bar{u}_j$ and the failure time instant k_j are unknown. More general failures (1.6) or (1.8) can be similarly addressed (see Chapters 2 through 5). We consider the failure compensation problem for the case that any up to $m - 1$ of the total m actuators may fail during system operation. The multi-output case (see Chapter 5) can also be addressed in discrete time.

The basic assumption for actuator failure compensation is

(A8.0)) The system (8.1) is so designed that for any up to $m - 1$ actuator failures, the remaining actuators can still achieve a desired control objective when $\mu = 0$ and $d(k) = 0$.

The task of robust adaptive failure compensation control is to adjust the remaining controls to achieve the desired control objective when there are up to $m - 1$ actuator failures whose parameters, together with the plant parameters, are all unknown, for the case when $\mu = 0$ and $d(k) = 0$, and to ensure closed-loop signal boundedness when $\mu \neq 0$ is small and $d(k) \neq 0$ and small tracking errors when $\mu \neq 0$ and $d(k) \neq 0$ are small.

Let $v(k) = [v_1, \ldots, v_m]^T$ be the applied control input vector. Similar to (1.3), in the presence of actuator failures modeled in (8.5), the system input vector $u(k)$ can be expressed as

$$u(k) = v(k) + \sigma(\bar{u} - v(k)) \qquad (8.6)$$

with

$$\bar{u} = [\bar{u}_1, \bar{u}_2, \ldots, \bar{u}_m]^T, \ \sigma = \mathrm{diag}\{\sigma_1, \sigma_2, \ldots, \sigma_m\} \qquad (8.7)$$

$$\sigma_i = \begin{cases} 1 & \text{if the } i\text{th actuator fails, i.e., } u_i = \bar{u}_i \\ 0 & \text{otherwise.} \end{cases} \qquad (8.8)$$

The control objective is to design a feedback control $v(t)$ for the plant (8.1) unknown with actuator failures (8.5) unknown, under Assumption (A8.0), such that despite the control error $u - v = \sigma(\bar{u} - v)$, all closed-loop signals are bounded, and the plant output $y(k)$ asymptotically tracks a given reference output $y_m(k)$ when $\mu = 0$ and $d(k) = 0$, where

$$y_m(k) = W_m(z)[r](k), \ W_m(z) = \frac{1}{P_m(z)} \qquad (8.9)$$

where $P_m(z)$ is a stable monic polynomial of degree n^*, and $r(k)$ is a bounded reference signal. Robustness of system performance requires closed-loop signal boundedness for $\mu \neq 0$ small and $d(k) \neq 0$ and small tracking errors for $\mu \neq 0$ and $d(k) \neq 0$ small. A model reference adaptive control approach [118], [125] is used for meeting the control objective and the robustness requirement.

8.2 Plant-Model Output Matching

We first show that there is a nominal plant-model matching controller when there is no unmodeled dynamics $\mu\Delta_m(z)$, $\mu\Delta_a(z)$ and no disturbance $d(k)$. Considering the failure model (8.5), we use the controller structure

$$v_1(k) = \cdots = v_m(k) = v_0(k) \overset{\triangle}{=} v_0^*(k)$$
$$= \theta_1^{*T}\omega_1(k) + \theta_2^{*T}\omega_2(k) + \theta_{20}^*y(k) + \theta_3^*r(k) + \theta_4^* \qquad (8.10)$$

where $\theta_1^* \in R^{n-1}$, $\theta_2^* \in R^{n-1}$, $\theta_{20}^* \in R^n$, and $\theta_3^* \in R$ are to be defined for plant-model output matching, $\theta_4^* \in R$ is to be chosen for compensation of the actuation error $u - v = \sigma(\bar{u} - v)$, and

$$\omega_1(k) = \frac{a(z)}{\Lambda(z)}[v_0](k), \quad \omega_2(k) = \frac{a(z)}{\Lambda(z)}[y](k) \tag{8.11}$$

with $a(z) = [1, z, \ldots, z^{n-2}]^T$, and $\Lambda(z)$ is a monic stable polynomial of degree $n-1$ (e.g., $\Lambda(z) = z^{n-1}$). This equal-actuation scheme is suitable for actuators of the same type, such as a group of ailerons.

Suppose that at time instant k there are p failed actuators, that is, $u_j(k) = \bar{u}_j$, $j = j_1, \ldots, j_p$. Then, with the chosen control signals $v_i(k) = v_0(k) = v_0^*(k)$ as in (8.10), we can express the plant (8.1) with $\mu = 0$ and $d(k) = 0$ as

$$y(k) = G_u(z)[v_0](k) + \bar{y}(k) \tag{8.12}$$

where

$$G_u(z) = \sum_{j \neq j_1, \ldots, j_p} \frac{k_{pj} Z_j(z)}{P(z)} \triangleq \frac{k_p Z_a(z)}{P(z)}, \quad j = 1, \ldots, m \tag{8.13}$$

for some scalar k_p and monic polynomial $Z_a(z)$ and

$$\bar{y}(k) = \sum_{j = j_1, \ldots j_p} \frac{k_{pj} Z_j(z)}{P(z)}[\bar{u}_j](k). \tag{8.14}$$

Note that with different failure patterns, $Z_a(z)$ and k_p are also different, which results in different (piecewise constant) matching parameters.

For the existence of the plant-model output matching controller, we assume that for all failure patterns,

(A8.1a) $Z_a(z)$ has degree $n - n^*$, i.e., $\sum_{j \neq j_1, \ldots j_p} k_{pj} Z_j(z)/P(z)$, $p \in \{0, 1, \ldots, m - 1\}$, have the same relative degree n^*, and
(A8.1b) $Z_a(z)$ is stable, i.e., $\sum_{j \neq j_1, \ldots j_p} k_{pj} Z_j(z)/P(z)$, $p \in \{0, 1, \ldots, m - 1\}$, are minimum phase.

Introducing

$$F_1(z) = \theta_1^{*T} \frac{a(z)}{\Lambda(z)}, \quad F_2(z) = \theta_2^{*T} \frac{a(z)}{\Lambda(z)} + \theta_{20}^* \tag{8.15}$$

we express the control signal $v_0(k) = v_0^*(k)$ as

$$v_0(k) = (F_1(z) + F_2(z)G_u(z))[v_0](t) + F_2(z)[\bar{y}](k) + \theta_3^* r(k) + \theta_4^* \tag{8.16}$$

which leads to the closed-loop system

$$y(k) = G(z)(1 - F_1(z) - F_2(z)G_u(z))^{-1}[F_2(z)[\bar{y}] + \theta_3^* r + \theta_4^*](k) + \bar{y}(k). \quad (8.17)$$

Considering

$$(1 - F_1(z) - F_2(z)G_u(z))^{-1} = \frac{\Lambda(z)P(z)}{P_0(z)} \quad (8.18)$$

where

$$P_0(z) = \Lambda(z)P(z) - \theta_1^* a(z)P(z) - \theta_2^{*T} a(z)k_p Z_a(z) - \Lambda(z)\theta_{20}^* k_p Z_a(z) \quad (8.19)$$

and determining the parameters θ_1^*, θ_2^*, θ_{20}^*, and θ_3^* from

$$\theta_1^{*T} a(z)P(z) + (\theta_2^{*T} a(z) + \theta_{20}^* \Lambda))k_p Z_a(z)$$
$$= \Lambda(z)(P(z) - k_p \theta_3^* Z_a(z) P_m(z)) \quad (8.20)$$

with $\theta_3^* = k_p^{-1}$, we have

$$(1 - F_1(z) - F_2(z)G_u(z))^{-1} = \frac{P(z)}{Z_a(z)P_m(z)}. \quad (8.21)$$

Substituting (8.13) and (8.21) into (8.17), we have

$$y(k) = \frac{1}{\theta_3^* P_m(z)}[F_2(z)[\bar{y}] + \theta_3^* r + \theta_4^* + \theta_3^* P_m(z)[\bar{y}]](k) \quad (8.22)$$

Using (8.20), we have

$$F_2(z) + \theta_3^* P_m(z) = \frac{\theta_{20}^{*T} a(z) + \theta_{20}^* \Lambda(z) + \theta_3^* P_m(z)\Lambda(z)}{\Lambda(z)}$$
$$= \frac{P(z)}{k_p Z_a(z)} - \frac{\theta_1^{*T} a(z)P(z)}{\Lambda(z)k_p(z)Z_a(z)}. \quad (8.23)$$

Finally, we obtain the closed-loop system expression

$$y(k) = \frac{1}{P_m(z)}[r](k) + \frac{1}{\theta_3^* P_m(z)}$$
$$\cdot \left[\frac{\Lambda(z) - \theta_1^{*T} a(z)}{\Lambda(z)k_p Z_a(z)} \sum_{j=j_1,\dots,j_p} k_{pj} Z_j(z)[\bar{u}_j] + \theta_4^* \right](k). \quad (8.24)$$

Since $P_m(z)$, $\Lambda(z)$, and $Z_a(z)$ are stable, there exists a constant θ_4^* such that

$$f_p(k) = \frac{1}{\theta_3^* P_m(z)} \left[\frac{\Lambda(z) - \theta_1^{*T} a(z)}{\Lambda(z)k_p Z_a(z)} \sum_{j=j_1,\dots,j_p} k_{pj} Z_j(z)[\bar{u}_j] + \theta_4^* \right](k) \quad (8.25)$$

converge to zero exponentially as $k \to \infty$, so that $\lim_{k \to \infty}(y(k) - y_m(k)) = 0$ exponentially (note that the Lemma 3.2.1 can be used to derive plant-model matching for the case of varying failure parameters; see Section 3.2).

Similar to that in Chapter 2, we let $\{k_i, k_i + 1, \ldots, k_{i+1} - 1\}$, $i = 0, 1, \ldots, m_0$, with $k_0 = 0$, be the time instant sets on which the actuator failure pattern is fixed, that is, actuators only fail at time k_i, $i = 1, \ldots, m_0$. Since there are m actuators, at least one of them does not fail; we have $m_0 < m$ and $k_{m_0+1} = \infty$. Then, at time k_j, $j = 1, \ldots, m_0$, the plant-model matching parameters θ_1^*, θ_2^*, θ_{20}^*, θ_3^*, and θ_4^* change their values such that

$$\theta_1^* = \theta_{1(i)}^*, \ \theta_2^* = \theta_{2(i)}^*, \ \theta_{20}^* = \theta_{20(i)}^*, \ \theta_3^* = \theta_{3(i)}^*, \ \theta_4^* = \theta_{4(i)}^* \tag{8.26}$$

that is, the plant-model matching parameters θ_1^*, θ_2^*, θ_{20}^*, θ_3^*, and θ_4^* are piecewise constant parameters, because the plant has different characterizations under different actuator failure conditions.

8.3 Adaptive Control Design

Now we develop an adaptive control scheme for the plant (8.1) with unknown plant parameters, and with unknown actuator failures (8.5) when $\mu = 0$ and $d(k) = 0$. In this case, the parameters θ_1^*, θ_2^*, θ_{20}^*, θ_3^*, and θ_4^* are unknown. We use the adaptive version of the controller (8.10) as

$$v_1(k) = \cdots = v_m(k) \triangleq v_0(k)$$
$$= \theta_1^T \omega_1(k) + \theta_2^T \omega_2(k) + \theta_{20} y(k) + \theta_3 r(k) + \theta_4 \tag{8.27}$$

where $\theta_1(k) \in R^{n-1}$, $\theta_2(k) \in R^{n-1}$, $\theta_{20}(t) \in R$, $\theta_3(k) \in R$, and $\theta_4(k) \in R$ are the estimates of the unknown parameters θ_1^*, θ_2^*, θ_{20}^*, θ_3^*, and θ_4^*.

To derive an error equation, we define

$$\theta^* = [\theta_1^{*T}, \theta_2^{*T}, \theta_{20}^*, \theta_3^*, \theta_4^*]^T \in R^{2n+1} \tag{8.28}$$
$$\theta(k) = [\theta_1^T(k), \theta_2^T(k), \theta_{20}(k), \theta_3(k), \theta_4(k)]^T \in R^{2n+1} \tag{8.29}$$
$$\omega(k) = [\omega_1^T(k), \omega_2^T(k), y(k), r(k), 1]^T \in R^{2n+1} \tag{8.30}$$
$$\tilde{\theta}(k) = \theta(k) - \theta^*. \tag{8.31}$$

Ignoring exponentially decaying terms, we have the tracking error equation

$$e(k) = y(k) - y_m(k) = \frac{k_p}{P_m(z)}[\tilde{\theta}^T \omega](k). \tag{8.32}$$

Introducing the auxiliary signals

$$\zeta(k) = W_m(z)[\omega](k) \tag{8.33}$$
$$\zeta(k) = \theta^T(k)\zeta(k) - W_m(z)[\theta^T \omega](k) \tag{8.34}$$
$$\epsilon(k) = e(k) + \rho(k)\xi(k) \tag{8.35}$$

where $\rho(k)$ is the estimate of $\rho^* = 1/\theta_3^*$. Then, we choose the adaptive laws

$$\theta(k+1) - \theta(k) = -\frac{\text{sign}[k_p]\Gamma\zeta(k)\epsilon(k)}{m^2(k)} \quad (8.36)$$

$$\rho(k+1) - \rho(k) = -\frac{\gamma\xi(k)\epsilon(k)}{m^2(k)} \quad (8.37)$$

where

$$m^2(k) = 1 + \zeta^T(k)\zeta(k) + \xi^2(k). \quad (8.38)$$

and Γ and γ are constant adaptation gains, which satisfy

$$0 < \Gamma = \Gamma^T < \frac{\gamma_\theta}{k_p^0} I_{2n+1}, \ 0 < \gamma_\theta < 2, \ 0 < \gamma < 2 \quad (8.39)$$

To implement (8.36), we need the following assumption

(A8.2) $\text{sign}[k_p]$, the sign of k_p, is known, and $|k_p| \leq k_p^0$ for some known constant $k_p^0 > 0$, where $k_p = \rho^* = 1/\theta_3^*$.

To analyze the stability and tracking performance of the adaptive control system, we define the positive definite function

$$V(\tilde{\theta}, \tilde{\rho}) = |\rho^*|\tilde{\theta}^T \Gamma^{-1}\tilde{\theta} + \gamma^{-1}\tilde{\rho}^2, \ \tilde{\theta} = \theta - \theta^*, \ \tilde{\rho} = \rho - \rho^*. \quad (8.40)$$

From (8.32)–(8.35), we obtain

$$\epsilon(k) = \rho^*\tilde{\theta}^T(k)\zeta(k) + \tilde{\rho}(k)\xi(k) + \epsilon_p(k) \quad (8.41)$$

where

$$\begin{aligned}\epsilon_p(k) = \ &\rho^*(\theta^{*T}(k)\zeta(k) - W_m(z)[\theta^{*T}\omega](k)) \\ &+ W_m(z)\rho^*[\tilde{\theta}^T\omega](k) - \rho^* W_m(z)[\tilde{\theta}^T\omega](k). \end{aligned} \quad (8.42)$$

Then, ignoring the effect of the exponentially decaying terms $\epsilon_p(k)$ in (8.41), the time increment of $V(\tilde{\theta}, \tilde{\rho})$ along the trajectories (8.36) and (8.37) is

$$\begin{aligned}&V(\tilde{\theta}(k+1), \tilde{\rho}(k+1)) - V(\tilde{\theta}(k), \tilde{\rho}(k)) \\ &= -\left(2 - \frac{|k_p|\zeta^T(k)\Gamma\zeta(k) + \gamma\xi^2(k)}{m^2(k)}\right)\frac{\epsilon^2(k)}{m^2(k)} \\ &\leq -(2 - \gamma_m)\frac{\epsilon^2(k)}{m^2(k)} \leq 0\end{aligned} \quad (8.43)$$

for $k \in \{k_i, k_i + 1, \ldots, k_{i+1} - 2\}$, $i = 0, 1, \ldots, m_0$, for some $m_0 < m$, where $\gamma_m = \max\{\gamma_\theta, \gamma\} < 2$. Hence, over each time set $\{k_i, k_i + 1, \ldots, k_{i+1} - 1\}$, $i = 0, 1, \ldots, m_0$, with $k_0 = 0$, $V(\tilde{\theta}(k), \tilde{\rho}(k))$ as a function of k is nonincreasing.

After a finite number m_0 of finite parameter jumps, that is, for $k > k_{m_0}$, $V(\tilde{\theta}(k), \tilde{\rho}(k)) \leq V(\tilde{\theta}(0), \tilde{\rho}(0)) + \Delta V$, where ΔV represents the total jumping value of $V(\tilde{\theta}(k), \tilde{\rho}(k))$ over these sets, so that $\theta(k) \in L^\infty$, $\rho(k) \in L^\infty$.

From (8.43) it follows that

$$\sum_{k=0}^{\infty} \frac{\epsilon^2(k)}{m^2(k)} \leq \frac{1}{2-\gamma_m}\left(V(\tilde{\theta}(0), \tilde{\rho}(0)) - V(\tilde{\theta}(\infty), \tilde{\rho}(\infty)) + \sum_{i=1}^{m_0} \Delta V_i\right) \quad (8.44)$$

where $\sum_{i=1}^{m_0} \Delta V_i$ denotes the sum of finite parameter jumps. From (8.44) we have $\epsilon(k)/m(k) \in L^2$; from (8.41) $\epsilon(k)/m(k) \in L^\infty$; and from (8.36) we have $\theta(k+1) - \theta(k) \in L^2$. In summary, we have

Lemma 8.3.1. *The adaptive controller (8.27), with the adaptive laws (8.36) and (8.37), applied to the system (8.1) with actuator failures (8.5), guarantees that $\theta(k)$, $\rho(k) \in L^\infty$, $\epsilon(k)/m(k) \in L^2 \cap L^\infty$, and $\theta(k+1) - \theta(k) \in L^2$.*

Based on this result, using the analysis method in [125], we can prove the following closed-loop stability and asymptotic tracking properties.

Theorem 8.3.1. *The adaptive controller (8.27), with the adaptive law (8.36)–(8.37), applied to the system (8.1) with actuator failures (8.5), guarantees that all closed-loop signals are bounded and the tracking error $e(k) = y(k) - y_m(k)$ goes to zero as k goes to infinity.*

8.4 Robust Adaptive Failure Compensation

The above adaptive control scheme was designed for $\mu = 0$ and $d(k) = 0$ and may not ensure stability in the presence of unmodeled dynamics $\mu \Delta_m(z)$, $\mu \Delta_a(z)$ for $\mu \neq 0$ and output disturbance $d(k) \neq 0$. In this section, we derive robust adaptive laws that ensure robust stability in the presence of $\mu \Delta_m(z)$, $\mu \Delta_a(z)$, and $d(k)$, under the following assumptions:

(A8.3a) $\Delta_{aj}(z)$ and $\Delta_{mj}(z)/P_m(z), j = 1, \ldots, m$, are strictly proper and rational transfer functions;

(A8.3b) $W_m(qz)\Delta_m(qz)(z+1)$ and $\Delta_a(qz)(z+1)$ are stable with a finite gain independent of μ for some constant $q \in (0,1)$, i.e., the impulse function $h_m(k)$, $h_a(k)$ of $W_m(qz)\Delta_m(qz)(z+1)$ and $\Delta_a(qz)(z+1)$ satisfy

$$\sum_{k=0}^{\infty} |h_m(k)| < c < \infty, \quad \sum_{k=0}^{\infty} |h_a(k)| < c < \infty \quad (8.45)$$

for some constant $c > 0$ independent of μ; and

(A8.3c) $d(k) \in L^\infty$, that is, $d(k)$ is bounded.

8.4.1 Robustness of Plant-Model Matching

We first examine robustness of plant-model output matching with nominal parameters. For robust non-adaptive control with $G_0(z)$ known, we can relax Assumptions (A8.3a) and (A8.3b) as

(A8.3d) $\Delta_{aj}(z)$ and $\Delta_{mj}(z)/P_m(z), j = 1, \ldots, m$, are proper and rational;

(A8.3e) for some constant $c > 0$ independent of μ, the impulse functions $h_{m0}(k)$, $h_{a0}(k)$ of $W_m(z)\Delta_m(z)$, $\Delta_a(z)$ satisfy

$$\sum_{k=0}^{\infty} |h_{m0}(k)| < c < \infty, \ \sum_{k=0}^{\infty} |h_{a0}(k)| < c < \infty. \tag{8.46}$$

Ignoring the exponentially decaying effect of the initial condition and finite parameter jumps, and defining $F_1(z)$ and $F_2(z)$ as (8.15), from (8.10) and (8.20) with $\theta_1^*, \theta_2^*, \theta_{20}^*, \theta_3^*$ and θ_4^*, we have

$$y(k) - y_m(k) = \frac{k_p}{P_m(z)}(1 - F_1(z))[\mu(\Delta_m(z) + G_u^{-1}(z)\Delta_a(z))][v](k)$$
$$+ \frac{k_p}{P_m(z)}(1 - F_1(z))G_u^{-1}(z)[d](k) \tag{8.47}$$

$$(1 - F_1(z))[v](k) = F_2(z)[y - y_m](k) + (\frac{F_2(z)}{P_m(z)} + \theta_3^*)[r](k). \tag{8.48}$$

Using (8.2), (8.20), (8.47), and (8.48), we have

$$\left(1 - \mu\frac{k_p}{P_m(z)}(\Delta_m(z) + G_u^{-1}(z)\Delta_a(z))F_2(z)\right)[y - y_m](k)$$
$$= \mu\frac{k_p}{P_m(z)}(\Delta_m(z) + G_u^{-1}(z)\Delta_a(z))\left(\frac{F_2(z)}{P_m(z)} + \theta_3^*\right)[r](k)$$
$$+ \frac{k_p}{P_m(z)}(1 - F_1(z))G_u^{-1}(z)[d](k) \tag{8.49}$$

$$\left(1 - \mu\frac{k_p}{P_m(z)}(\Delta_m(z) + G_u^{-1}(z)\Delta_a(z))F_2(z)\right)[v](k)$$
$$= G_u^{-1}(z)\frac{1}{P_m(z)}[r + k_pF_2(z)[d]](k). \tag{8.50}$$

Since $\frac{1}{P_m(z)}(\Delta_m(z)+G_u^{-1}(z)\Delta_a(z))$ is stable and proper, $(1-\mu\frac{k_p}{P_m(z)}(\Delta_m(z)+G_u^{-1}(z)\Delta_a(z))F_2(z))^{-1}$ has a stable and proper inverse if $\mu \in [0, \mu_0)$ for

$$\mu_0 = \frac{1}{\sup_{\omega \in [0,2\pi]}|\frac{k_p}{P_m(z)}(\Delta_m(z) + G_u^{-1}(z)\Delta_a(z))F_2(z)|_{z=e^{j\omega}}} > 0 \tag{8.51}$$

whose existence is ensured by Assumptions (A8.3d) and (A8.3e). Then, all closed-loop system signals are bounded and the tracking error $e(k) = y(k) - y_m(k)$ converges exponentially to the residual set

$$S_0 = \{e : |e| \leq b_1 \mu \bar{r} + b_2 \bar{d}\} \tag{8.52}$$

for some $b_i > 0, i = 1, 2$, where $\bar{r}, \bar{d}$ are the upper bounds for $|r(k)|, |d(k)|$.

8.4.2 Robust Adaptive Laws

For robust adaptive control, we need to update the parameters of (8.27) with an adaptive law that is robust when $\mu \neq 0$ and $d(k) \neq 0$, with unknown plant parameters and actuator failures, that is, $\theta_1^*, \theta_2^*, \theta_{20}^*, \theta_3^*$, and θ_4^* are unknown. To derive an error equation, we use $\theta^*, \theta(k), \omega(k)$, and $\tilde{\theta}(k)$ in (8.28)–(8.31) and, ignoring exponentially decaying terms, obtain

$$e(k) = y(k) - y_m(k) = \frac{k_p}{P_m(z)}[\tilde{\theta}^T \omega](k) + \mu \Delta(z)[v](k) + H(z)[d](k) \tag{8.53}$$

where

$$\Delta(z) = k_p(1 - F_1(z))\frac{1}{P_m(z)}\Delta_m(z) + \left(1 + \frac{k_p}{P_m(z)}F_2(z)\right)\Delta_a(z) \tag{8.54}$$

$$H(z) = 1 + \frac{k_p}{P_m(z)}F_2(z). \tag{8.55}$$

We then generate a new normalizing signal $m(k)$ from

$$m(k+1) = (1 - \delta_0)m(k) + \delta_1(|u(k)| + |y(k)| + 1) \tag{8.56}$$

where $m(0) > 0$, $q < 1 - \delta_0 < 1$ with q defined in Assumption (A8.3b).

To handle the unmodeled dynamics $\mu \Delta_m(z), \mu \Delta_a(z)$ and the disturbance $d(k)$, as in [57], [118], we introduce the switching signals $\sigma_\theta(k)$ and $\sigma_\rho(k)$:

$$\sigma_\theta(k) = \begin{cases} 0 & \text{if } \|\theta(k)\|_2 < 2M_1 \\ \sigma_0 & \text{if } \|\theta(k)\|_2 \geq 2M_1 \end{cases} \tag{8.57}$$

$$\sigma_\rho(k) = \begin{cases} 0 & \text{if} |\rho(k)| < 2M_2 \\ \sigma_0 & \text{if } |\rho(k)| \geq 2M_2 \end{cases} \tag{8.58}$$

where $0 < \sigma_0 < (1 - \gamma_m)/2$, $\gamma_m = \max\{\gamma_\theta k_p^0, \gamma\} < 1$, and $M_1 > \|\theta^*\|_2$ (with $\|\cdot\|_2$ being the Euclidean vector norm), $M_2 > |\rho^*|$. To implement $\sigma_\theta(k)$ and $\sigma_\rho(k)$, we need the knowledge of parameter bounds M_1 and M_2.

Introducing the auxiliary signals

$$\zeta(k) = W_m[\omega](k) \tag{8.59}$$

$$\xi(k) = \theta^T(k)\zeta(k) - W_m(z)[\theta^T\omega](k) \tag{8.60}$$

$$\epsilon(k) = e(k) + \rho(k)\xi(k) \tag{8.61}$$

where $\rho(k)$ is the estimate of $\rho^* = 1/\theta_3^*$, we choose the robust adaptive laws

$$\theta(k+1) = \theta(k) - \frac{\text{sign}[k_p]\Gamma\zeta(k)\epsilon(k)}{m^2(k)} - \sigma_\theta(k)\theta(k) \tag{8.62}$$

$$\rho(k+1) = \rho(k) - \frac{\gamma\xi(k)\epsilon(k)}{m^2(k)} - \sigma_\rho(k)\rho(k) \tag{8.63}$$

and the adaptation gains Γ and γ are constant and satisfy

$$0 < \Gamma = \Gamma^T < \frac{\gamma_\theta}{k_p^0}I_{2n+1},\ 0 < \gamma_\theta < \frac{1}{k_p^0},\ 0 < \gamma < 1,\ k_p^0 \ge |k_p|. \tag{8.64}$$

In this case, the estimation error equation now is

$$\epsilon(k) = \rho^*\tilde{\theta}^T(k)\zeta(k) + \tilde{\rho}(k)\xi(k) + \mu\eta(k) + \bar{d}(k) \tag{8.65}$$

where

$$\eta(k) = \Delta(z)[u](k),\ \bar{d}(k) = H(z)[d](k). \tag{8.66}$$

The time increment of $V(\tilde{\theta}(k), \tilde{\rho}(k))$ in (8.40) along (8.62) and (8.63) is

$$V(\tilde{\theta}(k+1), \tilde{\rho}(k+1)) - V(\tilde{\theta}(k)), \tilde{\rho}(k)$$

$$= -\left(2 - \frac{|k_p|\zeta^T(k)\Gamma\zeta(k) + \gamma\xi^2(k)}{m^2(k)}\right)\frac{\epsilon^2(k)}{m^2(k)} + \frac{2\epsilon(k)(\mu\eta(k) + \bar{d}(k))}{m^2(k)}$$

$$+2k_p\sigma_\theta(k)\theta^T(k)\frac{\epsilon(k)\zeta(k)}{m^2(k)} + 2\sigma_\rho(k)\rho(k)\frac{\epsilon(k)\xi(k)}{m^2(k)}$$

$$+|k_p|\sigma_\theta^2(k)\theta^T(k)\gamma_\theta^{-1}\tilde{\theta}(k) + \sigma_\rho^2(k)\rho(k)\gamma^{-1}\rho(k)$$

$$-2|k_p|\sigma_\theta(k)\theta^T(k)\gamma_\theta^{-1}\tilde{\theta}(k) - 2\sigma_\rho(k)\rho(k)\gamma^{-1}\rho(k)$$

$$\le -\frac{\gamma_m}{2}\frac{\epsilon^2(k)}{m^2(k)} + \frac{4}{2-\gamma_m}\frac{\mu^2\eta^2(k) + \bar{d}^2(k)}{m^2(k)} - (1-\gamma_m)\frac{\epsilon^2(k)}{m^2(k)}$$

$$+2k_p\sigma_\theta(k)\theta^T(k)\frac{\epsilon(k)\zeta(k)}{m^2(k)} + 2\sigma_\rho(k)\rho(k)\frac{\epsilon(k)\xi(k)}{m^2(k)}$$

$$+|k_p|\sigma_\theta^2(k)\theta^T(k)\gamma_\theta^{-1}\tilde{\theta}(k) + \sigma_\rho^2(k)\rho(k)\gamma^{-1}\rho(k)$$

$$-2|k_p|\sigma_\theta(k)\theta^T(k)\gamma_\theta^{-1}\tilde{\theta}(k) - 2\sigma_\rho(k)\rho(k)\gamma^{-1}\rho(k)$$

$$\le -\frac{\gamma_m}{2}\frac{\epsilon^2(k)}{m^2(k)} + \frac{4}{2-\gamma_m}\frac{\mu^2\eta^2(k) + \bar{d}^2(k)}{m^2(k)}$$

$$-\frac{|k_p|k_p^0\sigma_\theta(k)}{2}\theta^T(k)\theta(k) - \frac{\sigma_\rho(k)}{2}\rho^2(k). \tag{8.67}$$

From the definition of $\sigma_\theta(k)$, $\sigma_\rho(k)$, we see the term $\sigma_\theta(k)\theta^T(k)\theta(k)$ grows unbounded if $\theta(k)$ grows unbounded, and similarly the term $\sigma_\rho(k)\rho^2(k)$ is unbounded, and from Assumptions (A8.3a), (A8.3b), and (A8.3c), it can be shown that $(\mu^2\eta^2(k) + \bar{d}^2(k))/m^2(k)$ is bounded [57], [118], [125]. Hence $V(\tilde{\theta}(k), \tilde{\rho}(k))$ is bounded, that is, $\theta(k) \in L^\infty$, $\rho(k) \in L^\infty$, and so is $\epsilon(k)/m(k) \in L^\infty$.

Based on this result, using the analysis method in [125], we can prove the following robustness properties of the closed-loop system.

Theorem 8.4.1. *There exists a $\mu^* > 0$ such that for all $\mu \in [0, \mu^*)$, the controller (8.27) with the adaptive laws (8.62) and (8.63), applied to the system (8.1), ensures closed-loop signal boundedness and mean output tracking:*

$$\int_{t_1}^{t_2} (y(t) - y_m(t))^2(t)\, dt \le c_1 + c_2 \int_{t_1}^{t_2} \bar{d}^2(t)\, dt + c_3\mu^2 \tag{8.68}$$

for some constant $c_i > 0$, $i = 1, 2, 3$, and any $t_2 > t_1 \ge 0$.

8.5 Boeing 747 Lateral Control Simulation

We apply the adaptive control scheme consisting of (8.27) and (8.33)–(8.37) to a linearized Boeing 747 aircraft lateral dynamics model whose original continuous-time model is from [37] (see Section 4.4).

Plant Model. The linearized discrete-time model of the lateral dynamics of Boeing 747 with one augmented actuation vector can be described as

$$\begin{aligned} x(k+1) &= Ax(k) + Bu(k),\; y(k) = Cx(k) \\ x(k) &= [\beta, y_r, p, \phi]^T,\; B = [b_1, b_2] \end{aligned} \tag{8.69}$$

where β is the side-slip angle, y_r is the yaw rate, p is the roll rate, ϕ is the roll angle, y is the plant output, which is the yaw rate in this case, and u is the control input vector, which contains two control signals $u = [u_1, u_2]^T$ to represent two rudder servos: δ_{r1}, δ_{r2}, from a two-piece rudder for achieving compensation in the presence of actuator failures.

With sampling time $T = 0.1$ sec, using the continuous-time data [37] (see Section 4.4), we have the dynamics matrices in discrete time as

$$A = \begin{bmatrix} 0.9902 & -0.0985 & 0.0082 & 0.0041 \\ 0.0597 & 0.9855 & -0.0028 & 0.0001 \\ -0.2956 & 0.0525 & 0.9533 & -0.0006 \\ -0.0147 & 0.0104 & 0.0977 & 1.000 \end{bmatrix} \tag{8.70}$$

$$b_1 = \begin{bmatrix} 0.0031 \\ -0.0472 \\ 0.0137 \\ 0.0005 \end{bmatrix}, \quad b_2 = \begin{bmatrix} 0.0036 \\ -0.0497 \\ 0.0182 \\ 0.0007 \end{bmatrix} \tag{8.71}$$

$$C = \begin{bmatrix} 0 & 1 & 0 & 0 \end{bmatrix} \tag{8.72}$$

where b_2 is augmented for studying actuator failure compensation.

The control objective is to design an output feedback control vector $v = [v_1, v_2]^T$ applied to the plant (8.69) with actuator failures (8.5), to ensure that all closed-loop signals are bounded and the plant output (yaw rate) tracks a reference signal y_m in the presence of actuator failures.

Simulation Results. For the plant (8.69)–(8.72) with two inputs u_1, u_2, Assumptions (A8.1a) and (A8.1b) are satisfied: $(C, A, b_i), i = 1, 2$, are observable, controllable, minimum phase, and with relative degree 1, which is sufficient for stable plant-model matching with one actuator failure, and so is $(C, A, b_1 + b_2)$, which is for the no-failure case.

For simulation, we choose the reference model as $W_m(z) = 1/z$. In this study, we simulate the case of one actuator failure: $u_2(t) = 0.04$ (rad), $t \geq 60$ (sec). The simulation parameters are $\Gamma = 3I$, $\gamma = 1$, $x(0) = [0, 0, 0, 0]^T$, $y_m(0) = 0$, $y(0) = 0.01$, $\theta(0) = 0.9[-0.1700, 0.1571, 0.7575, 0.2839, -5.5475, 0.6237, 25.8161, -21.1864, 0]^T$, and $\rho(0) = 0$.

The simulation results, including the plant output $y(t)$, reference output $y_m(t)$, tracking error $e(t) = y(t) - y_m(t)$, and control input $v_1(t) = v_2(t)$, are shown in Figure 8.1 for the reference input $r(t) = 0.02(1 - e^{-0.01t})$ and in Figure 8.2 for $r(t) = 0.01 \sin(0.01t)$.

The system responses are as expected: At the time instant when one of the actuators fails, there is a transient response in the tracking errors, and as the time goes on, the output tracking errors become smaller, while all signals in the adaptive control system are bounded.

In summary, in this chapter, we study the adaptive actuator failure compensation problem for discrete-time systems. The robustness problem of the actuator failure compensation scheme with respect to unmodeled dynamics and output disturbance is also addressed. Desired failure compensation performance is achievable in discrete time, by an output feedback adaptive actuator failure compensation scheme, when unmodeled dynamics and output disturbance are absent. To ensure robustness of system performance, robust adaptive laws can be used to handle stable unmodeled dynamics and bounded disturbances, based on robust adaptive control theory.

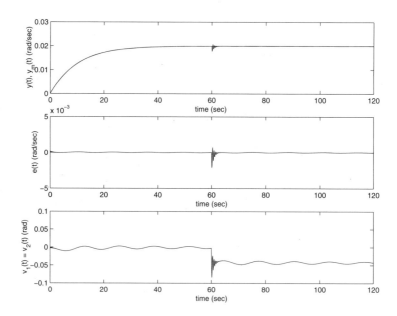

Fig. 8.1. System response for $r(t) = 0.02(1 - e^{-0.01t})$.

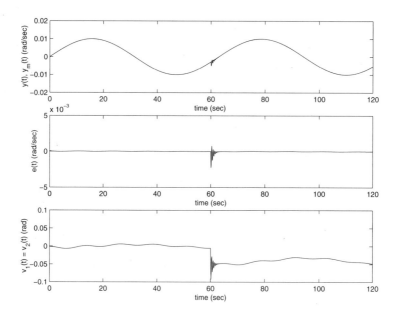

Fig. 8.2. System response for $r(t) = 0.01 \sin(0.01t)$.

Chapter 9

Failure Compensation for Nonlinear Systems

The adaptive actuator failure compensation problem has been addressed so far for linear systems with actuator failures characterized by inputs stuck at some values fixed or variant not influenced by control action. Both the failure patterns and the failure parameters including failure values and failure time are assumed unknown. However, most real-life systems are nonlinear dynamic systems, for example, aircraft flight control systems. Actuator failure compensation for nonlinear systems is an area of major interest. In the following chapters we further our research to several classes of nonlinear systems, whose dynamics can be transformed into some canonical forms that are subject to actuator failures. A key issue of direct adaptive control is the parametrization of the nonlinear system with actuator failures and the applied feedback control scheme for a given control objective under certain matching conditions. Based on a suitable parametrization, an adaptive scheme can be used to update the controller parameters for the case when the plant parameters and actuator failure parameters are both unknown.

This chapter gives an introduction to adaptive actuator failure compensation for nonlinear systems. In Section 9.1, we formulate the problem of compensation control for nonlinear systems. In Section 9.2, we present an adaptive actuator failure compensation control scheme for the class of feedback linearizable nonlinear systems. In Section 9.3, we present a different adaptive design based on different design conditions. Both designs guarantee closed-loop stability and asymptotic tracking in the presence of unknown system parameters and actuator failure parameters.

9.1 Problem Formulation

To investigate the actuator failure compensation control problem for nonlinear systems, we consider the nonlinear plant

$$\dot{x} = f(x) + \sum_{i=1}^{m} g_i(x)u_i, \ y = h(x) \qquad (9.1)$$

where $x \in R^n$ is the state, $y \in R$ is the output, and $u_i \in R$ is the input, which may fail during operation, and $f(x) \in R^n$, $g_i(x) \in R^n$, and $h(x) \in R$ are smooth.

As modeled in (1.2), the actuator failures are characterized as

$$u_i(t) = \bar{u}_i, \ t \geq t_i, \ i \in \{1, 2, \ldots, m\} \tag{9.2}$$

where the constant value $\bar{u}_i$ and the failure time instant t_i are unknown. For the plant (9.1) with the actuator failures (9.2), as in (1.3), the input vector $u = [u_1, u_2, \ldots, u_m]^T \in R^m$ can be expressed as

$$u = v(t) + \sigma(\bar{u} - v(t)) \tag{9.3}$$

where $v(t) = [v_1(t), v_2(t) \ldots, v_m(t)]^T$ is the applied control to be designed,

$$
\begin{aligned}
\bar{u} &= [\bar{u}_1, \bar{u}_2, \ldots, \bar{u}_m]^T \\
\sigma &= \text{diag}\{\sigma_1, \sigma_2, \ldots, \sigma_m\} \\
\sigma_i &= \begin{cases} 1 & \text{if the } i\text{th actuator fails, i.e., } u_i = \bar{u}_i \\ 0 & \text{otherwise} \end{cases}
\end{aligned} \tag{9.4}
$$

and the plant can be rewritten as

$$
\begin{aligned}
\dot{x} &= f(x) + g(x)\sigma\bar{u} + g(x)(I - \sigma)v \\
y &= h(x)
\end{aligned} \tag{9.5}
$$

where $g(x) = [g_1(x), g_2(x), \ldots, g_m(x)] \in R^{n \times m}$, for a certain failure pattern σ. It can be seen that when failures take place, not only will the applied control be checked, but also uncertainties will be brought into the system. A desirable control design should be able to handle uncertainties caused by system parameter and structural changes caused by all possible actuator failures.

The control objective is to use a state or output feedback control law for the plant (9.1) with the actuator failures (9.2) or a more general one given in (1.6) or (1.7) to ensure that all signals in the closed-loop system are bounded and the plant output $y(t)$ asymptotically tracks a given reference signal $y_m(t)$ generated from a reference system.

To accomplish the task, as stated in Section 1.3.3, the following basic assumption for the actuator failure compensation problem is needed:

(A9.0) The plant (9.1) is so constructed that for any up to $m - 1$ actuator failures, the remaining actuators can still achieve the above control objective if the knowledge of the plant dynamics and actuator failures is available for control design.

Clearly, this assumption relies on the state-of-the-art of nonlinear adaptive control, that is, the existing design tools for adaptive control of the plant (9.1) without actuator failures. Moreover, the actuator failures need to be handled. In the following chapters, we investigate the classes of nonlinear systems for which such design tools are available for actuator failure-free cases. Our main effort is to develop adaptive actuator failure compensation schemes for such adaptive feedback controllers applicable in the presence of unknown actuator failures. Three typical classes of nonlinear systems to be considered are feedback linearizable systems, parametric-strict-feedback systems, and output-feedback systems. In this chapter, we investigate the actuator failure compensation problem for systems that are feedback linearizable. In the next section, we present a solution to this problem based on a basic structure condition on the system dynamics. A different solution will be derived in Section 9.3 based on a different design condition.

9.2 Design for Feedback Linearizable Systems

We first develop a solution to adaptive actuator failure compensation control of systems that are feedback linearizable and with actuator failures.

Structure Condition. Considering the plant (9.1), we assume that

(A9.2a) $g_i(x) \in \text{span}\{g_0(x)\}$, $g_0(x) \in R^n$, for $i = 1, 2, \ldots, m$, and the nominal system

$$\dot{x} = f(x) + g_0(x)u_0, \; y = h(x) \tag{9.6}$$

where $u_0 \in R$, is feedback linearizable with relative degree ρ.

Diffeomorphism. We first introduce the concept of diffeomorphism. A continuously differentiable vector function $T(x) \in R^n$ for $x \in R^n$ is a diffeomorphism of U onto V for some open subsets U and V of R^n if
(i) $T(U) = V$,
(ii) T is one-to-one, and
(iii) the inverse function T^{-1} from V to U is continuously differentiable.

If $\bar{y} = T(\bar{x})$ and $\frac{\partial T}{\partial x}|_{x=\bar{x}}$ is nonsingular, then there exist open sets U containing $\bar{x}$ and V containing $\bar{y}$ such that T is a diffeomorphism of U onto V [138]. Since the nominal system (9.6) is feedback linearizable with index ρ, there exists a diffeomorphism $[\xi, \eta]^T = T(x) = [T_c(x), T_z(x)]^T$, $\xi \in R^\rho$ and $\eta \in R^\gamma$ ($\eta = T(z(x)$ may not be unique), $\rho + \gamma = n$, defined as [60]

$$\xi = T_c(x) = \begin{bmatrix} T_1(x) \\ T_2(x) \\ \vdots \\ T_\rho(x) \end{bmatrix} = \begin{bmatrix} h(x) \\ L_{f(x)}h(x) \\ \vdots \\ L_{f(x)}^{\rho-1}h(x) \end{bmatrix}$$

$$\eta = T_z(x) = \begin{bmatrix} T_{\rho+1}(x) \\ T_{\rho+2}(x) \\ \vdots \\ T_n(x) \end{bmatrix} \tag{9.7}$$

where $L_{f(x)}h(x)$ is the Lie derivative of $h(x)$ along $f(x)$ (defined as $L_f h = \frac{\partial h}{\partial x} f = \frac{\partial h}{\partial x_1} f_1 + \cdots + \frac{\partial h}{\partial x_n} f_n$), to transform system (9.6) into

$$\begin{aligned} \dot{\xi} &= A\xi + B(\varphi(\xi,\eta) + \beta_0(\xi,\eta)u_0) \\ \dot{\eta} &= \psi(\xi,\eta) \\ y &= C\xi \end{aligned} \tag{9.8}$$

where

$$A = \begin{bmatrix} 0 & 1 & 0 & \cdots & 0 & 0 \\ 0 & 0 & 1 & \cdots & 0 & 0 \\ & & \cdots & \cdots & & \\ 0 & 0 & \cdots & 0 & 0 & 1 \\ 0 & 0 & \cdots & 0 & 0 & 0 \end{bmatrix} \in R^{\rho\times\rho}, \; B = \begin{bmatrix} 0 \\ 0 \\ \vdots \\ 0 \\ 1 \end{bmatrix} \in R^{\rho\times 1}$$

$$C = \begin{bmatrix} 1 & 0 & \cdots & \cdots & 0 & 0 \end{bmatrix} \in R^{1\times\rho} \tag{9.9}$$

$$\varphi(\xi,\eta) = L_{f(x)}^\rho h(x), \; \beta_0(\xi,\eta) = L_{g_0(x)}L_{f(x)}^{\rho-1}h(x) \neq 0. \tag{9.10}$$

In this expression, $\dot{\eta} = \psi(\xi,\eta)$ is the *zero dynamics* subsystem. The system (9.8) consists of a cascade of integrators from input to output and a zero dynamic subsystem that is unobservable after exact cancellation of η in the output y by a feedback linearization control input $u_0 = (1/\beta_0(\xi,\eta))(w_0 - \varphi(\xi,\eta))$ for a linear feedback control law w_0 based on ξ.

From Assumption (A9.2a), it follows that with the diffeomorphism (9.7), the plant (9.1) can be transformed as

$$\begin{aligned} \dot{\xi} &= A\xi + B(\varphi(\xi,\eta) + \beta^T(\xi,\eta)u) \\ \dot{\eta} &= \psi(\xi,\eta) \\ y &= C\xi \end{aligned} \tag{9.11}$$

where

$$\begin{aligned} \beta(\xi,\eta) &= [\beta_1(\xi,\eta),\ldots,\beta_m(\xi,\eta)]^T \\ &= [L_{g_1(x)}L_{f(x)}^{\rho-1}h(x),\ldots,L_{g_m(x)}L_{f(x)}^{\rho-1}h(x)]^T. \end{aligned} \tag{9.12}$$

With the actual input described as that in (9.3) in the presence of actuator failures, we rewrite the plant (9.11) as

$$
\begin{aligned}
\dot{\xi} &= A\xi + B(\varphi(\xi,\eta) + \beta^T(\xi,\eta)\sigma\bar{u} + \beta^T(\xi,\eta)(I-\sigma)v) \\
\dot{\eta} &= \psi(\xi,\eta) \\
y &= C\xi.
\end{aligned}
\tag{9.13}
$$

Remark 9.2.1. From (9.11), we see that under Assumption (A9.2a) the zero dynamics of the plant (9.1) are the same as those of the nominal system (9.6) and will not change in the presence of actuator failures. In other words, actuator failures neither change the structure of the zero dynamics nor enter the zero dynamics as additional inputs. □

Remark 9.2.2. In order to explain the significance of Assumption (A9.2a) to the diffeomorphism (9.7), we examine the linear case to understand why no input enters the zero dynamics. Consider a linear system $\dot{x} = A_0 x + b_1 u_1 + b_2 u_2, y = cx$ with two inputs u_1 and u_2 and one output y, where (A_0, b_1) and (A_0, b_2) are both controllable. In the linear case, Assumption (A9.2a) indicates that b_1 and b_2 are parallel to each other, that is, $b_1 = kb_2$, where k is a nonzero constant. Obviously, (c, A_0, b_1) and (c, A_0, b_2) have the same relative degree and same zeros under Assumption (A9.2a).

For the system $\dot{x} = A_0 x + b_1 u_1, y = cx$ with relative degree ρ, a linear transformation $z = T_1 x$ can be found to lead to the control canonical form

$$
\dot{z} = A_0 z + b_c u_1, \; y = c_c z
\tag{9.14}
$$

where

$$
A_c = \begin{bmatrix}
0 & 1 & 0 & \cdots & 0 & 0 \\
0 & 0 & 1 & \cdots & 0 & 0 \\
 & & \cdots & \cdots & & \\
0 & 0 & \cdots & 0 & 0 & 1 \\
-a_0 & -a_1 & \cdots & -a_{n-3} & -a_{n-2} & -a_{n-1}
\end{bmatrix}, \; b_c = \begin{bmatrix} 0 \\ 0 \\ \vdots \\ 0 \\ 1 \end{bmatrix}
$$

$$
c_c = \begin{bmatrix} c_0 & c_1 & \cdots & c_{n-\rho-1} & c_{n-\rho} & 0 & \cdots & 0 \end{bmatrix}.
\tag{9.15}
$$

Now let us derive the feedback linearizable form for the linear system (c, A_0, b_1). For the first ρ coordinates, we choose

$$
\begin{aligned}
\xi_i &= c_c A_c^{i-1} z, \; i = 1, 2, \ldots, \rho \\
\eta_i &= z_i, \; i = 1, 2, \ldots, r, \; r = n - \rho
\end{aligned}
\tag{9.16}
\tag{9.17}
$$

to have the system transformed into the feedback linearizable form [60]:

$$\dot{\xi} = A\xi + Bu_1$$
$$\dot{\eta} = Q\eta + R\xi$$
$$y = C\xi \tag{9.18}$$

where A, B, and C are defined in (9.9) and the eigenvalues of Q in the zero dynamics are exactly the zeros of (c, A_0, b_1). In fact, the transformation $[\xi, \eta]^T = T_2(z)$ is still linear, that is, $T_2(z) = T_2 z$, where

$$T_2 = \begin{bmatrix} C \\ CA \\ \vdots \\ CA^{\rho-1} \\ P \end{bmatrix}, \quad P = \begin{bmatrix} 1 & 0 & \cdots & 0 & \cdots & 0 \\ 0 & 1 & \cdots & 0 & \cdots & 0 \\ & & \cdots & \cdots & & \\ 0 & 0 & \cdots & 1 & 0 & 0 \end{bmatrix} \in R^{r \times n}. \tag{9.19}$$

Since $b_1 = kb_2$ under Assumption (A9.2a), the diffeomorphism $[\xi, \eta]^T = T(x) = T_2 T_1 x$ also transforms the original system (c, A_0, b_1, b_2) into

$$\dot{\xi} = A\xi + B(u_1 + ku_2)$$
$$\dot{\eta} = Q\eta + R\xi$$
$$y = C\xi \tag{9.20}$$

with no additional input in the zero dynamics, and the eigenvalues of Q are the zeros of (c, A_0, b_1) or (c, A_0, b_2) (ξ is now also influenced by u_2). $\qquad\square$

Stable Zero Dynamics Assumption. With the control objective that the output y is required to track the output $y_m(t)$ of a reference system $W_m(s)$, that is, $y_m(t) = W_m(s)[r](t)$, where $W_m(s) = 1/P_m(s)$ is the transfer function of the reference system and $P_m(s) = s^\rho + a_\rho s^{\rho-1} + \cdots + a_2 s + a_1$ is a stable polynomial, the states η in the zero dynamics system $\dot{\eta} = \psi(\xi, \eta)$ actually will be cancelled and become unobservable in the output. Therefore, for the stability of the closed-loop system, we assume that

(A9.3a) The system (9.6) is minimum phase, that is, the zero dynamics $\dot{\eta} = \psi(\xi, \eta)$ are *input-to-state stable* (ISS).

As a definition, a nonlinear system $\dot{x} = f(t, x, u)$ with an equilibrium point $x = x_e$ is said to be input-to-state stable if its solution $\tilde{x}(t) = x(t) - x_e$ satisfies $\|\tilde{x}(t)\| \le \beta(\|x(t_0) - x_e\|, t - t_0) + \gamma(\sup_{t_0 \le \tau \le t} \|u(\tau)\|)$ for any $x(t_0) \in R^n$ and any bounded input $u(t) \in R^m$ with some class-$\mathcal{K}$ functions γ (i.e., $\gamma(0) = 0$, $\gamma(r) > 0$ for $r > 0$, and $\gamma(\cdot)$ is strictly increasing) and β (with respect to $\|x(t_0) - x_e\|$), while β is decreasing with respect to $t - t_0$ and $\lim_{t \to \infty} \beta(\|x(t_0) - x_e\|, t - t_0) = 0$ [67].

Reference Model. A state variable model for the reference system $W_m(s)$ with relative degree ρ is

$$\dot{\xi}_m = A_m \xi_m + B_m r$$
$$y_m = C_m \xi_m \qquad (9.21)$$

where $\xi_m = [y_m, \dot{y}_m, \ddot{y}_m, \ldots, y_m^{(\rho-1)}]^T$, $r \in R$ is the reference input signal, which is bounded and piecewise continuous, and

$$A_m = \begin{bmatrix} 0 & 1 & 0 & \cdots & 0 & 0 \\ 0 & 0 & 1 & \cdots & 0 & 0 \\ & & \cdots & \cdots & & \\ 0 & 0 & \cdots & 0 & 0 & 1 \\ -a_1 & -a_2 & \cdots & -a_{\rho-2} & -a_{\rho-1} & -a_\rho \end{bmatrix} \in R^{\rho \times \rho}$$

$$B_m = \begin{bmatrix} 0 \\ 0 \\ \vdots \\ 0 \\ 1 \end{bmatrix} \in R^{\rho \times 1}$$

$$C_m = \begin{bmatrix} 1 & 0 & \cdots & \cdots & 0 & 0 \end{bmatrix} \in R^{1 \times \rho}. \qquad (9.22)$$

With the canonical form (9.13) of the plant after transformation and the state model of the reference system (9.21), which has the same relative degree ρ as the nonlinear plant (9.13), we solve the problem of partial state tracking of the reference system instead, which leads to output tracking $y(t) \to y_m(t)$.

Controller Structure. We first develop a nominal controller structure, assuming the knowledge of the actuator failures (both the failure pattern σ and the failure value $\bar{u}$) at time t, by choosing the controller

$$v_i = \frac{1}{\beta_i(\xi, \eta)} \left(k_{1,i}^T \xi + k_{2,i}(r - \varphi(\xi, \eta)) + k_{3,i}^T \beta(\xi, \eta) \right), \quad i = 1, 2, \ldots, m \quad (9.23)$$

where $k_{1,i} \in R^\rho$, $k_{2,i} \in R$, $k_{3,i} \in R^m$, such that

$$[k_{1,1}, k_{1,2}, \ldots, k_{1,m}](I - \sigma)[1, 1, \ldots, 1]^T = -a$$

$$[k_{2,1}, k_{2,2}, \ldots, k_{2,m}](I - \sigma)[1, 1, \ldots, 1]^T = 1$$

$$[k_{3,1}, k_{3,2}, \ldots, k_{3,m}](I - \sigma)[1, 1, \ldots, 1]^T = -\sigma\bar{u} \qquad (9.24)$$

with $a = [a_1, a_2, \ldots, a_\rho]^T$. It is easy to verify that for any actuator failure pattern there always exist $k_{1,i}$, $k_{2,i}$, and $k_{3,i}$ to satisfy the equations in (9.24). With the controller (9.23), the closed-loop system becomes

$$\dot{\xi} = A_m \xi + Br$$
$$\dot{\eta} = \psi(\xi, \eta)$$
$$y = C\xi \tag{9.25}$$

where A_m is given in (9.22). Since η is unobservable in the output y with the applied control and $B = B_m$, $C = C_m$, $y(t) = W_m(s)[r](t)$, which means that $y(t)$ will exponentially track $y_m(t) = W_m(s)[r](t)$. With the choice of the state model (9.21) for the reference system, $\xi = [y, \dot{y}, \ldots, y^{(\rho-1)}]^T$ will exponentially track $\xi_m = [y_m, \dot{y}_m, \ldots, y_m^{(\rho-1)}]^T$.

To handle actuator failure uncertainties, we develop an adaptive version of the controller (9.23) as

$$v_i = \frac{1}{\beta_i(\xi, \eta)} \left(\hat{k}_{1,i}^T(t)\xi(t) + \hat{k}_{2,i}(t)(r(t) - \varphi(\xi, \eta)) + \hat{k}_{3,i}^T(t)\beta(\xi, \eta) \right) \tag{9.26}$$

for $i = 1, 2, \ldots, m$, where $\hat{k}_{1,i}(t)$, $\hat{k}_{2,i}(t)$, and $\hat{k}_{3,i}(t)$ are the estimates of $k_{1,i}$, $k_{2,i}$, and $k_{3,i}$ that satisfy (9.24).

Adaptive Laws. To derive adaptive laws for $\hat{k}_{1,i}$, $\hat{k}_{2,i}$, and $\hat{k}_{3,i}$, $i = 1, 2, \ldots, m$, we define the parameter vectors and their errors:

$$\kappa_i = [k_{1,i}^T, k_{2,i}, k_{3,i}^T]^T$$
$$\hat{\kappa}_i = [\hat{k}_{1,i}^T, \hat{k}_{2,i}, \hat{k}_{3,i}^T]^T$$
$$\tilde{\kappa}_i = \hat{\kappa}_i - \kappa_i, \ i = 1, 2, \ldots, m \tag{9.27}$$

and introduce the signal

$$\omega(t) = [\xi^T(t), r(t) - \varphi(\xi, \eta), \beta^T(\xi, \eta)]^T. \tag{9.28}$$

Hence, we design the update law for $\hat{\kappa}_i$ as

$$\dot{\hat{\kappa}}_i = -e^T P B_m \Gamma_i \omega, \ i = 1, 2, \ldots, m \tag{9.29}$$

where $e = \xi - \xi_m$ is the tracking error, $\Gamma_i = \Gamma_i^T > 0$, and P is a positive definite symmetric matrix that satisfies the Lyapunov equation

$$P A_m + A_m^T P = -Q. \tag{9.30}$$

The adaptive scheme has the following desired properties.

Theorem 9.2.1. *With the ISS of the zero dynamics (A9.3a), the controller (9.26) with the adaptive laws (9.29) ensures the boundedness of the closed-loop signals and the asymptotic output tracking of the reference signal.*

Proof: Supposing that one or more than one actuator fails at time instants t_j, $j = 1, 2, \ldots, q$, $1 \leq q \leq m - 1$, and at time $t \in (t_{j-1}, t_j)$ there are p failed actuators in the system, that is, $u_i = \bar{u}_i$, $i = i_1, i_2, \ldots, i_p$, $\{i_1, i_2, \ldots, i_p\} \subset \{1, 2, \ldots, m\}$, $0 \leq p \leq m - 1$, the adaptive scheme can be analyzed by using the positive definite function defined on the interval (t_{j-1}, t_j), on which the actuator failure pattern σ is unchanged:

$$V_{\sigma_{j-1}}(t) = \frac{1}{2}e^T P e + \frac{1}{2} \sum_{i \neq i_1, \ldots, i_p} \tilde{\kappa}_i^T \Gamma_i^{-1} \tilde{\kappa}_i. \tag{9.31}$$

The time derivative of $V_\sigma(t)$ along the trajectories of (9.29) is

$$\dot{V}_{\sigma_{j-1}} = -e^T Q e \leq 0, \ t \in (t_{j-1}, t_j). \tag{9.32}$$

At $t = t_j$, $j = 1, 2, \ldots, q$, when there are actuators failing, some of the terms $\tilde{\kappa}_i^T \Gamma_i^{-1} \tilde{\kappa}_i$, $i \neq i_1, i_2, \ldots, i_p$, will drop from $V_{\sigma_{j-1}}(t)$ so that a new $V_{\sigma_j}(t)$ is formed on the next interval (t_j, t_{j+1}), and the remaining $\tilde{\kappa}_i$ will have a jumping because of the change of the value κ_i. From (9.32) it follows that

$$V_{\sigma_{j-1}}(t_j^-) < V_{\sigma_{j-1}}(t_{j-1}^+) \tag{9.33}$$

which indicates that $\tilde{\kappa}_i(t)$ and $e(t)$ are bounded on the interval (t_{j-1}, t_j) if $V_{\sigma_{j-1}}(t_{j-1}^+)$ is finite. Since the change of κ_i is finite, it in turn implies that $V_{\sigma_j}(t_j^+)$ is also finite, that is, for the next interval (t_j, t_{j+1}), the initial value of the function $V_{\sigma_j}(t)$ is finite. Due to the finite value of $V_{\sigma_0}(t_0)$, it turns out that the initial value for the last function $V_{\sigma_q}(t)$ defined on the (t_q, ∞) is finite so that we can conclude that $\xi(t)$ and $\kappa_i(t)$ are bounded and $e(t) \in L^2$.

Under the assumption that the zero dynamics system $\dot{\eta} = \psi(\xi, \eta)$ with input ξ is input-to-state stable, we conclude that η is bounded so that $\dot{e}(t)$ is bounded, which implies that $\lim_{t \to \infty} e(t) = 0$. Therefore, the closed-loop stability and asymptotic tracking $\lim_{t \to \infty}(y(t) - y_m(t)) = 0$ are ensured. ∇

For the case when the plant dynamics functions $f(x)$ and $g_i(x)$ contain unknown parameters, the above adaptive actuator failure compensation scheme can be redesigned to include an adaptive estimation algorithm that updates the estimates of unknown plant parameters for adaptive control [60], [71].

9.3 An Alternative Design

We now develop an alternative adaptive actuator failure compensation controller based on the following design condition, which is different from the condition (A9.2a) in Section 9.2, relaxed in the sense that the actuation vectors $g_i(x)$ and $g_j(x)$ do not need to be parallel to each other.

Structure Condition. For the nonlinear plant (9.1), we assume

(A9.2b) The system $\dot{x} = f(x) + g_i(x)u_i$, $y = h(x)$ has relative degree ρ, for $i = 1, 2, \ldots, m$.

This assumption means the plant (9.1) has the same relative degree for each actuation function $g_i(x)$, $i = 1, 2, \ldots, m$, that is,

$$
\begin{aligned}
L_{g_i(x)}L_{f(x)}^j h(x) &= 0, \quad j = 0, 1, \ldots, \rho - 2, \\
L_{g_i(x)}L_{f(x)}^{\rho-1} h(x) &\neq 0,
\end{aligned}
\qquad i = 1, 2, \ldots, m.
\qquad (9.34)
$$

This does not require that $g_i(x)$ and $g_j(x)$ are parallel to each other.

Actuation Scheme. We choose the proportional-actuation scheme as

$$
v_i = b_i(x)v_0, \quad i = 1, 2, \ldots, m
\qquad (9.35)
$$

v_0 is a control input to be designed, $b_i(x)$ is a scalar function to be chosen such that for $\forall x \in R^n$, $\forall \{i_1, i_2, \ldots, i_p\} \subset \{1, 2, \ldots, m\}$

$$
\sum_{i \neq i_1, \ldots, i_p} b_i(x) L_{g_i(x)} L_{f(x)}^{\rho-1} h(x) \neq 0.
\qquad (9.36)
$$

There always exist such $b_i(x)$, $i = 1, 2, \ldots, m$, to satisfy (9.36); for example, $b_i(x) = 1/L_{g_i(x)}L_{f(x)}^{\rho-1} h(x)$, with which $\sum_{i \neq i_1, \ldots, i_p} b_i(x) L_{g_i(x)} L_{f(x)}^{\rho-1} h(x) \geq 1$.

Diffeomorphism. With Assumption (A9.2b) and the actuation scheme (9.35), there exists a diffeomorphism $[\xi, \eta]^T = T(x) = [T_c(x), T_z(x)]^T$, $\xi \in R^\rho$ and $\eta \in R^\gamma$, $\rho + \gamma = n$, constructed as

$$
\xi = T_c(x) = \begin{bmatrix} T_1(x) \\ T_2(x) \\ \vdots \\ T_\rho(x) \end{bmatrix} = \begin{bmatrix} h(x) \\ L_{f(x)}h(x) \\ \vdots \\ L_{f(x)}^{\rho-1} h(x) \end{bmatrix}
$$

$$
\eta = T_z(x) = \begin{bmatrix} T_{\rho+1}(x) \\ T_{\rho+2}(x) \\ \vdots \\ T_n(x) \end{bmatrix}
\qquad (9.37)
$$

where $T_z(x)$ exists but is not unique [60], such that $[T(x)/\partial x]$ is nonsingular for $\forall x \in R^n$. This diffeomorphism is chosen to develop an adaptive compensation control scheme, that is, the variable $\eta(t)$ will be used for feedback control. Such a chosen diffeomorphism transforms the system (9.5) into

$$\dot{\xi} = A\xi + B(\varphi(\xi,\eta) + \beta^T(\xi,\eta)\kappa_1 + \bar{\beta}^T(\xi,\eta)\kappa_2 v_0)$$
$$\dot{\eta} = \psi(\xi,\eta) + \varDelta(\xi,\eta)\kappa_1 + \bar{\varDelta}(\xi,\eta)\kappa_2 v_0$$
$$y = C\xi \tag{9.38}$$

where A, B, and C are the same as those given in (9.9), and

$$\kappa_1 = \sigma\bar{u}, \quad \kappa_2 = (I - \sigma)[1, 1, \ldots, 1]^T \tag{9.39}$$

which can be treated as some unknown constant vectors,

$$\varphi(\xi,\eta) = L^\rho_{f(x)} h(x)$$
$$\beta(\xi,\eta) = [L_{g_1(x)} L^{\rho-1}_{f(x)} h(x), \ldots, L_{g_m(x)} L^{\rho-1}_{f(x)} h(x)]^T$$
$$\bar{\beta}(\xi,\eta) = [b_1(x) L_{g_1(x)} L^{\rho-1}_{f(x)} h(x), \ldots, b_m(x) L_{g_m(x)} L^{\rho-1}_{f(x)} h(x)]^T$$
$$\psi(\xi,\eta) = \frac{\partial T_z(x)}{\partial x} f(x)$$
$$\varDelta(\xi,\eta) = \frac{\partial T_z(x)}{\partial x} g(x)$$
$$\bar{\varDelta}(\xi,\eta) = \frac{\partial T_z(x)}{\partial x} [b_1(x)g_1(x), \ldots, b_m(x)g_m(x)]^T. \tag{9.40}$$

One choice of $T_z(x)$ may be made based on the plant model (9.5) with the actuation scheme (9.35), in the case of no failure and under the condition

$$\frac{\partial T_z(x)}{\partial x} \sum_{i=1}^m b_i(x)g_i(x) = 0 \tag{9.41}$$

so that $\bar{\varDelta}(\xi,\eta)\kappa_2 = 0$ for the no-failure case.

   With a feedback control signal v_0 for the output tracking of a given reference signal, $\dot{\eta} = \psi(\xi,\eta) + \varDelta(\xi,\eta)\kappa_1 + \bar{\varDelta}(\xi,\eta)\kappa_2 v_0$ becomes the zero dynamics of the closed-loop system, which are dependent on the actuator failures.

Remark 9.3.1. For a chosen $T_z(x)$, the function $\bar{\varDelta}(\xi,\eta)\kappa_2$ in (9.38) may be nonzero for some failure patterns, that is, the zero dynamics explicitly depends on the control signal v_0, which may lead to difficulty in analyzing closed-loop system stability. To obtain a zero dynamics system independent of control input, for each failure pattern σ, we introduce a diffeomorphism $[\xi, \bar{\eta}]^T = T_\sigma(x) = [T_c(x), T_{z,\sigma}(x)]^T$, where $T_{z,\sigma}(x)$ is dependent on the failure pattern σ, for which we use $T_{z,\sigma}(x)$ to indicate a change of coordinates based on a failure pattern. For the diffeomorphism, $T_{z,\sigma}(x)$ is chosen to satisfy

$$\frac{\partial T_{z,\sigma}(x)}{\partial x} [b_1(x)g_1(x), \ldots, b_m(x)g_m(x)]^T \kappa_2 = 0. \tag{9.42}$$

This equation has a solution $T_{z,\sigma(x)}$, because from the actuation scheme (9.35) with $b_i(x)$ being chosen to satisfy (9.36), it follows that $[b_1(x)g_1(x),\ldots,$ $b_m(x)g_m(x)]^T \kappa_2 \neq 0$, for $\forall x \in R^n$ and any κ_2 associated with the failure pattern σ, that is, $\mathrm{span}\{[b_1(x)g_1(x),\ldots,b_m(x)g_m(x)]^T\kappa_2\}$ is a one-dimensional, nonsingular distribution [60]. For the defined $T_c(x)$, in addition to (9.42), $T_{z,\sigma}(x)$ can be chosen to make $T_\sigma(x)$ a diffeomorphism [60]. With such a diffeomorphism $T_\sigma(x)$, the resulting zero dynamics system is represented by

$$\dot{\bar{\eta}} = \psi_\sigma(\xi, \bar{\eta}) + \Delta_\sigma(\xi, \bar{\eta})\kappa_1 \tag{9.43}$$

where

$$\psi_\sigma(\xi, \bar{\eta}) = \frac{\partial T_{z,\sigma}(x)}{\partial x} f_0(x), \quad \Delta_\sigma(\xi, \bar{\eta}) = \frac{\partial T_{z,\sigma}(x)}{\partial x} g(x) \tag{9.44}$$

which are dependent on the failure pattern σ.

The zero dynamics system (9.43), which characterizes the internal property of the system zero dynamics without being involved with the control signal v_0, is equivalent to the zero dynamics in (9.38) in the sense that there is an invertible transformation from $[\xi, \bar{\eta}]^T$ to $[\xi, \eta]^T$: $[\xi, \eta]^T = T(T_\sigma^{-1}(\xi, \bar{\eta}))$, where both $T(x)$ and $T_\sigma(x)$ are well-defined diffeomorphisms.   □

Remark 9.3.2. When some actuators fail during operation, the failure pattern σ will change, which means that κ_1 and κ_2 are not fixed constant vectors and will change when failures occur. Since the times of changing are finite (less than m), adaptive design automatically adjusts their estimates from one steady value to another. On the other hand, the changes of the failure pattern σ will influence the controlled plant significantly and in turn the stability of the closed-loop system. At the beginning, we mentioned that actuator failures not only bring disturbance into systems, but also alter the structure of systems. In view of (9.43), actuator failures influence the zero dynamics in two ways. One comes from κ_1 clearly. The other is through $T_{z,\sigma}(x)$, which will also change with the occurrence of actuator failures and reshape the whole structure of the zero dynamics.   □

Stable Zero Dynamics Assumption. To make $y(t)$ track the reference signal $y_m(t)$ from the reference system $W_m(s) = 1/P_m(s)$ whose state model is defined in (9.21), we make the following assumption for the stability of the adaptive compensation design:

(A9.3b) The system (9.38) is minimum phase for all possible failure patterns, that is, the zero dynamics (9.43) are *input-to-state stable* (ISS) with respect to ξ as the input for all possible σ (and κ_1).

Control Law. We design the adaptive controller

$$v_i = b_i(x)v_0, \ i = 1, 2, \ldots, m$$

$$v_0 = \frac{1}{\bar{\beta}^T \hat{\kappa}_2}[r - a^T \xi - \varphi - \beta^T \hat{\kappa}_1] \tag{9.45}$$

where $a = [a_1, a_2, \ldots, a_\rho]^T$, and $\hat{\kappa}_1$ and $\hat{\kappa}_2$ are the estimates of κ_1 and κ_2.

Adaptive Laws. In order to avoid singularity, we will introduce a projection into the adaptive law design. Define

$$\mathrm{proj}_{[x^T y, \lambda]}(f(t)) = \begin{cases} f(t) & \text{if } x^T y > \lambda \\ f(t) & \text{if } x^T y = \lambda \text{ and } x^T \dot{y} \geq 0 \\ 0 & \text{if } x^T y = \lambda \text{ and } x^T \dot{y} < 0 \end{cases} \tag{9.46}$$

where $\lambda > 0$ is a constant.

Hence a lower bound on $|\bar{\beta}^T(\xi, \eta)\kappa_2|$ is needed, that is, $|\bar{\beta}^T(\xi, \eta)\kappa_2| > \lambda > 0$ with a known constant λ, and the sign of $\bar{\beta}^T \kappa_2$ is known. In fact, by choosing appropriate $b_i(x)$ to satisfy (9.36), the lower bound condition can always be guaranteed in this case.

Choose the update laws for $\hat{\kappa}_1$ and $\hat{\kappa}_2$ as

$$\dot{\hat{\kappa}}_1 = e^T P B_m \Gamma_1 \beta$$

$$\dot{\hat{\kappa}}_2 = \mathrm{proj}_{[\mathrm{sign}[\bar{\beta}^T \kappa_2]\bar{\beta}^T \hat{\kappa}_2, \lambda]}(e^T P B_m \Gamma_2 \bar{\beta} v_0) \tag{9.47}$$

where $e = \xi - \xi_m$ is the tracking error, P is a positive definite symmetric matrix that satisfies the Lyapunov equation

$$P A_m + A_m^T P = -Q \tag{9.48}$$

and $\Gamma_1 = \Gamma_1^T > 0$, $\Gamma_2 = \Gamma_2^T > 0$ are the adaptation gain matrices.

Remark 9.3.3. In this design, the adaptive compensation scheme is based on the actuation scheme (9.35) so that the control inputs for all the alive actuators are related to each other. However, in Section 9.2, with a more restrictive structure condition (because Assumption (A9.2a) always implies Assumption (A9.2b)), the design for each actuator is independent. □

With the adaptive scheme (9.45) and (9.47), the closed-loop system has the following desired properties.

Theorem 9.3.1. *With the ISS zero dynamics (A9.3b), the controller (9.45) with adaptive laws (9.47) ensures the boundedness of the closed-loop signals and the asymptotic output tracking of the reference signal.*

Proof: This adaptive compensation scheme presented in (9.45) and (9.47) can be analyzed by using the positive definite function

$$V(e, \tilde{\kappa}_1, \tilde{\kappa}_2) = \frac{1}{2} e^T P e + \frac{1}{2} \tilde{\kappa}_1^T \Gamma_1^{-1} \tilde{\kappa}_1 + \frac{1}{2} \tilde{\kappa}_2^T \Gamma_2^{-1} \tilde{\kappa}_2 \tag{9.49}$$

whose time derivative of along the trajectories of (9.45) is

$$\dot{V} = -e^T Q e - \delta(t) \le -e^T Q e \le 0, \ t \ne t_j, \ j = 1, 2, \ldots, q \tag{9.50}$$

where $\delta(t) = \tilde{\kappa}_2^T \left(e^T P B_m \Gamma_2 \bar{\beta} v_0 - \text{proj}_{[\text{sign}[\bar{\beta}^T \kappa_2] \bar{\beta}^T \hat{\kappa}_2, \lambda]} (e^T P B_m \Gamma_2 \bar{\beta} v_0) \right) \ge 0$ introduced by the projection, and t_j, $j = 1, 2, \ldots, q$, $1 \le q \le m - 1$, are the time instants when one or more than one of the actuators possibly fail. At $t = t_j$, the parameters κ_1 and κ_2 defined in (9.39) change their values, causing a finite jumping in them and $V(e, \tilde{\kappa}_1, \tilde{\kappa}_2)$ as well. It follows from (9.50) that $V(e, \tilde{\kappa}_1, \tilde{\kappa}_2) \in L^\infty$ and $e(t) \in L^2$. Then we conclude that $\xi(t)$, $\hat{\kappa}_1(t)$, and $\hat{\kappa}_2(t)$ are bounded. Furthermore, $\bar{\eta}(t)$ is bounded because of the ISS of the zero dynamics (9.43). In view of Remark 9.3.1, we have that $\eta(t)$ is also bounded, which implies that $\dot{e}$ is bounded so that $e(t)$ goes to zero when t goes to infinity. Hence, the closed-loop stability and asymptotic tracking: $\lim_{t \to \infty} (y(t) - y_m(t)) = 0$ can be established. ∇

The case when the functions $f(x)$ and $g_i(x)$ have unknown parameters can be treated by combining this adaptive actuator failure compensation design and an adaptive control design for unknown plant parameters [60], [71].

9.4 Issues for Nonlinear Dynamics

In summary, we have formulated and discussed the actuator failure compensation problem for nonlinear systems. The adaptive compensation control of feedback linearizable systems for actuator failures is investigated and two adaptive schemes are designed based on different structure conditions. The closed-loop system stability and asymptotic tracking are proved for the proposed adaptive actuator failure compensation control schemes.

Plant Parameter Uncertainties. So far, we have only considered actuator failure compensation for the class of feedback linearizable systems with actuator failure uncertainties but without parameter uncertainties. It is important to develop adaptive control schemes to handle parameter uncertainties in addition to actuator failure uncertainties, as parameter uncertainties usually exist in many real-life systems such as aircraft flight control systems, power systems, robotic systems, and so on.

As an illustrative example, we give a nonlinear aircraft wing model based on the state-space form of the wing-rock model with ailerons modeled as first-order actuator dynamics [71], which is in the *parametric-strict-feedback form*. With three augmented actuation parameters (b_1, b_2, and b_3) for actuator failure compensation study, the wing model can be described as

$$\dot{\phi} = p$$
$$\dot{p} = \theta_1 + \theta_2\phi + \theta_3 p + \theta_4|\phi|p + \theta_5|p|p + \delta$$
$$\dot{\delta} = -\frac{1}{\tau}\delta + \frac{b_1}{\tau}u_1 + \frac{b_2}{\tau}u_2 + \frac{b_3}{\tau}u_3 \qquad (9.51)$$

where u_1, u_2, and u_3 are the inputs, τ is the aileron time constant, and θ_i, $i = 1, 2, \ldots, 5$, and b_j, $j = 1, 2, 3$, are unknown constant parameters. An adaptive compensation controller is needed for the model (9.2) to achieve the output tracking objective $\phi(t) - \phi_m(t) \to 0$, which means that the wing rock is suppressed despite the presence of unexpected actuator failures.

To handle parameter uncertainties in nonlinear dynamic systems, two powerful methods can be applied, that is, adaptive feedback linearization [109] and adaptive backstepping [71].

Adaptive Feedback Linearization. Adaptive feedback linearization is based on the feedback linearization design [60], with adaptive laws for estimating the unknown parameters (which are related to the system parameters) in a nominal control law (see [109]). Such an adaptive design usually leads to a large dimension of parameter estimates when the relative degrees of the controlled systems are larger than 1. This is due to the complete feedback linearization procedure that let all terms related to uncertainties and nonlinearities to the last equation (for example, the ρth differential equation if the relative degree is ρ), while the control input in this equation is to cancel the nonlinearities and to handle the uncertainties.

For example, consider a feedback linearizable nonlinear system with m actuators and l unknown parameters θ_i, $i = 1, 2, \ldots, l$, described by

$$\dot{x} = f_0(x) + \sum_{i=1}^{l} \theta_i f_i(x) + g(x)u$$
$$y = h(x) \qquad (9.52)$$

where $f_0(x) \in R^n$, $f_i(x) \in R^n$, $i = 1, 2, \ldots, l$, and $g(x) = [g_1(x), g_2(x), \ldots, g_m(x)] \in R^{n \times m}$ are known functions, $u \in R^m$. Assume that the system has the same relative degree for each actuation function $g_j(x)$, $j = 1, 2, \ldots, m$. With the knowledge of θ_i, $i = 1, 2, \ldots, l$, the control scheme proposed in this chapter can be applied for achieving tracking objective by considering that $f(x) = f_0(x) + \sum_{i=1}^{l} \theta_i f_i(x)$ for the system (9.1). Since θ_i, $i = 1, 2, \ldots, l$, are unknown, we now have uncertainties in $\varphi(\xi, \eta)$ and $\beta(\xi, \eta)$ for both (9.11) and (9.38). With complete feedback linearization, the uncertainties are not only those unknown parameters themselves but also a set of combinations of them in the functions $\varphi(\xi, \eta)$ and $\beta(\xi, \eta)$. Writing $\varphi(\xi, \eta)$ and $\beta(\xi, \eta)$ as

$$\begin{aligned}
\varphi(\xi, \eta) &= \theta_\varphi^T \bar{\varphi}(\xi, \eta), \\
\beta(\xi, \eta) &= \theta_\beta^T \bar{\beta}(\xi, \eta)
\end{aligned} \tag{9.53}$$

where $\bar{\varphi}(\xi, \eta)$ and $\bar{\beta}(\xi, \eta)$ are matrices of known functions obtained from $f_0(x)$, $f_i(x)$, $i = 1, 2, \ldots, l$, and $g(x)$, we notice that the dimensions of the vectors θ_φ and θ_β are both very large and will increase significantly with the relative degree. For instance, for a system with relative degree ρ and l unknown parameters, the dimension of θ_φ could be at least $\sum_{i=1}^{\rho} l^i$. With consideration of actuator failures, the total number of the parameter estimates to be updated would be even larger.

Adaptive Backstepping. While the design based on feedback linearization is direct to derive and easy to understand but with the cost of over-parametrization, another powerful design tool is the backstepping technique for which the system may not be completely linearized with the presence of unknown parameters. As compared with a feedback linearization design, a backstepping design preserves some useful nonlinearities of the system so that the control design is more flexible.

Backstepping can be made adaptive to handle system parameter uncertainties. With tuning functions [71], the problem of over-parametrization can be solved, which means for the above system (9.52), only l unknown parameters have to be estimated via adaptive update laws.

Backstepping is a design method using a recursive procedure to generate a desired control signal step by step from the first step related to the control task such as output tracking. At each step, based on a positive definite function that becomes a part of the Lyapunov function to be used for stability analysis, a stabilization function is obtained that will be considered as a new task for the next step. The last stabilization function becomes the controller to guarantee the control objective, which is exactly the first task. Therefore, for backstepping design, the nonlinearities with unknown parameters as $\theta_i f_i(x)$ in (9.52) are not pushed to the last equation as the feedback linearization design does so. The nonlinearities are kept in each differential equation such that each equation is treated as a first-order system with the corresponding task. Combined with tuning functions, the backstepping design minimizes the dynamic order of the adaptive controller, which may reduce the cost for implementing the control scheme in real-life systems.

Output Feedback Control. In case that the system states are not measured, the actuator failure compensation controller is required to be designed based on output feedback, which is unlike the linear case, where the problem can be solved similar to the one using state feedback combined with an observer to estimate the unmeasured states. For nonlinear systems, in general, even if we have a desired observer that can estimate the states within an exponentially decaying error, we may not guarantee the stability of the closed-loop system if simply replacing the state in the state feedback controller with the estimated state from the observer. Moreover, since we are studying the actuator failure problem for which actuator failures may change the structure of the system, the uncertainties have to be considered for both the control design and the observer design. It is more difficult to develop an output feedback design for nonlinear systems than linear systems, because of the conversion of a state feedback design to an output feedback design.

For output feedback adaptive control, the basic issue is the parametrization of a state observer and a control law. An adaptive observer is needed to accommodate the additional effect caused by uncertain actuator failures, while a parametrized adaptive controller will be designed based on the parametrized adaptive observer with the parameter estimates to be updated online by some adaptive laws. Hence, for output feedback design, a main issue is how to design such an observer, combined with which a suitable adaptive controller can be derived. Thus, to have a desired nominal observer with the knowledge of actuator failures such that the adaptive version of it still can capture the unmeasured states fast enough to be used for the control design is the key issue in output feedback compensation control.

Output feedback control schemes have been developed mostly for the class of output-feedback systems in which only the system nonlinearities are functions of the measured system output. Our control objective is to design an output feedback controller such that the system output asymptotically tracks a prescribed reference signal, and at the same time all closed-loop signals are guaranteed to be bounded, in the presence of unexpected actuator failures. Adaptive backstepping can be applied to solve the actuator failure compensation problem for output-feedback systems with output feedback. A nominal observer constructed with some filters can estimate the state ideally in the sense that estimation error converges to zero exponentially when implemented with true parameters of system and actuator failures. When implemented with the parameter estimates, the adaptive observer still provides desired estimation of the states, with which an adaptive control scheme can be developed so as to ensure the stability and output tracking objective.

The output-feedback canonical form characterizes a large class of nonlinear systems for which nonlinearities are dependent on the output. However, there are more situations when the system nonlinearities depend on other state variables. For example, for aircraft flight control systems, due to the complexity of the dynamics of an aircraft, nonlinearities are dependent not only on the output but also on some state variables.

Consider the dynamics of the angle of attack from a rigid-body longitudinal hypersonic aircraft model, described by

$$
\begin{aligned}
\dot{x}_1 &= x_2 + \psi_1(x_3, x_4, y) \\
\dot{x}_2 &= \psi_2(x_2, x_4, y) + b_1 x_4^2 u_1 + b_2 x_4^2 u_2 \\
\dot{x}_3 &= \psi_3(x_3, x_4, y) \\
\dot{x}_4 &= \psi_4(x_3, x_4, y) \\
y &= x_1
\end{aligned}
\tag{9.54}
$$

where x_1 is the angle of attack, x_2 is the pitch rate, which is unmeasured, x_3 is the flight-path angle, x_4 is the velocity, $u_1(t)$ and $u_2(t)$ are the elevator segment deflection angles, b_1 and b_2 are unknown constants with known signs, and $\psi_1(x_3, y)$, $\psi_2(x_2, y)$, $\psi_3(x_3, y)$, and $\psi_4(x_3, y)$ are known nonlinear functions. This state-dependent nonlinear system is beyond the output-feedback form. For output feedback adaptive actuator failure compensation control, a complete backstepping solution is given for the output-feedback systems. We also extend the developed backstepping design to systems with state-dependent nonlinearities by giving a general framework for solving the actuator failure compensation problem with output feedback.

Next, adaptive compensation schemes for parametric-strict-feedback systems and output-feedback systems will be developed in Chapter 10 and Chapter 11, respectively, to achieve output tracking control objective in the presence of unknown actuator failures as well as unknown system parameters, under different system structures and design conditions. Detailed stability analysis and extensive simulation results on aircraft systems are given to illustrate the effectiveness of our failure compensation control schemes.

Chapter 10

State Feedback Designs for Nonlinear Systems

In this chapter, we further the study of adaptive actuator failure compensation control to nonlinear systems with parameter uncertainties. We fisrt develop adaptive control solutions to *parametric-strict-feedback* systems without zero dynamics in the presence of actuator failures in Section 10.1. The actuator failure compensation problem is characterized in terms of two actuation models, and output matching conditions for the two actuation models are derived. Adaptive compensation control schemes are developed for parametric-strict-feedback systems without zero dynamics based on matching designs for two different actuation models. Closed-loop signal boundedness and asymptotic output tracking are ensured by the developed adaptive control schemes. Simulation results for the control of an aircraft wing-rock model with augmented actuators verifies the desired performance of our adaptive designs.

The actuator failure compensation problem is studied for the *parametric-strict-feedback* systems with zero dynamics in Section 10.2. Actuator failure compensation control of the parametric-strict-feedback systems is investigated in Sections 10.2.1 and 10.2.2. Two adaptive state feedback control schemes based on two structure conditions are presented and the closed-loop system is guaranteed to achieve desired control objective for both control designs. A related robustness issue is addressed in Section 10.2.3. In Section 10.2.4, the nonlinear longitudinal dynamic model of a twin otter aircraft is studied and an adaptive control scheme is applied to the nonlinear aircraft model for output tracking in the presence of unknown parameters and unexpected actuator failures. Simulation results demonstrate the desired performance of the closed-loop system with adaptive actuator failure compensation.

10.1 Design for Systems Without Zero Dynamics

Consider the parametric-strict-feedback nonlinear plant with m actuators:

$$\dot{x}_i = x_{i+1} + \theta^T \varphi_i(x_1, \dots, x_i), \ i = 1, 2, \dots, n-1 \qquad (10.1)$$

$$\dot{x}_n = \varphi_0(x) + \theta^T \varphi_n(x) + \beta^T(x)u \tag{10.2}$$

$$y = x_1 \tag{10.3}$$

where $\theta \in R^p$ is the vector of unknown constant parameters, $\varphi_0 \in R$ and $\varphi_i \in R^p$, $i = 1, 2, \ldots, n$, are known smooth nonlinear functions, and $\beta(x) \in R^m$ is a vector of the partially known functions and has different forms based on different assumptions that are used for different control designs. The state variables x_i, $i = 1, 2, \ldots, n$, are measurable, $u(t) \in R^m$ is the input vector whose components may fail during the system operation, and $y(t)$ is the plant output. The actuator failures considered in this chapter are modeled as

$$u_i(t) = \bar{u}_i, \ t \geq t_i, \ i \in \{1, 2, \ldots, m\} \tag{10.4}$$

where the constant value $\bar{u}_i$ and the failure time instant t_i are unknown. The system (10.1)–(10.3) meets the basic assumption for actuator failure compensation that for any up to $m - 1$ actuator failures, the remaining actuators can still achieve a desired stabilization and output tracking control objective, when implemented with known parameters.

As given in (1.3), due to actuator failures, the input vector $u(t)$ is

$$u = v(t) + \sigma(\bar{u} - v(t)) \tag{10.5}$$

where $v(t)$ is the vector of applied control inputs, $\bar{u} = [\bar{u}_1, \bar{u}_2, \ldots, \bar{u}_m]^T$ is a constant vector with $\bar{u}_i$, $i \in \{1, 2, \ldots, m\}$, some unknown values of the up to $m - 1$ actuator failures, and

$$\sigma = \mathrm{diag}\{\sigma_1, \sigma_2, \ldots, \sigma_m\}$$
$$\sigma_i = \begin{cases} 1 & \text{if the } i\text{th actuator fails, i.e., } u_i = \bar{u}_i \\ 0 & \text{otherwise.} \end{cases} \tag{10.6}$$

We should note that for the adaptive actuator failure problems solved in this chapter, both the plant parameters and failure paramenters are unknown. To formulate the actuator failure compensation problems, next we introduce two actuation models for the system (10.1)–(10.3).

Actuation Model I. The first actuation model is with

$$\beta^T(x) = [\beta_1(x), \beta_2(x), \ldots, \beta_m(x)]. \tag{10.7}$$

The actuator failure compensation scheme for actuation model (10.7) will be developed with the help of the solution to the nominal plant:

$$\dot{x}_i = x_{i+1} + \theta^T \varphi_i(x_1, \ldots, x_i), \; i = 1, 2, \ldots, n-1 \qquad (10.8)$$

$$\dot{x}_n = \varphi_0(x) + \theta^T \varphi_n(x) + \beta_0(x) u_0 \qquad (10.9)$$

where $\beta_0 \in R$ is a known smooth nonlinear function of x in R^n, and $\beta_0(x) \neq 0$ for all $x \in R^n$. To avoid a control singularity, we assume that

(A10.1a) $|\beta_0(x)| > \bar{\beta} > 0$, for some constant $\bar{\beta}$, $\forall x \in R^n$.

Based on the basic assumption for actuator failure compensation, when all but one actuator fail, that is, $u_j(t) = \bar{u}_j = 0$, $j = 1, 2, \ldots, i-1, i+1, \ldots, m$, the expression (10.2) with (10.7) becomes (10.9), with

$$\beta_i u_i = \beta_0 u_0. \qquad (10.10)$$

To meet the output tracking objective, we need the assumption:

(A10.2a) There exist constant scalars k^*_{s1i}, $i = 1, 2, \ldots, m$, such that

$$k^*_{s1i} \beta_i(x) = \beta_0(x). \qquad (10.11)$$

Since $\beta_0(x) \neq 0, \forall x \in R^n$, $\beta_i(x) \neq 0, \forall x \in R^n, i = 1, 2, \ldots, m$. For adaptive control, k^*_{s1i}, $i = 1, 2, \ldots, m$, are not known, while the signs of k^*_{s1i}, $i = 1, 2, \ldots, m$, are needed for the design of stable parameter adaptive update laws, so we need the assumption:

(A10.3a) The sign of the parameter k^*_{s1i} in (10.11), which is represented by $\text{sign}[k^*_{s1i}]$, is known for $i = 1, 2, \ldots, m$.

Actuation Model II. The second actuation model to consider is with

$$\beta(x) = [b_1 \beta_1(x), b_2 \beta_2(x), \ldots, b_m \beta_m(x)]^T \qquad (10.12)$$

where $\beta_i(x) \in R$, $i = 1, 2, \ldots, m$, are known smooth nonlinear functions of x with $\beta_i(x) \neq 0$, $\forall x \in R^n$, and b_i, $i = 1, 2, \ldots, m$, are unknown constants. To simplify the expression, we define

$$u_a(t) = [\beta_1(x(t))u_1(t), \beta_2(x(t))u_2(t), \ldots, \beta_m(x(t))u_m(t)]^T$$
$$b = [b_1, b_2, \ldots, b_m]^T$$
$$\beta_a(x) = [\beta_1(x), \beta_2(x), \ldots, \beta_m(x)]^T \qquad (10.13)$$

to rewrite (10.2) as

$$\dot{x}_n = \varphi_0(x) + \theta^T \varphi_n(x) + b^T u_a. \qquad (10.14)$$

In the presence of actuator failures, the input vector $u_a(t)$ is described as

$$u_a(t) = v_a(t) + \sigma(\bar{u}_a - v_a(t)) \tag{10.15}$$

where $v_a(t) = [\beta_1(x(t))v_1(t), \beta_2(x(t))v_2(t), \ldots, \beta_m(x(t))v_m(t)]^T$ with the control inputs $v_i(t)$, $i = 1, 2, \ldots, m$, to be designed,

$$\bar{u}_a = [\beta_1(x)\bar{u}_1, \beta_2(x)\bar{u}_2, \ldots, \beta_m(x)\bar{u}_m]^T \tag{10.16}$$

with $\bar{u}_i$, $i \in \{1, 2, \ldots, m\}$, some unknown constant actuator failure values, and the diagonal matrix σ in (10.5) indicates the failure pattern.

The actuator failure compensation scheme for actuation model (10.12) will be developed with the help of the solution to the nominal plant

$$\dot{x}_i = x_{i+1} + \theta^T \varphi_i(x_1, \ldots, x_i), \ i = 1, 2, \ldots, n - 1 \tag{10.17}$$
$$\dot{x}_n = \varphi_0(x) + \theta^T \varphi_n(x) + u_0. \tag{10.18}$$

Based on backstepping [71], a state feedback control u_0 can be designed for closed-loop stability and output tracking, under the assumption

(A10.1b) $|\beta_i(x)| > \bar{\beta}_i > 0$, for some constant $\bar{\beta}_i$, $i = 1, 2, \ldots, m$.

Based on the basic assumption for actuator failure compensation, when all but one actuator fail, that is, $u_j(t) = \bar{u}_j = 0$, $j = 1, 2, \ldots, i-1, i+1, \ldots, m$, (10.2) with (10.12) and (10.13) becomes (10.18), with

$$b_i \beta_i u_i = b_i u_{ai} = u_0. \tag{10.19}$$

To meet the output tracking objective, we need the assumption:

(A10.2b) There exist constant scalars k^*_{s1i}, $i = 1, \ldots, m$, such that

$$b_i k^*_{s1i} = 1 \tag{10.20}$$

which means that $b_i \neq 0$, and $k^*_{s1i} = \frac{1}{b_i}$, $i = 1, \ldots, m$.

For adaptive control, we only need to know the signs of k^*_{s1i}, $i = 1, 2, \ldots, m$, that is, we assume

(A10.3b) The sign of the parameter k^*_{s1i} in (10.20), which is represented by sign$[k^*_{s1i}]$, is known for $i = 1, 2, \ldots, m$.

The control objective is to design a state feedback control $v(t)$ for the nonlinear plant (10.1)–(10.3) with the actuator failure model (10.4), to ensure that all closed-loop signals are bounded and the output $y(t)$ tracks a reference signal $y_m(t)$ despite the uncertainties in actuator failures (up to $m-1$ failures with unknown failure time and parameters), where the reference signal $y_m(t)$ and its first n derivatives, $\dot{y}_m$, $\ddot{y}_m$, ... , $y_m^{(n)}$, are known and bounded.

Two solutions to this problem will be proposed respectively for the two actuation models. We will propose output matching conditions in the presence of actuator failures in Section 10.1.1 and develop adaptive compensation schemes by employing the backstepping technique for state feedback control design in Section 10.1.2. Stability analysis for the closed-loop systems with the two adaptive control schemes will also be given.

10.1.1 Output Matching Design

Two different matching schemes are proposed in this section for the system (10.1)–(10.3) with the actuator failure model (10.4) based on the actuation model (10.7) and actuation model (10.12).

Output Matching Design I. For the system (10.1)–(10.3) with actuator failures (10.5), and Assumption (A10.2a), we derive output matching conditions for the actuation model (10.7), based on the controller structure

$$v(t) = k_1^* u_0(t) + k_2^* \tag{10.21}$$

where $k_1^* \in R^m$, and $k_2^* \in R^m$ is used to compensate for actuator failures. Supposing that at time t there are p failed actuators, that is, $u_j(t) = \bar{u}_j$, $j = j_1, \ldots, j_p$, $1 \leq p \leq m - 1$, from (10.6) and (10.21), we have

$$\beta^T u = \beta^T(v + \sigma(\bar{u} - v)) = \beta^T(I - \sigma)k_1^* u_0 + \beta^T(I - \sigma)k_2^* + \beta^T \sigma \bar{u}. \tag{10.22}$$

To match $\beta^T u$ to $\beta_0 u_0$, we need

$$\beta^T(I - \sigma)k_1^* = \sum_{i \neq j_1, \ldots, j_p} k_{1i}^* \beta_i = \sum_{i \neq j_1, \ldots, j_p} \frac{k_{1i}^*}{k_{s1i}^*} \beta_0 = \beta_0 \tag{10.23}$$

that is, the matching condition for k_1^* is

$$\sum_{i \neq j_1, \ldots, j_p} \frac{k_{1i}^*}{k_{s1i}^*} = 1. \tag{10.24}$$

In addition, we choose k_2^* to make

$$\begin{aligned}
\beta^T(I - \sigma)k_2^* + \beta^T \sigma \bar{u} &= \sum_{i \neq j_1, \ldots, j_p} k_{2i}^* \beta_i + \sum_{i = j_1, \ldots, j_p} \beta_i \bar{u}_i \\
&= \sum_{i \neq j_1, \ldots, j_p} \frac{k_{2i}^*}{k_{s1i}^*} \beta_0 + \sum_{i = j_1, \ldots, j_p} \frac{\bar{u}_i}{k_{s1i}^*} \beta_0 = 0
\end{aligned} \tag{10.25}$$

so that the matching condition for k_2^* is

$$\sum_{i\neq j_1,\ldots,j_p} \frac{k_{2i}^*}{k_{s1i}^*} + \sum_{i=j_1,\ldots,j_p} \frac{\bar{u}_i}{k_{s1i}^*} = 0. \tag{10.26}$$

The choice of k_{1i}^* and k_{2i}^*, $i = j_1,\ldots,j_p$, is irrelevant to the closed-loop system and may be

$$k_{1i}^* = 0, k_{2i}^* = 0, i = j_1,\ldots,j_p. \tag{10.27}$$

Output Matching Design II. For the system (10.1)–(10.3) with actuator failures (10.15) and Assumption (A10.2b), we derive matching conditions for the actuation model (10.12), based on the controller structure

$$v_a(t) = k_1^* u_0(t) + K_2^* \beta_a(x) \tag{10.28}$$

where $k_1^* \in R^m$ and $K_2^* = [k_{21}^*, k_{22}^*, \ldots, k_{2m}^*]^T \in R^{m\times m}$, and $K_2^* \beta_a(x)$ is used to compensate for the actuation error caused by actuator failures.

Suppose that at time t there are p failed actuators, that is, $u_j(t) = \bar{u}_j$ or $u_{aj}(t) = \bar{u}_{aj}$, $j = j_1, j_2, \ldots, j_p$, $1 \leq p \leq m-1$. Then from (10.15) and (10.28), we obtain

$$b^T u_a = b^T(v_a + \sigma(\bar{u}_a - v_a)) = b^T(I-\sigma)k_1^* u_0 + b^T(I-\sigma)K_2^* \beta_a + b^T \sigma \bar{u}_a. \tag{10.29}$$

To match $b^T u_a$ to u_0, we need

$$b^T(I-\sigma)k_1^* = \sum_{i\neq j_1,\ldots,j_p} k_{1i}^* b_i = 1 \tag{10.30}$$

that is, the matching condition for k_1^* is

$$\sum_{i\neq j_1,\ldots,j_p} \frac{k_{1i}^*}{k_{s1i}^*} = 1. \tag{10.31}$$

The gain matrix $K_2^* = [k_{21}^*, k_{22}^*, \ldots, k_{2m}^*]^T$ is chosen to make

$$b^T(I-\sigma)K_2^* \beta_a + b^T \sigma \bar{u}_a$$

$$= \sum_{i\neq j_1,\ldots,j_p} b_i k_{2i}^{*T} \beta_a + \sum_{i=j_1,\ldots,j_p} b_i \beta_i \bar{u}_i$$

$$= \sum_{l=1}^{m} \sum_{i\neq j_1,\ldots,j_p} b_i k_{2il}^* \beta_l + \sum_{l=j_1,\ldots,j_p} b_l \bar{u}_l \beta_l$$

$$= \sum_{l\neq j_1,\ldots,j_p} \sum_{i\neq j_1,\ldots,j_p} b_i k_{2il}^* \beta_l$$

$$+ \sum_{l=j_1,\ldots,j_p} \left(\sum_{i\neq j_1,\ldots,j_p} b_i k_{2il}^* + b_l \bar{u}_l \right) \beta_l = 0 \tag{10.32}$$

where $k_{2i}^* = [k_{2i1}^*, k_{2i2}^*, \ldots, k_{2im}^*]^T$. From (10.12) we know that there at least exist some k_{2i}^*, $i \neq j_1, j_2, \ldots, j_p$, to have

$$\sum_{i \neq j_1, \ldots, j_p} b_i k_{2il}^* + b_l \bar{u}_l = 0, \; l = j_1, j_2, \ldots, j_p \tag{10.33}$$

and hence $k_{2il}^* = 0$ for $l \neq j_1, j_2, \ldots, j_p$, $i \neq j_1, j_2, \ldots, j_p$. It follows from (10.12) and Assumption (A10.2b) that the matching condition for K_2^* is

$$\sum_{i \neq j_1, \ldots, j_p} \frac{k_{2il}^*}{k_{s1i}^*} + \frac{\bar{u}_l}{k_{s1l}^*} = 0, \; l = j_1, j_2, \ldots, j_p. \tag{10.34}$$

In this case, the choice of k_{1i}^* and $k_{2i}^* \in R^m$, $i = j_1, \ldots, j_p$, is also irrelevant to the closed-loop system and can be set to

$$k_{1i}^* = 0, k_{2i}^* = 0, i = j_1, \ldots, j_p. \tag{10.35}$$

10.1.2 Adaptive Actuator Failure Compensation

We now apply the backstepping technique [71] to derive two adaptive control laws for the system (10.1)–(10.3) with actuator failure model (10.4) based on the two actuation models mentioned above. The design has a procedure of n steps for the system, which has order n. At each step, an error variable z_i is defined and a stabilizing function α_i is designed to stabilize an ith-order subsystem with respect to a Lyapunov function V_i. The state feedback control law and the adaptive update laws are given at the last step. Since (10.1) is common, the first $n-1$ steps are the same for both models, which will be addressed at first, and then the adaptive designs derived from the nth step for the two actuation models are given with stability analysis.

First $n-1$ Steps of Backstepping. For the adaptive designs I and II, the first $n-1$ steps of the adaptive backstepping procedure are the same.

Step 1: Defining the tracking error $z_1 = x_1 - y_m$, where $y_m(t)$ is the reference signal, and $z_2 = x_2 - \alpha_1$, we have

$$\dot{z}_1 = \dot{x}_1 - \dot{y}_m = z_2 + \alpha_1 + \theta^T \varphi_1 - \dot{y}_m. \tag{10.36}$$

At this step, (10.36) is viewed as a first-order system to be stabilized by α_1 with respect to the Lyapunov function

$$V_1 = \frac{1}{2} z_1^2 + \frac{1}{2} (\hat{\theta} - \theta)^T \Gamma^{-1} (\hat{\theta} - \theta) \tag{10.37}$$

where $\Gamma = \Gamma^T > 0$, and $\hat{\theta}$ is the estimate of θ. The derivative of V_1 is

$$\dot{V}_1 = z_1(z_2 + \alpha_1 + \hat{\theta}^T \varphi_1 - \dot{y}_m) + (\hat{\theta} - \theta)^T \Gamma^{-1}(\dot{\hat{\theta}} - \Gamma z_1 \varphi_1). \tag{10.38}$$

Choosing the first stabilizing function as

$$\alpha_1 = -c_1 z_1 - \hat{\theta}^T \varphi_1 + \dot{y}_m \tag{10.39}$$

and defining the first tuning function as

$$\tau_1 = \Gamma z_1 \varphi_1 \tag{10.40}$$

we obtain

$$\dot{z}_1 = z_2 - c_1 z_1 - (\hat{\theta} - \theta)^T \varphi_1 \tag{10.41}$$

$$\dot{V}_1 = -c_1 z_1^2 + z_1 z_2 + (\hat{\theta} - \theta)^T \Gamma^{-1}(\dot{\hat{\theta}} - \tau_1). \tag{10.42}$$

The second term $z_1 z_2$ will be cancelled at the next step, and the update law for $\hat{\theta}$ as $\dot{\hat{\theta}}$ in (10.42) is to be given later on.

Step 2: Introducing $z_3 = x_3 - \alpha_2$, we have

$$\dot{z}_2 = \dot{x}_2 - \dot{\alpha}_1 = x_3 + \theta^T \varphi_2 - \frac{\partial \alpha_1}{\partial x_1}(x_2 + \theta^T \varphi_1)$$

$$-\frac{\partial \alpha_1}{\partial \hat{\theta}}\dot{\hat{\theta}} - \frac{\partial \alpha_1}{\partial r}\dot{y}_m - \frac{\partial \alpha_1}{\partial \dot{y}_m}\ddot{y}_m. \tag{10.43}$$

Here (10.41) and (10.43) is a second-order system to be stabilized by α_1 given in (10.39) and α_2 with respect to $V_2 = V_1 + \frac{1}{2}z_2^2$. The derivative of V_2 is

$$\dot{V}_2 = -c_1 z_1^2 + z_2[z_1 + z_3 + \alpha_2 - \frac{\partial \alpha_1}{\partial x_1}x_2 - \frac{\partial \alpha_1}{\partial \hat{\theta}}\dot{\hat{\theta}}$$

$$-\frac{\partial \alpha_1}{\partial r}\dot{y}_m - \frac{\partial \alpha_1}{\partial \dot{y}_m}\ddot{y}_m + \hat{\theta}^T(\varphi_2 - \frac{\partial \alpha_1}{\partial x_1}\varphi_1)]$$

$$+(\hat{\theta} - \theta)^T \Gamma^{-1}[\dot{\hat{\theta}} - \tau_1 - \Gamma z_2(\varphi_2 - \frac{\partial \alpha_1}{\partial x_1}\varphi_1)]. \tag{10.44}$$

Define the tuning function τ_2 as

$$\tau_2 = \tau_1 + \Gamma z_2 \left(\varphi_2 - \frac{\partial \alpha_1}{\partial x_1}\varphi_1 \right) \tag{10.45}$$

and choose the stabilizing function α_2 as

$$\alpha_2 = -z_1 - c_2 z_2 + \frac{\partial \alpha_1}{\partial x_1}x_2 + \frac{\partial \alpha_1}{\partial \hat{\theta}}\tau_2$$

$$+\frac{\partial \alpha_1}{\partial r}\dot{y}_m + \frac{\partial \alpha_1}{\partial \dot{y}_m}\ddot{y}_m - \hat{\theta}^T\left(\varphi_2 - \frac{\partial \alpha_1}{\partial x_1}\varphi_1 \right). \tag{10.46}$$

Hence we rewrite $\dot{V}_2$ as

$$\dot{V}_2 = -c_1 z_1^2 - c_2 z_2^2 + z_2 z_3 - z_2 \frac{\partial \alpha_1}{\partial \hat{\theta}} (\dot{\hat{\theta}} - \tau_2) + (\hat{\theta} - \theta)^T \Gamma^{-1} (\dot{\hat{\theta}} - \tau_2) \quad (10.47)$$

and $\dot{z}_2$ as

$$\dot{z}_2 = -z_1 - c_2 z_2 + z_3 - (\hat{\theta} - \theta)^T \left(\varphi_2 - \frac{\partial \alpha_1}{\partial x_1} \varphi_1 \right) - \frac{\partial \alpha_1}{\partial \hat{\theta}} (\dot{\hat{\theta}} - \tau_2). \quad (10.48)$$

Step i: Introducing $z_{i+1} = x_{i+1} - \alpha_i$ and $V_i = V_{i-1} + \frac{1}{2} z_i^2$, we have

$$\dot{z}_i = \dot{x}_i - \dot{\alpha}_{i-1} = x_{i+1} + \theta^T \varphi_i$$
$$- \sum_{k=1}^{i-1} \frac{\partial \alpha_{i-1}}{\partial x_k} (x_{k+1} + \theta^T \varphi_k) - \frac{\partial \alpha_{i-1}}{\partial \hat{\theta}} \dot{\hat{\theta}} - \sum_{k=1}^{i} \frac{\partial \alpha_{i-1}}{\partial y_m^{(k-1)}} y_m^{(k)}. \quad (10.49)$$

Now, the ith-order system $(z_1, z_2, \ldots, z_i)$ is to be stabilized by $\alpha_1, \alpha_2, \ldots, \alpha_i$ with respect to $V_i = V_{i-1} + \frac{1}{2} z_i^2$. The derivative of V_i is

$$\dot{V}_i = -\sum_{k=1}^{i-1} c_k z_k^2 - \left(\sum_{k=1}^{i-2} z_{k+1} \frac{\partial \alpha_k}{\partial \hat{\theta}} \right) (\dot{\hat{\theta}} - \tau_{i-1})$$
$$+ z_i \Big[z_{i-1} + z_{i+1} + \alpha_i - \sum_{k=1}^{i-1} \frac{\partial \alpha_{i-1}}{\partial x_k} x_{k+1} - \frac{\partial \alpha_{i-1}}{\partial \hat{\theta}} \dot{\hat{\theta}}$$
$$- \sum_{k=1}^{i} \frac{\partial \alpha_{i-1}}{\partial y_m^{(k-1)}} y_m^{(k)} + \hat{\theta}^T \left(\varphi_i - \sum_{k=1}^{i-1} \frac{\partial \alpha_{i-1}}{\partial x_k} \varphi_k \right) \Big]$$
$$+ (\hat{\theta} - \theta)^T \Gamma^{-1} [\dot{\hat{\theta}} - \Gamma \sum_{l=1}^{i} z_l \left(\varphi_l - \sum_{k=1}^{l-1} \frac{\partial \alpha_{l-1}}{\partial x_k} \varphi_k \right)]. \quad (10.50)$$

Defining the tuning function τ_i as

$$\tau_i = \tau_{i-1} + \Gamma z_i \left(\varphi_i - \sum_{k=1}^{i-1} \frac{\partial \alpha_{i-1}}{\partial x_k} \varphi_k \right) = \Gamma \sum_{l=1}^{i} z_l \left(\varphi_l - \sum_{k=1}^{l-1} \frac{\partial \alpha_{l-1}}{\partial x_k} \varphi_k \right) \quad (10.51)$$

and choosing the stabilizing function α_i as

$$\alpha_i = -z_{i-1} - c_i z_i + \sum_{k=1}^{i-1} \frac{\partial \alpha_{i-1}}{\partial x_k} x_{k+1} + \frac{\partial \alpha_{i-1}}{\partial \hat{\theta}} \tau_i + \sum_{k=1}^{i} \frac{\partial \alpha_{i-1}}{\partial y_m^{(k-1)}} y_m^{(k)}$$
$$- \left(\hat{\theta}^T - \sum_{k=1}^{i-2} z_{k+1} \frac{\partial \alpha_k}{\partial \hat{\theta}} \Gamma \right) \left(\varphi_i - \sum_{k=1}^{i-1} \frac{\partial \alpha_{i-1}}{\partial x_k} \varphi_k \right) \quad (10.52)$$

we rewrite $\dot{V}_i$ as

$$\dot{V}_i = -\sum_{k=1}^{i} c_k z_k^2 + z_i z_{i+1} - \sum_{k=1}^{i-1} z_{k+1} \frac{\partial \alpha_k}{\partial \hat{\theta}} (\dot{\hat{\theta}} - \tau_i) + (\hat{\theta} - \theta)^T \Gamma^{-1} (\dot{\hat{\theta}} - \tau_i) \quad (10.53)$$

and $\dot{z}_i$ as

$$\dot{z}_i = -z_{i-1} - c_i z_i + z_{i+1} - \frac{\partial \alpha_{i-1}}{\partial \hat{\theta}} (\dot{\hat{\theta}} - \tau_i) - (\hat{\theta} - \theta)^T \left(\varphi_i - \sum_{k=1}^{i-1} \frac{\partial \alpha_{i-1}}{\partial x_k} \varphi_k \right)$$

$$+ \left(\sum_{k=1}^{i-2} z_{k+1} \frac{\partial \alpha_k}{\partial \hat{\theta}} \Gamma \right) \left(\varphi_i - \sum_{k=1}^{i-1} \frac{\partial \alpha_{i-1}}{\partial x_k} \varphi_k \right). \quad (10.54)$$

The nth Step of Backstepping. The last step (the nth step) of the adaptive backstepping procedure is different for the adaptive designs I and II.

Adaptive design I. For the actuation model (10.7), we use the following controller structure:

$$v(t) = k_1 u_0(t) + k_2 \quad (10.55)$$

where $k_1 \in R^m$ and $k_2 \in R^m$ are the estimates of k_1^* and k_2^*. Define the parameter errors as

$$\tilde{k}_{1i} = k_{1i} - k_{1i}^*, \ \tilde{k}_{2i} = k_{2i} - k_{2i}^*. \quad (10.56)$$

Suppose there are p failed actuators, that is, $u_j(t) = \bar{u}_j, \ j = j_1, j_2, \ldots, j_p,$ $1 \le p \le m - 1$. From (10.6), (10.11), (10.23), (10.25), (10.55), and (10.56), we rewrite (10.2) as

$$\dot{x}_n = \varphi_0(x) + \theta^T \varphi_n(x) + \sum_{i \ne j_1, \ldots, j_p} k_{1i} \beta_i u_0$$

$$+ \sum_{i \ne j_1, \ldots, j_p} k_{2i} \beta_i + \sum_{i = j_1, \ldots, j_p} \beta_i \bar{u}_i$$

$$= \varphi_0(x) + \theta^T \varphi_n(x) + \beta_0(x) u_0$$

$$+ \sum_{i \ne j_1, \ldots, j_p} \tilde{k}_{1i} \beta_i u_0 + \sum_{i \ne j_1, \ldots, j_p} \tilde{k}_{2i} \beta_i$$

$$= \varphi_0(x) + \theta^T \varphi_n(x) + \beta_0(x) u_0$$

$$+ \sum_{i \ne j_1, \ldots, j_p} \frac{1}{k_{s1i}^*} \tilde{k}_{1i} \beta_0 u_0 + \sum_{i \ne j_1, \ldots, j_p} \frac{1}{k_{s1i}^*} \tilde{k}_{2i} \beta_0. \quad (10.57)$$

Applying the following nth step of the backstepping method to the dynamics presented in (10.57) for design I, an adaptive scheme will be developed to ensure that the output $y(t)$ of the system with the actuator failure model described in (10.1)–(10.3) and (10.4) will track the reference signal $y_m(t)$.

Step n: Introducing

$$V_n = V_{n-1} + \frac{1}{2}z_n^2 + \sum_{i \neq j_1, \dots, j_p} \frac{1}{2|k_{s1i}^*|}\Gamma_{1i}\tilde{k}_{1i}^2 + \sum_{i \neq j_1, \dots, j_p} \frac{1}{2|k_{s1i}^*|}\Gamma_{2i}\tilde{k}_{2i}^2 \quad (10.58)$$

where $\Gamma_{1i} > 0 \in R$ and $\Gamma_{2i} > 0 \in R$, we have

$$
\begin{aligned}
\dot{V}_n =\ & -\sum_{k=1}^{n-1} c_k z_k^2 + \left(\sum_{k=1}^{n-2} z_{k+1}\frac{\partial \alpha_k}{\partial \hat{\theta}}\right)(\tau_{n-1} - \dot{\hat{\theta}}) \\
& + z_n[z_{n-1} + \beta_0 u_0 + \varphi_0 - \sum_{k=1}^{n-1}\frac{\partial \alpha_{n-1}}{\partial x_k}x_{k+1} - \frac{\partial \alpha_{n-1}}{\partial \hat{\theta}}\dot{\hat{\theta}} \\
& - \sum_{k=1}^{n}\frac{\partial \alpha_{n-1}}{\partial y_m^{(k-1)}}y_m^{(k)} + \hat{\theta}^T(\varphi_n - \sum_{k=1}^{n-1}\frac{\partial \alpha_{n-1}}{\partial x_k}\varphi_k)] \\
& + (\hat{\theta} - \theta)^T \Gamma^{-1}[\dot{\hat{\theta}} - \Gamma\sum_{l=1}^{n} z_l(\varphi_l - \sum_{k=1}^{l-1}\frac{\partial \alpha_{l-1}}{\partial x_k}\varphi_k)] \\
& + z_n\left(\sum_{i \neq j_1, \dots, j_p}\frac{1}{k_{s1i}^*}\tilde{k}_{1i}\beta_0 u_0 + \sum_{i \neq j_1, \dots, j_p}\frac{1}{k_{s1i}^*}\tilde{k}_{2i}\beta_0\right) \\
& + \sum_{i \neq j_1, \dots, j_p}\frac{1}{|k_{s1i}^*|}\Gamma_{1i}\tilde{k}_{1i}\dot{\tilde{k}}_{1i} + \sum_{i \neq j_1, \dots, j_p}\frac{1}{|k_{s1i}^*|}\Gamma_{2i}\tilde{k}_{2i}\dot{\tilde{k}}_{2i}. \quad (10.59)
\end{aligned}
$$

Note that $\dot{\tilde{k}}_{1i} = \dot{k}_{1i}$ and $\dot{\tilde{k}}_{2i} = \dot{k}_{2i}$. At this step, we can design the adaptive update law for $\hat{\theta}$ as

$$
\begin{aligned}
\dot{\hat{\theta}} = \tau_n &= \tau_{n-1} + \Gamma z_n\left(\varphi_n - \sum_{k=1}^{n-1}\frac{\partial \alpha_{n-1}}{\partial x_k}\varphi_k\right) \\
&= \Gamma\sum_{l=1}^{n} z_l\left(\varphi_l - \sum_{k=1}^{l-1}\frac{\partial \alpha_{l-1}}{\partial x_k}\varphi_k\right) \quad (10.60)
\end{aligned}
$$

and choose the control u_0 as

$$
\begin{aligned}
u_0 =\ & \frac{1}{\beta_0}[-z_{n-1} - c_n z_n - \varphi_0 + \sum_{k=1}^{n-1}\frac{\partial \alpha_{n-1}}{\partial x_k}x_{k+1} \\
& + \frac{\partial \alpha_{n-1}}{\partial \hat{\theta}}\dot{\hat{\theta}} + \sum_{k=1}^{n}\frac{\partial \alpha_{n-1}}{\partial y_m^{(k-1)}}y_m^{(k)} \\
& - \left(\hat{\theta}^T - \sum_{k=1}^{n-2} z_{k+1}\frac{\partial \alpha_k}{\partial \hat{\theta}}\Gamma\right)\left(\varphi_n - \sum_{k=1}^{n-1}\frac{\partial \alpha_{n-1}}{\partial x_k}\varphi_k\right)]. \quad (10.61)
\end{aligned}
$$

For the resulting closed-loop system, we have

$$\dot{V}_n = -\sum_{k=1}^{n} c_k z_k^2 + \sum_{i \neq j_1,\dots,j_p} \frac{1}{|k_{s1i}^*|} \Gamma_{1i} \tilde{k}_{1i}(\dot{k}_{1i} + \text{sign}[k_{s1i}^*]\Gamma_{1i}^{-1}\beta_0 z_n u_0)$$

$$+ \sum_{i \neq j_1,\dots,j_p} \frac{1}{|k_{s1i}^*|} \Gamma_{2i} \tilde{k}_{2i}(\dot{k}_{2i} + \text{sign}[k_{s1i}^*]\Gamma_{2i}^{-1}\beta_0 z_n). \tag{10.62}$$

Therefore, the adaptive laws for k_{1i} and k_{2i}, $i \neq j_1, j_2, \dots, j_p$, are chosen as

$$\dot{k}_{1i} = -\text{sign}[k_{s1i}^*]\Gamma_{1i}^{-1}\beta_0 z_n u_0, \; i \neq j_1, j_2, \dots, j_p \tag{10.63}$$

$$\dot{k}_{2i} = -\text{sign}[k_{s1i}^*]\Gamma_{2i}^{-1}\beta_0 z_n, \; i \neq j_1, j_2, \dots, j_p. \tag{10.64}$$

Since the failures are unknown, we choose the above adaptive laws for all k_{1i} and k_{2i}, $i = 1, 2, \dots, m$, that is,

$$\dot{k}_{1i} = -\text{sign}[k_{s1i}^*]\Gamma_{1i}^{-1}\beta_0 z_n u_0, \; i = 1, 2, \dots, m \tag{10.65}$$

$$\dot{k}_{2i} = -\text{sign}[k_{s1i}^*]\Gamma_{2i}^{-1}\beta_0 z_n, \; i = 1, 2, \dots, m. \tag{10.66}$$

The closed-loop system with the adaptive design for actuation model I has the following properties.

Theorem 10.1.1. *The adaptive backstepping control design with the parameters update law (10.60) and the controller structure (10.55) consisting of and the state feedback law (10.61) and the adaptive laws (10.65) and (10.66), applied to the system (10.1)–(10.3) with the actuator failure model (10.4) for actuation model I (10.7) based on the Assumption (A10.2a), ensures that all closed-loop signals are bounded and $\lim_{t\to\infty}(y(t) - y_m(t)) = 0$.*

Proof: In view of (10.62), (10.63), and (10.64), we have

$$\dot{V}_n = -\sum_{k=1}^{n} c_k z_k^2 \leq 0. \tag{10.67}$$

It follows from (10.67) that $\hat{\theta} \in L^\infty$ and $z_k \in L^\infty$, $k = 1, 2, \dots, n$. Since $z_1 = x_1 - r$ is bounded, we have $x_1 \in L^\infty$ and hence $\varphi_1(x_1) \in L^\infty$. It follows from (10.39) that $\alpha_1 \in L^\infty$. By $z_2 = x_2 - \alpha_1$, we have $x_2 \in L^\infty$ and hence $\varphi_2(x_1, x_2) \in L^\infty$. It follows from (10.46) that $\alpha_2 \in L^\infty$. Continuing in the same way, we prove that $\alpha_i \in L^\infty$, $i = 1, 2, \dots, n - 1$, $x_i \in L^\infty$, $i = 1, 2, \dots, n$. Then according to (10.61), we have $u_0 \in L^\infty$. From (10.67), we also have $k_{1i} \in L^\infty$ and $k_{2i} \in L^\infty$, $i \neq j_1, j_2, \dots, j_p$. In addition, from (10.65) and (10.66) it follows that

$$k_{1j}(t) = k_{1j}(0) - \text{sign}[k_{s1j}^*]\Gamma_{1j}^{-1}\text{sign}[k_{s1i}^*]\Gamma_{1i}(k_{1i}(0) - k_{1i}(t))$$

$$= k_{1j}(0) - \text{sign}[k_{s1j}^*]\text{sign}[k_{s1i}^*](k_{1i}(0) - k_{1i}(t)) \tag{10.68}$$

for $j = j_1, j_2, \ldots, j_p$, $\forall i \neq j_1, j_2, \ldots, j_p$. Since $k_{1i}(t)$, $i \neq j_1, j_2, \ldots, j_p$, are bounded, it follows that $k_{1j}(t)$ is bounded for all $j = j_1, j_2, \ldots, j_p$. Similarly, k_{2j} is bounded for $j = j_1, j_2, \ldots, j_p$. From (10.55), we have the boundedness of $v(t)$ so that all the closed-loop signals are bounded.

From (10.67), actually we have $z_k \in L^2$, $k = 1, 2, \ldots, n$, and from (10.36) and the boundedness of the closed-loop signals, we have $\dot{z}_1 \in L^\infty$. It follows that $\lim_{t \to \infty} z_1(t) = 0$, which means that $\lim_{t \to \infty}(y(t) - y_m(t)) = 0$. ∇

Adaptive design II. For the actuation model (10.12), we use the following controller structure:

$$v_a(t) = k_1 u_0(t) + K_2 \beta_a(x) \tag{10.69}$$

where $k_1 \in R^m$, $K_2 = [k_{21}, k_{22}, \ldots, k_{2m}]^T \in R^{m \times m}$ are the estimates of k_1^* and K_2^*, and $\beta_a(x) = [\beta_1(x), \beta_2(x), \ldots, \beta_m(x)]^T$. Since $\beta_i(x)$ is known, the actual control input for each $v_i(t)$ in $v(t)$ is

$$v_i(t) = \frac{1}{\beta_i(x)} v_{ai}(t), \; i = 1, 2, \ldots, m. \tag{10.70}$$

Define the parameter errors

$$\tilde{k}_{1i} = k_{1i} - k_{1i}^*, \; \tilde{k}_{2i} = k_{2i} - k_{2i}^* \tag{10.71}$$

Suppose there are p failed actuators, that is, $u_j(t) = \bar{u}_j$ or $u_{aj}(t) = \bar{u}_{aj}$, $j = j_1, \ldots, j_p$, $1 \le p \le m - 1$. From (10.14), (10.15), (10.20), (10.30), (10.32), (10.69), and (10.71), we rewrite (10.2) as

$$
\begin{aligned}
\dot{x}_n &= \varphi_0(x) + \theta^T \varphi_n(x) \\
&\quad + \sum_{i \neq j_1, \ldots, j_p} b_i \beta_i v_i + \sum_{i = j_1, \ldots, j_p} b_i \beta_i \bar{u}_i \\
&= \varphi_0(x) + \theta^T \varphi_n(x) + \sum_{i \neq j_1, \ldots, j_p} b_i k_{1i} u_0 \\
&\quad + \sum_{i \neq j_1, \ldots, j_p} b_i k_{2i}^T \beta_a + \sum_{i = j_1, \ldots, j_p} b_i \beta_i \bar{u}_i \\
&= \varphi_0(x) + \theta^T \varphi_n(x) + u_0 \\
&\quad + \sum_{i \neq j_1, \ldots, j_p} \frac{1}{k_{s1i}^*} \tilde{k}_{1i} u_0 + \sum_{i \neq j_1, \ldots, j_p} \frac{1}{k_{s1i}^*} \tilde{k}_{2i}^T \beta_a. \tag{10.72}
\end{aligned}
$$

Applying the following nth step of the backstepping method to the dynamics presented in (10.70) for design II, an adaptive scheme will be developed to ensure that the output $y(t)$ of the system with the actuator failure model described in (10.1)–(10.3) and (10.4) will track the reference signal $y_m(t)$.

Step n: Introducing

$$V_n = V_{n-1} + \frac{1}{2}z_n^2 + \sum_{i \neq j_1,\ldots,j_p} \frac{1}{2|k_{s1i}^*|}\Gamma_{1i}\tilde{k}_{1i}^2$$

$$+ \sum_{i \neq j_1,\ldots,j_p} \frac{1}{2|k_{s1i}^*|}\tilde{k}_{2i}^T\Gamma_{2i}\tilde{k}_{2i} \tag{10.73}$$

where $\Gamma_{1i} > 0$ in R and $\Gamma_{2i} > 0$ in $R^{m \times m}$, we have

$$\dot{V}_n = -\sum_{k=1}^{n-1} c_k z_k^2 + \left(\sum_{k=1}^{n-2} z_{k+1}\frac{\partial \alpha_k}{\partial \hat{\theta}}\right)(\tau_{n-1} - \dot{\hat{\theta}})$$

$$+ z_n\left[z_{n-1} + u_0 + \varphi_0 - \sum_{k=1}^{n-1}\frac{\partial \alpha_{n-1}}{\partial x_k}x_{k+1} - \frac{\partial \alpha_{n-1}}{\partial \hat{\theta}}\dot{\hat{\theta}}\right.$$

$$\left. -\sum_{k=1}^{n}\frac{\partial \alpha_{n-1}}{\partial y_m^{(k-1)}}y_m^{(k)} + \hat{\theta}^T\left(\varphi_n - \sum_{k=1}^{n-1}\frac{\partial \alpha_{n-1}}{\partial x_k}\varphi_k\right)\right]$$

$$+ (\hat{\theta} - \theta)^T\Gamma^{-1}[\dot{\hat{\theta}} - \Gamma\sum_{l=1}^{n}z_l\left(\varphi_l - \sum_{k=1}^{l-1}\frac{\partial \alpha_{l-1}}{\partial x_k}\varphi_k\right)]$$

$$+ z_n\left(\sum_{i \neq j_1,\ldots,j_p} \frac{1}{k_{s1i}^*}\tilde{k}_{1i}u_0 + \sum_{i \neq j_1,\ldots,j_p} \frac{1}{k_{s1i}^*}\tilde{k}_{2i}^T\beta_a\right)$$

$$+ \sum_{i \neq j_1,\ldots,j_p} \frac{1}{|k_{s1i}^*|}\Gamma_{1i}\tilde{k}_{1i}\dot{k}_{1i} + \sum_{i \neq j_1,\ldots,j_p} \frac{1}{|k_{s1i}^*|}\tilde{k}_{2i}^T\Gamma_{2i}\dot{k}_{2i}. \tag{10.74}$$

Note that $\dot{\tilde{k}}_{1i} = \dot{k}_{1i}$ and $\dot{\tilde{k}}_{2i} = \dot{k}_{2i}$. Therefore, we can design the adaptive update law for $\hat{\theta}$ as

$$\dot{\hat{\theta}} = \tau_n = \tau_{n-1} + \Gamma z_n\left(\varphi_n - \sum_{k=1}^{n-1}\frac{\partial \alpha_{n-1}}{\partial x_k}\varphi_k\right)$$

$$= \Gamma\sum_{l=1}^{n}z_l\left(\varphi_l - \sum_{k=1}^{l-1}\frac{\partial \alpha_{l-1}}{\partial x_k}\varphi_k\right) \tag{10.75}$$

and choose the control u_0 as

$$u_0 = -z_{n-1} - c_n z_n - \varphi_0 + \sum_{k=1}^{n-1}\frac{\partial \alpha_{n-1}}{\partial x_k}x_{k+1} + \frac{\partial \alpha_{n-1}}{\partial \hat{\theta}}\dot{\hat{\theta}} + \sum_{k=1}^{n}\frac{\partial \alpha_{n-1}}{\partial y_m^{(k-1)}}y_m^{(k)}$$

$$- \left(\hat{\theta}^T - \sum_{k=1}^{n-2}z_{k+1}\frac{\partial \alpha_k}{\partial \hat{\theta}}\Gamma\right)\left(\varphi_n - \sum_{k=1}^{n-1}\frac{\partial \alpha_{n-1}}{\partial x_k}\varphi_k\right). \tag{10.76}$$

For the resulting closed-loop system, we have

$$
\dot{V}_n = -\sum_{k=1}^{n} c_k z_k^2 + \sum_{i \neq j_1,\dots,j_p} \frac{1}{|k_{s1i}^*|} \Gamma_{1i} \tilde{k}_{1i} (\dot{k}_{1i} + \text{sign}[k_{s1i}^*] \Gamma_{1i}^{-1} z_n u_0)
$$
$$
+ \sum_{i \neq j_1,\dots,j_p} \frac{1}{|k_{s1i}^*|} \tilde{k}_{2i}^T \Gamma_{2i} (\dot{k}_{2i} + \text{sign}[k_{s1i}^*] \Gamma_{2i}^{-1} z_n \beta_a). \tag{10.77}
$$

Therefore, the adaptive laws for k_{1i} and k_{2i}, $i \neq j_1, j_2, \dots, j_p$, are chosen as

$$
\dot{k}_{1i} = -\text{sign}[k_{s1i}^*] \Gamma_{1i}^{-1} z_n u_0, \ i \neq j_1, j_2, \dots, j_p \tag{10.78}
$$
$$
\dot{k}_{2i} = -\text{sign}[k_{s1i}^*] \Gamma_{2i}^{-1} z_n \beta_a, \ i \neq j_1, j_2, \dots, j_p. \tag{10.79}
$$

Since the failures are unknown, we choose the above adaptive laws for all k_{1i} and k_{2i}, $i = 1, 2, \dots, m$, that is,

$$
\dot{k}_{1i} = -\text{sign}[k_{s1i}^*] \Gamma_{1i}^{-1} z_n u_0, \ i = 1, 2, \dots, m \tag{10.80}
$$
$$
\dot{k}_{2i} = -\text{sign}[k_{s1i}^*] \Gamma_{2i}^{-1} z_n \beta_a, \ i = 1, 2, \dots, m. \tag{10.81}
$$

The closed-loop system with the adaptive design for actuation model II has the following properties.

Theorem 10.1.2. *The adaptive backstepping control design with the parameters update law (10.75) and the controller structure (10.69) and (10.70) consisting of the state feedback law (10.76) and the adaptive laws (10.80) and (10.81), applied to the system (10.1)–(10.3) with the actuator failure model (10.4) for actuation model II (10.12) based on Assumption (A10.2b), ensures that all closed-loop signals are bounded and the output tracking error $y(t) - y_m(t)$ goes to zero as t goes to infinity.*

Proof: From (10.77), (10.78), and (10.79), we have

$$
\dot{V}_n = -\sum_{k=1}^{n} c_k z_k^2 \leq 0. \tag{10.82}
$$

It follows from (10.82) that $\hat{\theta} \in L^\infty$ and $z_k \in L^\infty$, $k = 1, 2, \dots, n$. Since $z_1 = x_1 - r$ is bounded, we have $x_1 \in L^\infty$ and hence $\varphi_1(x_1) \in L^\infty$. It follows from (10.39) that $\alpha_1 \in L^\infty$. By $z_2 = x_2 - \alpha_1$, we have $x_2 \in L^\infty$ and hence $\varphi_2(x_1, x_2) \in L^\infty$. It follows from (10.46) that $\alpha_2 \in L^\infty$. Continuing in the same way, we prove that $\alpha_i \in L^\infty$, $i = 1, 2, \dots, n-1$, $x_i \in L^\infty$, $i = 1, 2, \dots, n$. Then according to (10.74), we have $u_0 \in L^\infty$. From (10.82) we also have $k_{1i} \in L^\infty$ and $k_{2i} \in L^\infty$, $i \neq j_1, j_2, \dots, j_p$. In addition, from (10.80) and (10.81) it follows that

$$k_{1j}(t) = k_{1j}(0) - \text{sign}[k^*_{s1j}]\Gamma^{-1}_{1j}\text{sign}[k^*_{s1i}]\Gamma_{1i}(k_{1i}(0) - k_{1i}(t))$$

$$= k_{1j}(0) - \text{sign}[k^*_{s1j}]\text{sign}[k^*_{s1i}](k_{1i}(0) - k_{1i}(t)) \qquad (10.83)$$

for $j = j_1, \ldots, j_p$, $\forall i \neq j_1, \ldots, j_p$. Since $k_{1i}(t)$, $i \neq j_1, \ldots, j_p$, are bounded, it follows that $k_{1j}(t)$ is bounded for all $j = j_1, \ldots, j_p$. Similarly, according to

$$k_{2j}(t) = k_{2j}(0) - \text{sign}[k^*_{s1j}]\Gamma^{-1}_{2j}\text{sign}[k^*_{s1i}]\Gamma_{2i}(k_{2i}(0) - k_{2i}(t)) \qquad (10.84)$$

for $j = j_1, j_2, \ldots, j_p$, $\forall i \neq j_1, j_2, \ldots, j_p$, $k_{2j} \in R^m$ is bounded for $j = j_1, j_2, \ldots, j_p$. From (10.69) and (10.70) we have the boundedness of $v(t)$ so that all the closed-loop signals are bounded.

From (10.82) actually we have $z_k \in L^2$, $k = 1, 2, \ldots, n$, and from (10.36) and the boundedness of the closed-loop signals, we have $\dot{z}_1 \in L^\infty$. It follows that $\lim_{t \to \infty} z_1(t) = 0$, which means that the output tracking error $y(t) - y_m(t)$ goes to zero as t goes to infinity. ∇

10.1.3 Wing Rock Control of an Aircraft Model

We apply the adaptive actuator failure compensation control design to a nonlinear aircraft wing model with unknown actuator failures. The model we consider here is based on the state-space form of the F-18 HARV-like wing-rock model with ailerons modeled as a first-order actuator dynamics [71], which is in the parametric-strict-feedback form (also see Section 9.4). With three augmented actuation parameters (b_1, b_2, b_3) for actuator failure compensation study, the aircraft wing model is described as

$$\dot{\phi} = p$$
$$\dot{p} = \theta_1 + \theta_2\phi + \theta_3p + \theta_4|\phi|p + \theta_5|p|p + \delta$$
$$\dot{\delta} = -\frac{1}{\tau}\delta + \frac{b_1}{\tau}u_1 + \frac{b_2}{\tau}u_2 + \frac{b_3}{\tau}u_3 \qquad (10.85)$$

where u_1, u_2, and u_3 are the inputs with the unit taken as rad, which are used to control the aileron, τ is the aileron time constant, and θ_i, $i = 1, 2, \ldots, 5$, and b_j, $j = 1, 2, 3$, are unknown constant parameters. Considering the plant (10.85) as actuation model I given in (10.7), we now apply the adaptive feedback compensation scheme with the controller (10.55) and parameter update law (10.60) for the model to achieve the output tracking objective $\phi(t) - \phi_m(t) \to 0$, with the actuator failures

$$u_1(t) = \begin{cases} v_1(t) & \text{for } t \in [0, 10] \\ \bar{u}_1 & \text{for } t \in [10, \infty), \end{cases}$$

$$u_2(t) = \begin{cases} v_2(t) & \text{for } t \in [0, 20] \\ \bar{u}_2 & \text{for } t \in [20, \infty) \end{cases} \qquad (10.86)$$

while $u_3(t) = v_3(t)$, $t \in [0, \infty)$. This means that the wing rock is suppressed despite the presence of actuator failures. For simulation, we choose the parameters as $c_i = 5.1$, $i = 1, 2, 3$, $\Gamma = 0.1I$, and $\Gamma_{1j} = 1.05$, $\Gamma_{2j} = 10.5$, $j = 1, 2, 3$, and the initial condition as $\hat{\theta}_i(0) = 1.35\theta_i$, $i = 1, 2, \ldots, 5$, with the true value: $\theta_1 = 0$, $\theta_2 = -26.67$, $\theta_3 = 0.76485$, $\theta_4 = -2.9225$, and $\theta_5 = 0$, $\phi(0) = 0.4$, $p(0) = \delta(0) = 0$, $k_1(0) = [0.8\ 1.2\ 1.5]^T$, $k_2(0) = [-2\ 0\ 1]^T$. Figure 10.1 shows the output tracking error $e(t) = \phi(t) - \phi_m(t)$ for an exponentially decaying reference trajectory $\phi_m(t)$ governed by $(s + 10)(s^2 + 4s + 24.25)\phi_m(s) = 0$ when the adaptive controller (10.55) acts through the aileron with $b_1 = 0.75$, $b_2 = 0.3$, and $b_3 = 0.45$ as well as $\tau = 1/15$ in the presence of the actuator failures (10.86) where $\bar{u}_1 = 3$ and $\bar{u}_2 = 2$, and the three control inputs are shown in Figure 10.2. With the aileron time constant τ, the deflection angle δ is in a reasonable interval during operation.

In summary, for nonlinear systems in the parametric-strict-feedback form without zero dynamics, we have characterized the systems with actuator failures in terms of two actuation models and derived the matching conditions for actuator failure compensation. Two adaptive compensation schemes have been developed with the desired controllers and parameter update laws. The effectiveness of the adaptive control scheme applied to an aircraft wing model with actuator failures has been verified by simulation.

10.2 Design for Systems with Zero Dynamics

To investigate the actuator failure compensation control problem for nonlinear systems, we consider the nonlinear plant

$$\dot{x} = f_0(x) + \sum_{i=1}^{l} \theta_i f_i(x) + \sum_{j=1}^{m} \mu_j g_j(x) u_j$$
$$y = h(x) \tag{10.87}$$

where $x \in R^n$ is the state, $y \in R$ is the output, and $u_j \in R$, $j = 1, 2, \ldots, m$, are the plant inputs, which may fail during operation, $f_i(x) \in R^n$, $i = 0, 1, \ldots, l$, $g_j(x) \in R^n$, $j = 1, 2, \ldots, m$, and $h(x) \in R$ are smooth functions, θ_i, $i = 1, 2, \ldots, l$, and μ_j, $j = 1, 2, \ldots, m$, are unknown constant parameters. For the actuator failure model (10.4), we rewrite the plant (10.87) as

$$\dot{x} = f_0(x) + F(x)\theta + g(x)\mu\sigma\bar{u} + g(x)\mu(I - \sigma)v$$
$$y = h(x) \tag{10.88}$$

for a certain failure pattern σ, where

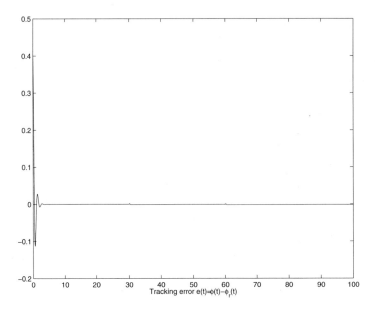

Fig. 10.1. Output tacking error $(u_1 = 3$ at $t = 10$, $u_2 = 2$ at $t = 20)$.

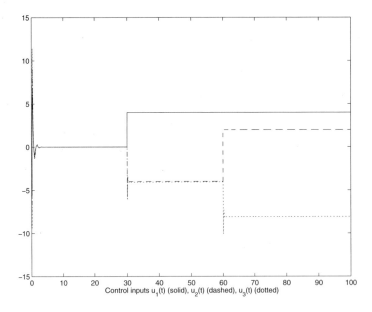

Fig. 10.2. Control inputs$(u_1 = 3$ at $t = 10$, $u_2 = 2$ at $t = 20)$.

$$F(x) = [f_1(x), f_2(x), \ldots, f_l(x)] \in R^{n \times l} \qquad (10.89)$$

$$\theta = [\theta_1, \theta_2, \ldots, \theta_l]^T \in R^l \qquad (10.90)$$

$$g(x) = [g_1(x), g_2(x), \ldots, g_m(x)] \in R^{n \times m} \qquad (10.91)$$

$$\mu = \text{diag}\{\mu_1, \mu_2, \ldots, \mu_m\} \in R^{m \times m}. \qquad (10.92)$$

The control objective is to use state feedback design for the plant (10.87) with up to $m - 1$ failures of u_j, $j = 1, 2, \ldots, m$, to ensure that all signals in the closed-loop system are bounded and the plant output $y(t)$ asymptotically tracks a prescribed reference signal $y_m(t)$.

Unlike that addressed in Section 10.1, in this actuator failure compensation problem, the nonlinear plant (10.87) may contain zero dynamics.

10.2.1 An Adaptive Failure Compensation Design

To meet the basic assumption for failure compensation, some system structure conditions are needed for achieving the output tracking when there are up to $m - 1$ actuator failures. In this subsection, we first develop an adaptive actuator failure compensation scheme under the following condition.

Structure Condition. For the plant (10.87), we assume that

(A10.4a) $g_j(x) \in \text{span}\{g_0(x)\}$, $g_0(x) \in R^n$, for $j = 1, 2, \ldots, m$, and the nominal system

$$\dot{x} = f_0(x) + F(x)\theta + g_0(x)u_0, \ y = h(x) \qquad (10.93)$$

where $u_0 \in R$, is transformable into the parametric-strict-feedback form (10.8) and (10.9) with relative degree ρ.

Remark 10.2.1. For the nominal system (10.93) to be transformable into the parametric-strict-feedback form with relative degree ρ, through a parameter-independent transformation, we need the differential geometric conditions

$$L_{g_0(x)} L_{f_0(x)}^k h(x) = 0, \ k = 0, 1, \ldots, \rho - 2,$$

$$L_{g_0(x)} L_{f_0(x)}^{\rho-1} h(x) \neq 0 \qquad (10.94)$$

$$\xi_{k+1} = L_{f_0(x)}^k h(x),$$

$$L_{f_i(x)} L_{f_0(x)}^k h(x) = \varphi_{k+1,i}(\xi_1, \ldots, \xi_{k+1}) \qquad (10.95)$$

for $i = 1, 2, \ldots, l$, $k = 0, 1, \ldots, \rho - 2$ [71]. The condition (10.94) implies that such a nonlinear system as $\dot{x} = f_0(x) + g_0(x)u_0$, $y = h(x)$ is feedback linearizable with relative degree ρ.   □

Diffeomorphism. Under Assumption (A10.4a), there exists a diffeomorphism $[\xi, \eta]^T = T(x) = [T_c(x), T_z(x)]^T$, with

$$\xi = T_c(x) = \begin{bmatrix} T_1(x) \\ T_2(x) \\ \vdots \\ T_\rho(x) \end{bmatrix} = \begin{bmatrix} h(x) \\ L_{f_0(x)}h(x) \\ \vdots \\ L_{f_0(x)}^{\rho-1}h(x) \end{bmatrix}$$

$$\eta = T_z(x) = \begin{bmatrix} T_{\rho+1}(x) \\ T_{\rho+2}(x) \\ \vdots \\ T_n(x) \end{bmatrix} \tag{10.96}$$

to transform the system (10.93) into the parametric-strict-feedback form

$$\begin{aligned}
\dot{\xi}_1 &= \xi_2 + \varphi_1^T(\xi_1)\theta \\
\dot{\xi}_2 &= \xi_3 + \varphi_2^T(\xi_1, \xi_2)\theta \\
&\cdots \\
\dot{\xi}_{\rho-1} &= \xi_\rho + \varphi_{\rho-1}^T(\xi_1, \ldots, \xi_{\rho-1})\theta \\
\dot{\xi}_\rho &= \varphi_0(\xi, \eta) + \varphi_\rho^T(\xi, \eta)\theta + \beta_0(\xi, \eta)u_0 \\
\dot{\eta} &= \psi(\xi, \eta) + \Psi(\xi, \eta)\theta \\
y &= \xi_1
\end{aligned} \tag{10.97}$$

where

$$\begin{aligned}
\varphi_0(\xi, \eta) &= L_{f_0(x)}^\rho h(x) \\
\varphi_k(\xi_1, \ldots, \xi_k) &= [L_{f_1(x)}L_{f_0(x)}^{k-1}h(x), \ldots, \\
&\qquad L_{f_l(x)}L_{f_0(x)}^{k-1}h(x)]^T, \ k = 1, \ldots, \rho - 2 \\
\varphi_\rho(\xi, \eta) &= [L_{f_1(x)}L_{f_0(x)}^{\rho-1}h(x), \ldots, L_{f_l(x)}L_{f_0(x)}^{\rho-1}h(x)]^T \\
\beta_0(\xi, \eta) &= L_{g_0(x)}L_{f_0(x)}^{\rho-1}h(x) \\
\psi(\xi, \eta) &= \frac{\partial T_z(x)}{\partial x}f_0(x) \\
\Psi(\xi, \eta) &= \frac{\partial T_z(x)}{\partial x}F(x)
\end{aligned} \tag{10.98}$$

and $\dot{\eta} = \psi(\xi, \eta) + \Psi(\xi, \eta)\theta$ is the zero dynamics of the nominal system, which is unobservable after cancellation of η in the output by a feedback control u_0.

From Assumption (A10.4a), it follows that with the diffeomorphism (10.96), the plant (10.87) can be transformed into

$$\dot{\xi}_i = \xi_{i+1} + \varphi_i^T(\xi_1, \ldots, \xi_i)\theta, \; i = 1, 2, \ldots, \rho - 1$$
$$\dot{\xi}_\rho = \varphi_0(\xi, \eta) + \varphi_\rho^T(\xi, \eta)\theta + \beta^T(\xi, \eta)\mu u$$
$$\dot{\eta} = \psi(\xi, \eta) + \Psi(\xi, \eta)\theta, \; y = \xi_1 \tag{10.99}$$

where

$$\beta(\xi, \eta) = [\beta_1(\xi, \eta), \ldots, \beta_m(\xi, \eta)]^T$$
$$= [L_{g_1(x)}L_{f_0(x)}^{\rho-1}h(x), \ldots, L_{g_m(x)}L_{f_0(x)}^{\rho-1}h(x)]^T \tag{10.100}$$

and μ is given in (10.92). With the actual input described as that in (10.89) in the presence of actuator failures, we rewrite the plant (10.99) as

$$\dot{\xi}_i = \xi_{i+1} + \varphi_i^T(\xi_1, \ldots, \xi_i)\theta, \; i = 1, 2, \ldots, \rho - 1$$
$$\dot{\xi}_\rho = \varphi_0(\xi, \eta) + \varphi_\rho^T(\xi, \eta)\theta + \beta^T(\xi, \eta)\mu\sigma\bar{u} + \beta^T(\xi, \eta)\mu(I - \sigma)v$$
$$\dot{\eta} = \psi(\xi, \eta) + \Psi(\xi, \eta)\theta, \; y = \xi_1. \tag{10.101}$$

Remark 10.2.2. Under Assumption (A10.4a), the zero dynamics of the plant (10.87) or (10.99) are the same as those of the nominal system (10.93), and $T_z(x)$ is determined by the nominal system such that the zero dynamics will not change in the presence of actuator failures. In other words, actuator failures neither change the structure of the zero dynamics nor enter the zero dynamics as additional inputs, which allows us to design control for each actuator u_i individually, $i = 1, 2, \ldots, m$. ☐

Stable Zero Dynamics Assumption. For the control objective that the output y is required to track a reference signal $y_m(t)$ whose first ρ derivatives are available with $y_m^{(\rho)}$ bounded and piecewise continuous, the states η in the zero dynamics actually will be cancelled and become unobservable in the output. Therefore, for the stability of the closed-loop system, we assume

(A10.5a) The nominal system (10.93) is minimum phase, that is, the zero dynamics $\dot{\eta} = \psi(\xi, \eta) + \Psi(\xi, \eta)\theta$ are *input-to-state stable* (ISS) [67] with respect to ξ as the input.

Backstepping Design for the Nominal System. With Assumption (A10.5a), a stable adaptive control scheme for the nominal system (10.93) can be obtained by applying the backstepping technique [71] such that the output $y(t)$ asymptotically tracks the reference signal $y_m(t)$.

The control signal u_0 is designed as

$$u_0 = \frac{1}{\beta_0(\xi, \eta)} v_0, \quad v_0 = \alpha_\rho - \varphi_0, \tag{10.102}$$

along with the adaptive law for updating the parameter estimate $\hat{\theta}$ of θ

$$\dot{\hat{\theta}} = \tau_\rho \tag{10.103}$$

with the error variables z_i, the stabilizing functions α_i, and the tuning functions τ_i, $i = 1, 2, \ldots, \rho$, defined by the recursive procedures:

$$z_1 = \xi_1 - y_m$$

$$\alpha_1 = -c_1 z_1 - \varphi_1^T \hat{\theta} + \dot{y}_m$$

$$\tau_1 = z_1 \Gamma \varphi_1$$

$$z_i = \xi_i - \alpha_{i-1}(\xi_1, \ldots, \xi_{i-1}, \hat{\theta}, y_m, \dot{y}_m, \ldots, y_m^{(i-1)}), \quad i = 2, 3, \ldots, \rho$$

$$\alpha_i = -c_i z_i - z_{i-1} - \varphi_i^T \hat{\theta} + \sum_{k=1}^{i-1} \frac{\partial \alpha_{i-1}}{\partial \xi_k} (\xi_{k+1} + \varphi_k^T \hat{\theta})$$

$$+ \frac{\partial \alpha_{i-1}}{\partial \hat{\theta}} \tau_{i-1} + \left(\sum_{k=1}^{i-1} z_{k+1} \frac{\partial \alpha_k}{\partial \hat{\theta}} \right) \Gamma \left(\varphi_i - \sum_{k=1}^{i-1} \frac{\partial \alpha_{i-1}}{\partial \xi_k} \varphi_k \right)$$

$$+ \sum_{k=1}^{i} \frac{\partial \alpha_{i-1}}{\partial y_m^{(k-1)}} y_m^{(k)}, \quad i = 2, 3, \ldots, \rho$$

$$\tau_i = \tau_{i-1} + z_i \Gamma \left(\varphi_i - \sum_{k=1}^{i-1} \frac{\partial \alpha_{i-1}}{\partial x_k} \varphi_k \right), \quad i = 2, 3, \ldots, \rho, \tag{10.104}$$

where $\Gamma = \Gamma^T > 0$ is the adaptation gain matrix and $c_i > 0$, $i = 1, 2, \ldots, \rho$, are some design constants.

Controller Structure. Before developing an adaptive failure compensation control scheme, we first present, assuming the knowledge of the uncertainty μ and the actuator failures, the ideal controller for the plant (10.101):

$$v_j = \frac{1}{\beta_j(\xi, \eta)} (k_{1,j} v_0 + k_{2,j}^T \beta(\xi, \eta)), \quad j = 1, 2, \ldots, m \tag{10.105}$$

where v_0 comes from the nominal control design given in (10.102), and $k_{1,j} \in R$ and $k_{2,j} \in R^m$ are some constant parameters chosen to satisfy

$$[k_{1,1}, k_{1,2}, \ldots, k_{1,m}] \mu (I - \sigma)[1, 1, \ldots, 1]^T = 1$$

$$[k_{2,1}, k_{2,2}, \ldots, k_{2,m}] (I - \sigma)[1, 1, \ldots, 1]^T = -\mu \sigma \bar{u}. \tag{10.106}$$

It can be proved that with the knowledge of μ and the actuator failures we always can find such $k_{1,j}$ and $k_{2,j}$ to satisfy (10.106).

With the controller (10.105), the closed-loop system behaves exactly the same as the nominal system with the controller (10.102), which means that $y(t)$ asymptotically tracks $y_m(t)$. Since the zero dynamics are identical to the nominal system and independent of actuator failures, under Assumption (A10.5a), closed-loop signal boundedness is ensured, too.

For adaptive control with σ, $\bar{u}$, and μ unknown, we develop an adaptive version of the feedback controller (10.105), that is,

$$v_j = \frac{1}{\beta_j(\xi, \eta)}(\hat{k}_{1,j}(t)v_0 + \hat{k}_{2,j}^T(t)\beta(\xi, \eta)), \ j = 1, 2, \ldots, m \qquad (10.107)$$

where $\hat{k}_{1,j}(t)$ and $\hat{k}_{2,j}(t)$ are the estimates of $k_{1,j}$ and $k_{2,j}$.

Adaptive Laws. To derive the adaptive laws for $\hat{k}_{1,j}(t)$ and $\hat{k}_{2,j}(t)$, the estimates of $k_{1,j}$ and $k_{2,j}$, $j = 1, 2, \ldots, m$, we define parameters and their errors according to the matching conditions (10.106),

$$\begin{aligned} \kappa_j &= [k_{1,j}, k_{2,j}^T]^T \\ \hat{\kappa}_j &= [\hat{k}_{1,j}, \hat{k}_{2,j}^T]^T \\ \tilde{\kappa}_j &= \hat{\kappa}_j - \kappa_j, \ j = 1, 2, \ldots, m \end{aligned} \qquad (10.108)$$

and introduce the signal

$$\omega(t) = [v_0, \beta^T(\xi, \eta)]^T. \qquad (10.109)$$

Then, we choose the adaptive update law for $\hat{\kappa}_i$ as

$$\dot{\hat{\kappa}}_j = -\text{sign}[\mu_j]z_\rho \Gamma_j \omega, \ j = 1, 2, \ldots, m \qquad (10.110)$$

where z_ρ is derived from the recursive procedures (10.104), $\Gamma_j = \Gamma_j^T > 0$, and $\text{sign}[\mu_j]$ is the sign of the unknown parameter μ_j, $j = 1, 2, \ldots, m$. To implement such adaptive laws we need to assume that

(A10.6) The signs of μ_j, $j = 1, 2, \ldots, m$, are known.

The closed-loop system has the following desired properties.

Theorem 10.2.1. *With the zero dynamics to be ISS (A10.5a), the adaptive controller (10.107) with the update laws (10.103) and (10.110) ensures the boundedness of the closed-loop signals and asymptotic output tracking, that is, $\lim_{t \to \infty}(y(t) - y_m(t)) = 0$.*

Proof: Suppose that one or more than one actuators fails at time instants t_i, $i = 1, 2, \ldots, q$, $1 \leq q \leq m - 1$, and at time $t \in (t_{i-1}, t_i)$ there are p failed actuators in the system, that is, $u_j = \bar{u}_j$, $j = j_1, j_2, \ldots, j_p$, $\{j_1, j_2, \ldots, j_p\} \subset \{1, 2, \ldots, m\}$, $0 \leq p \leq m - 1$. The adaptive scheme can be analyzed by using the positive definite function defined on the interval (t_{i-1}, t_i), on which the actuator failure pattern σ is unchanged:

$$V_{\sigma_{i-1}}(t) = \frac{1}{2} z^T(t) z(t) + \frac{1}{2} \tilde{\theta}^T(t) \Gamma^{-1} \tilde{\theta}(t)$$
$$+ \sum_{j \neq j_1, \ldots, j_p} \frac{|\mu_j|}{2} \tilde{\kappa}_j^T(t) \Gamma_j^{-1} \tilde{\kappa}_j(t) \qquad (10.111)$$

where $\tilde{\theta}(t) = \hat{\theta}(t) - \theta$ and $z(t) = [z_1(t), z_2(t), \ldots, z_\rho(t)]^T$ is a vector of the error variables defined in (10.104). The time derivative of $V_\sigma(t)$ along the trajectories of (10.103) and (10.110) is

$$\dot{V}_{\sigma_{i-1}} = - \sum_{k=1}^{\rho} c_k z_k^2(t) \leq 0, \ t \in (t_{i-1}, t_i) \qquad (10.112)$$

where $c_k > 0$ chosen for the control design. At $t = t_i$, $i = 1, 2, \ldots, q$, when there are actuators failing, some of the terms $\tilde{\kappa}_j^T(t) \Gamma_j^{-1} \tilde{\kappa}_j(t)$, $j \neq j_1, j_2, \ldots, j_p$, will be removed from the function $V_{\sigma_{i-1}}(t)$ to form the new $V_{\sigma_i}(t)$ on the next interval (t_i, t_{i+1}), and the remaining $\tilde{\kappa}_j(t)$ will have a jumping because of the change of the value κ_j. From (10.112) it follows that

$$V_{\sigma_{i-1}}(t_i^-) \leq V_{\sigma_{i-1}}(t_{i-1}^+) \qquad (10.113)$$

which indicates that $z(t)$, $\tilde{\theta}(t)$, and $\tilde{\kappa}_j(t)$ are bounded on the interval (t_{i-1}, t_i) if $V_{\sigma_{i-1}}(t_{i-1}^+)$ is finite. Since the change of κ_j is finite, it in turn implies that $V_{\sigma_i}(t_i^+)$ is also finite, that is, for the next interval (t_i, t_{i+1}), the initial of the function $V_{\sigma_i}(t)$ is finite. Due to the finite value of $V_{\sigma_0}(t_0)$, it turns out that the initial value for the last function $V_{\sigma_q}(t)$ defined on the (t_q, ∞) is finite such that we can conclude that $z(t)$, $\hat{\theta}(t)$, and $\hat{\kappa}_j(t)$, $j \neq j_1, j_2, \ldots, j_{p_q}$, are bounded and $z(t) \in L^2$. Since $z_1 = \xi_1 - y_m$, ξ_1 is bounded and hence α_1 is also bounded. With $z_2 = \xi_2 - \alpha_1$, $\xi_2 \in L^\infty$ and in turn $\alpha_2 \in L^\infty$. Continuing in the same way, we see that $\xi(t)$ is bounded. With the ISS of the zero dynamics, $\eta(t)$ is bounded with respect to $\xi(t)$ as the input. It follows that the control signals v_j, $j \neq j_1, j_2, \ldots, j_{p_q}$, are bounded and hence the closed-loop stability is established. Considering that $\dot{z} \in L^\infty$, we conclude that $z(t)$ goes to zero when t goes to infinity. With $z_1 = y - y_m$, the asymptotic tracking $\lim_{t \to \infty}(y(t) - y_m(t)) = 0$ is achieved. ∇

10.2.2 An Adaptive Design with Relaxed Conditions

We now develop an adaptive failure compensation controller, based on the following condition (weaker than condition (A10.4a) in Section 10.2.1).

Relative Degree Condition. For the plant (10.87), we assume

(A10.4b) The system $\dot{x} = f_0(x) + g_j(x)u_j$, $y = h(x)$ has the same relative degree ρ, for $j = 1, 2, \ldots, m$.

Remark 10.2.3. Assumption (A10.4b) implies that the plant (10.87) has the same relative degree for each actuation function $g_j(x)$, $j = 1, \ldots, m$, that is,

$$
\begin{aligned}
L_{g_j(x)}L_{f_0(x)}^k h(x) &= 0, \ k = 0, 1, \ldots, \rho - 2, \\
L_{g_j(x)}L_{f_0(x)}^{\rho-1} h(x) &\neq 0,
\end{aligned} \qquad j = 1, 2, \ldots, m. \qquad (10.114)
$$

This condition is similar to that for the linear case (see Chapter 3).   □

Actuation Scheme. Choose the proportional-actuation scheme as

$$
v_j = b_j(x)v_0, \ j = 1, 2, \ldots, m \qquad (10.115)
$$

where v_0 is a control input to be designed, $b_j(x)$ is a scalar function to be chosen such that for $\forall x \in R^n$ and $\forall \{j_1, j_2, \ldots, j_p\} \subset \{1, 2, \ldots, m\}$,

$$
\sum_{j \neq j_1, \ldots, j_p} \mu_j b_j(x) L_{g_j(x)} L_{f_0(x)}^{\rho-1} h(x) \neq 0. \qquad (10.116)
$$

With Assumption (A10.6), there exist such $b_j(x)$, $j = 1, 2, \ldots, m$, to satisfy (10.116), for example, $b_j(x) = \text{sign}[\mu_j](1/L_{g_j(x)}L_{f_0(x)}^{\rho-1}h(x))$ with which

$$
\sum_{j \neq j_1, \ldots, j_p} \mu_j b_j(x) L_{g_j(x)} L_{f_0(x)}^{\rho-1} h(x) \geq \min_{j \in \{1, 2, \ldots, m\}} |\mu_j| > 0. \qquad (10.117)
$$

Diffeomorphism. For plant (10.87) to be transformable into a parametric-strict-feedback form, we need the following differential geometric condition:

$$
\begin{aligned}
\xi_{k+1} &= L_{f_0(x)}^k h(x), \ k = 0, \ldots, \rho - 2 \\
L_{f_i(x)} L_{f_0(x)}^k h(x) &= \varphi_{k+1,i}(\xi_1, \ldots, \xi_{k+1}), \ i = 1, \ldots, l. \qquad (10.118)
\end{aligned}
$$

Based on Assumption (A10.4b), for the plant (2.1) to be transformable into parametric-strict-feedback form, with the actuation scheme (10.115), there exists a diffeomorphism $[\xi, \eta]^T = T(x) = [T_c(x), T_z(x)]^T$, where

$$\xi = T_c(x) = \begin{bmatrix} T_1(x) \\ T_2(x) \\ \vdots \\ T_\rho(x) \end{bmatrix} = \begin{bmatrix} h(x) \\ L_{f_0(x)}h(x) \\ \vdots \\ L_{f_0(x)}^{\rho-1}h(x) \end{bmatrix}$$

$$\eta = T_z(x) = \begin{bmatrix} T_{\rho+1}(x) \\ T_{\rho+2}(x) \\ \vdots \\ T_n(x) \end{bmatrix} \tag{10.119}$$

and $T_z(x)$ exists but is not unique [60], such that $[\frac{\partial T(x)}{\partial x}]$ is nonsingular for $\forall x \in R^n$, to transform the plant (10.91) with the presence of actuator failures into a canonical form consisting of a parametric-strict-feedback cascade subsystem and a zero dynamic subsystem:

$$\dot{\xi}_1 = \xi_2 + \varphi_1^T(\xi_1)\theta$$
$$\dot{\xi}_2 = \xi_3 + \varphi_2^T(\xi_1, \xi_2)\theta$$

$$\cdots$$

$$\dot{\xi}_{\rho-1} = \xi_\rho + \varphi_{\rho-1}^T(\xi_1, \dots, \xi_{\rho-1})\theta$$
$$\dot{\xi}_\rho = \varphi_0(\xi, \eta) + \varphi_\rho^T(\xi, \eta)\theta + \beta^T(\xi, \eta)\kappa_1 + \bar{\beta}^T(\xi, \eta)\kappa_2 v_0$$
$$y = \xi_1 \tag{10.120}$$
$$\dot{\eta} = \psi(\xi, \eta) + \Psi(\xi, \eta)\theta + \Delta(\xi, \eta)\kappa_1 + \bar{\Delta}(\xi, \eta)\kappa_2 v_0 \tag{10.121}$$

where $\xi = [\xi_1, \xi_2, \dots, \xi_\rho]^T \in R^\rho$, $\eta \in R^\gamma$, $\rho + \gamma = n$, $\kappa_1 = \mu\sigma\bar{u}$ and $\kappa_2 = \mu(I - \sigma)[1, 1, \dots, 1]^T$ treated as some unknown constant vectors, and

$$\varphi_0(\xi, \eta) = L_{f_0(x)}^\rho h(x)$$
$$\varphi_k(\xi_1, \dots, \xi_k) = [L_{f_1(x)}L_{f_0(x)}^{k-1}h(x), \dots,$$
$$L_{f_l(x)}L_{f_0(x)}^{k-1}h(x)]^T, \ k = 1, \dots, \rho - 2$$
$$\varphi_\rho(\xi, \eta) = [L_{f_1(x)}L_{f_0(x)}^{\rho-1}h(x), \dots, L_{f_l(x)}L_{f_0(x)}^{\rho-1}h(x)]^T$$
$$\beta(\xi, \eta) = [L_{g_1(x)}L_{f_0(x)}^{\rho-1}h(x), \dots, L_{g_m(x)}L_{f_0(x)}^{\rho-1}h(x)]^T$$
$$\bar{\beta}(\xi, \eta) = [b_1(x)L_{g_1(x)}L_{f_0(x)}^{\rho-1}h(x), \dots, b_m(x)L_{g_m(x)}L_{f_0(x)}^{\rho-1}h(x)]^T$$
$$\psi(\xi, \eta) = \frac{\partial T_z(x)}{\partial x}f_0(x)$$
$$\Psi(\xi, \eta) = \frac{\partial T_z(x)}{\partial x}F(x)$$
$$\Delta(\xi, \eta) = \frac{\partial T_z(x)}{\partial x}g(x)$$
$$\bar{\Delta}(\xi, \eta) = \frac{\partial T_z(x)}{\partial x}[b_1(x)g_1(x), \dots, b_m(x)g_m(x)]^T \tag{10.122}$$

and (10.121) is the zero dynamics subsystem. One choice of $T_z(x)$ may be made based on the plant model (10.87) with the actuation scheme (10.115), in the case of no failure and under the condition

$$\frac{\partial T_z(x)}{\partial x} \sum_{j=1}^{m} \mu_j b_j(x) g_j(x) = 0 \tag{10.123}$$

so that $\bar{\Delta}(\xi, \eta)\kappa_2 = 0$ for the no-failure case.

With a feedback control signal v_0, the subsystem (10.121) forms the zero dynamics of the closed-loop system, different from that in Section 10.2.1.

Remark 10.2.4. For a chosen $T_z(x)$, the function $\bar{\Delta}(\xi, \eta)\kappa_2$ in (10.121) may be nonzero for some failure patterns, that is, the zero dynamics (10.121) explicitly depends on the control signal v_0, which may lead to difficulty in analyzing closed-loop system stability. To obtain a zero dynamics system independent of control input, for each failure pattern σ, we introduce a diffeomorphism $[\xi, \bar{\eta}]^T = T_\sigma(x) = [T_c(x), T_{z,\sigma}(x)]^T$, where $T_{z,\sigma}(x)$ is dependent on the failure pattern σ and is chosen to satisfy

$$\frac{\partial T_{z,\sigma(x)}}{\partial x} [b_1(x)g_1(x), \ldots, b_m(x)g_m(x)]^T \kappa_2 = 0. \tag{10.124}$$

This equation has a solution $T_{z,\sigma(x)}$, because from the actuation scheme (10.115) with $b_j(x)$ being chosen to satisfy (10.116), it follows that $[b_1(x)g_1(x), \ldots, b_m(x)g_m(x)]^T \kappa_2 \neq 0$ for $\forall x \in R^n$ and any κ_2 associated with the failure pattern σ, that is, $\text{span}\{[b_1(x)g_1(x), \ldots, b_m(x)g_m(x)]^T \kappa_2\}$ is a one-dimensional, nonsingular distribution [60]. For the defined $T_c(x)$, in addition to (10.124), $T_{z,\sigma}(x)$ can be chosen to made $T_\sigma(x)$ a diffeomorphism [60]. With such a diffeomorphism $T_\sigma(x)$, the resulting zero dynamics system is

$$\dot{\bar{\eta}} = \psi_\sigma(\xi, \bar{\eta}) + \Psi_\sigma(\xi, \bar{\eta})\theta + \Delta_\sigma(\xi, \bar{\eta})\kappa_1 \tag{10.125}$$

where

$$\psi_\sigma(\xi, \bar{\eta}) = \frac{\partial T_{z,\sigma}(x)}{\partial x} f_0(x)$$

$$\Psi_\sigma(\xi, \bar{\eta}) = \frac{\partial T_{z,\sigma}(x)}{\partial x} F(x)$$

$$\Delta_\sigma(\xi, \bar{\eta}) = \frac{\partial T_{z,\sigma}(x)}{\partial x} g(x) \tag{10.126}$$

which are dependent on the failure pattern σ. The zero dynamics system (10.125), which characterizes the internal property of the system zero dynamics without being involved with the control signal v_0, is equivalent to the zero dynamics system (10.121) in the sense that there is an invertible transformation from $[\xi, \bar{\eta}]^T$ to $[\xi, \eta]^T$: $[\xi, \eta]^T = T(T_\sigma^{-1}(\xi, \bar{\eta}))$, where both $T(x)$ and $T_\sigma(x)$ are well-defined diffeomorphisms.   □

Stable Zero Dynamics Assumption. To make $y(t)$ track the reference signal $y_m(t)$, we need the following assumption for closed-loop stability:

(A10.5b) The system is minimum phase for all possible failure patterns, that is, the zero dynamics (10.125) are *input-to-state stable* (ISS) with respect to ξ as the input for all possible σ (and κ_1).

Adaptive Controller. We design the adaptive controller as

$$v_j = b_j(x)v_0, \ j = 1, 2, \ldots, m$$

$$v_0 = \frac{1}{\bar{\beta}^T \hat{\kappa}_2}(\alpha_\rho - \varphi_0 - \beta^T \hat{\kappa}_1) \qquad (10.127)$$

with the adaptive laws

$$\dot{\hat{\theta}} = \tau_\rho$$

$$\dot{\hat{\kappa}}_1 = z_\rho \Gamma_1 \beta$$

$$\dot{\hat{\kappa}}_2 = \text{proj}_{[\text{sign}[\bar{\beta}^T \kappa_2]\bar{\beta}^T \hat{\kappa}_2, \lambda]}(z_\rho \Gamma_2 \bar{\beta} v_0) \qquad (10.128)$$

where $\hat{\theta}$, $\hat{\kappa}_1$, and $\hat{\kappa}_2$ are the estimates of θ, κ_1, and κ_2, respectively,

$$\text{proj}_{[x^T y, \lambda]}(f(t)) = \begin{cases} f(t) & \text{if } x^T y > \lambda \\ f(t) & \text{if } x^T y = \lambda \text{ and } x^T \dot{y} \geq 0 \\ 0 & \text{if } x^T y = \lambda \text{ and } x^T \dot{y} < 0 \end{cases} \qquad (10.129)$$

with $\lambda > 0$ being a known constant,[1] and α_ρ, τ_ρ, and z_ρ are derived from the backstepping design procedures:

$$z_1 = \xi_1 - y_m$$

$$\alpha_1 = -c_1 z_1 - \varphi_1^T \hat{\theta} + \dot{y}_m$$

$$\tau_1 = z_1 \Gamma \varphi_1$$

$$z_i = \xi_i - \alpha_{i-1}(\xi_1, \ldots, \xi_{i-1}, \hat{\theta}, y_m, \dot{y}_m, \ldots, y_m^{(i-1)}), \ i = 2, 3, \ldots, \rho$$

$$\alpha_i = -c_i z_i - z_{i-1} - \varphi_i^T \hat{\theta}$$

$$+ \sum_{k=1}^{i-1} \frac{\partial \alpha_{i-1}}{\partial \xi_k}(\xi_{k+1} + \varphi_k^T \hat{\theta}) + \frac{\partial \alpha_{i-1}}{\partial \hat{\theta}} \tau_{i-1}$$

$$+ \left(\sum_{k=1}^{i-1} z_{k+1} \frac{\partial \alpha_k}{\partial \hat{\theta}} \right) \Gamma \left(\varphi_i - \sum_{k=1}^{i-1} \frac{\partial \alpha_{i-1}}{\partial \xi_k} \varphi_k \right)$$

---

[1] With Assumption (A10.6), (10.115), and (10.116), $|\bar{\beta}^T(\xi, \eta)\kappa_2| \geq \lambda$ is ensured and the sign of $\bar{\beta}^T \kappa_2$ is determined; for example, with the choice of $b_j(x) = \text{sign}|\mu_j|[L_{g_j(x)}L_{f_0(x)}^{\nu-1}h(x)]^{-1}$, $j = 1, 2, \ldots, m$, it follows that $|\bar{\beta}^T(\xi, \eta)\kappa_2| \geq \min_j\{|\mu_j|\}$ and $\text{sign}[\bar{\beta}^T \kappa_2] = 1$.

$$+ \sum_{k=1}^{i} \frac{\partial \alpha_{i-1}}{\partial y_m^{(k-1)}} y_m^{(k)}, \ i = 2, 3, \ldots, \rho$$

$$\tau_i = \tau_{i-1} + z_i \Gamma \left(\varphi_i - \sum_{k=1}^{i-1} \frac{\partial \alpha_{i-1}}{\partial x_k} \varphi_k \right), \ i = 2, 3, \ldots, \rho \quad (10.130)$$

where $\Gamma = \Gamma^T > 0$ and $\Gamma_i = \Gamma_i^T > 0$, $i = 1, 2$, are the adaptation gain matrices, $c_i > 0$, $i = 1, 2, \ldots, \rho$, are the design constants.

Remark 10.2.5. In this design, the adaptive compensation scheme is based on the actuation scheme (10.115) so that the control inputs for all the remaining actuators are related to each other. However, in Section 10.2.1, with a more restrictive structure condition (as Assumption (A10.4a) implies Assumption (A10.4b)), the design for each actuator is independently specified.   □

The adaptive control scheme (10.127)–(10.130) has the desired properties.

Theorem 10.2.2. *With the ISS zero dynamics (A10.5b), the controller (10.127) with the adaptive laws (10.128) ensures the boundedness of the closed loop signals and the asymptotic output tracking:* $\lim_{t \to \infty} (y(t) - y_m(t)) = 0$.

Proof: We analyze this adaptive scheme using the positive definite function

$$V(z, \tilde{\theta}, \tilde{\kappa}_1, \tilde{\kappa}_2) = \frac{1}{2} z^T(t) z(t) + \frac{1}{2} \tilde{\theta}^T(t) \Gamma^{-1} \tilde{\theta}(t)$$

$$+ \frac{1}{2} \tilde{\kappa}_1^T(t) \Gamma_1^{-1} \tilde{\kappa}_1(t) + \frac{1}{2} \tilde{\kappa}_2^T(t) \Gamma_2^{-1} \tilde{\kappa}_2(t) \quad (10.131)$$

where $\tilde{\theta}(t) = \hat{\theta}(t) - \theta$, $\tilde{\kappa}_1(t) = \hat{\kappa}_1(t) - \kappa_1$ and $\tilde{\kappa}_2(t) = \hat{\kappa}_2(t) - \kappa_2$. The time derivative of $V(z, \tilde{\theta}, \tilde{\kappa}_1, \tilde{\kappa}_2)$ along the trajectories of (10.122) is

$$\dot{V} = -\sum_{i=1}^{\rho} c_i z_i^2(t) - \delta(t) \leq -\sum_{i=1}^{\rho} c_i z_i^2(t) \leq 0, \ t \neq t_j, \ j = 1, 2, \ldots, q \quad (10.132)$$

where $\delta(t) = \tilde{\kappa}_2^T \left(z_\rho \bar{\beta} v_0 - \text{proj}_{[\text{sign}[\bar{\beta}^T \kappa_2] \bar{\beta}^T \hat{\kappa}_2, \lambda]} (z_\rho \bar{\beta} v_0) \right) \geq 0$ introduced by the projection, and t_j, $j = 1, 2, \ldots, q$, $1 \leq q \leq m - 1$, are the time instants when one or more than one of the actuators possibly fails. At $t = t_j$, the parameters κ_1 and κ_2 change their values, causing a finite jumping in them and $V(z, \tilde{\theta}, \tilde{\kappa}_1, \tilde{\kappa}_2)$ as well. It follows from (10.132) that $V(z, \tilde{\theta}, \tilde{\kappa}_1, \tilde{\kappa}_2) \in L^\infty$ and $z(t) \in L^2$. Then it can be concluded that $z(t)$, $\hat{\theta}(t)$, $\hat{\kappa}_1(t)$, and $\hat{\kappa}_2(t)$ are bounded. Since $z_1 = \xi_1 - y_m$, ξ_1 is bounded and hence α_1 is also bounded. With $z_2 = \xi_2 - \alpha_1$, $\xi_2 \in L^\infty$ and in turn $\alpha_2 \in L^\infty$. Continuing in the same way, we see that $\xi(t)$ is bounded. Furthermore, $\bar{\eta}(t)$ is bounded because of the ISS of the zero dynamics (10.125). In view of Remark 10.2.4, we have that $\eta(t)$ is also bounded, which implies that $\dot{z}$ is bounded so that $z(t)$ goes to zero when t goes to infinity. With $z_1 = y - y_m$, the asymptotic tracking $\lim_{t\to\infty}(y(t) - y_m(t)) = 0$ is achieved. ∇

10.2.3 Robustness

We now show that actuator failures cannot cause parameter drifting in our actuator failure compensation design. With the adaptive controller (10.107) and adaptive laws (10.103) and (10.110), the closed-loop system is

$$\dot{z} = A_{cl}(z,\eta,\hat{\theta})z + \Upsilon(z,\eta,\hat{\theta})\tilde{\theta} + e_\rho \sum_{j=1}^{m} \mu_j \bar{\omega}^T(z,\eta,\hat{\theta})\varsigma_j(t)$$

$$\dot{\eta} = \psi(z + \chi(z,\hat{\theta},y_m,\dot{y}_m,\ldots,y_m^{(\rho-1)}),\eta)$$
$$+\Psi(z + \chi(z,\hat{\theta},y_m,\dot{y}_m,\ldots,y_m^{(\rho-1)}),\eta)\theta$$

$$\dot{\hat{\theta}} = \Gamma\Upsilon(z,\eta,\hat{\theta})z$$

$$\dot{\tilde{\kappa}}_j = -\text{sign}[\mu_j]z_\rho\Gamma_j\bar{\omega}(z,\eta,\hat{\theta}), \quad j = 1,2,\ldots,m \qquad (10.133)$$

where

$$A_{cl}(z,\eta,\hat{\theta}) =$$

$$\begin{bmatrix}
-c_1 & 1 & 0 & \cdots & 0 \\
-1 & -c_2 & 1 - \frac{\partial\alpha_1}{\partial\hat{\theta}}\Gamma v_3 & \cdots & -\frac{\partial\alpha_1}{\partial\hat{\theta}}\Gamma v_\rho \\
0 & -1 + \frac{\partial\alpha_1}{\partial\hat{\theta}}\Gamma v_3 & \cdots & \cdots & -\frac{\partial\alpha_2}{\partial\hat{\theta}}\Gamma v_\rho \\
\vdots & \vdots & \ddots & \ddots & \vdots \\
0 & \frac{\partial\alpha_1}{\partial\hat{\theta}}\Gamma v_{\rho-1} & \cdots & \cdots & 1 - \frac{\partial\alpha_{\rho-2}}{\partial\hat{\theta}}\Gamma v_\rho \\
0 & \frac{\partial\alpha_1}{\partial\hat{\theta}}\Gamma v_\rho & \cdots & -1 + \frac{\partial\alpha_{\rho-2}}{\partial\hat{\theta}}\Gamma v_\rho & -c_n
\end{bmatrix}$$

$$\Upsilon(z,\eta,\hat{\theta}) = [v_1, v_2, \ldots, v_\rho] \qquad (10.134)$$

with $v_i = \varphi_i - \sum_{k=1}^{i-1}(\partial\alpha_{i-1}/\partial\xi_k)\varphi_k$, $i = 1,2,\ldots,\rho$, $e_\rho = [0,0,\ldots,1]^T \in R^\rho$, and

$$\chi(z,\hat{\theta},y_m,\dot{y}_m,\ldots,y_m^{(\rho-1)}) = [y_m,\alpha_1,\alpha_2,\ldots,\alpha_\rho]^T$$
$$\bar{\omega}(z,\eta,\hat{\theta}) = \omega(t) = [v_0,\beta^T]^T \qquad (10.135)$$

$\varsigma_j(t) = (1 - \sigma_j)\tilde{\kappa}_j(t)$ with σ_j defined in (10.90) and $\tilde{\kappa}_j$ defined in (10.108). It follows that $\varsigma_j(t)$, $j = 1,2,\ldots,m$, are some piecewise continuous signals that can be described as

$$\varsigma_j(t) = \tilde{\kappa}_j(t), \quad j \neq j_1, j_2, \ldots, j_p, \quad \forall t \geq 0 \qquad (10.136)$$

and for p failed actuators u_j, $j = j_1, j_2, \ldots, j_p$,

$$\varsigma_j(t) \;=\; \tilde{\kappa}_j(t), \text{ for } 0 \le t < t_j$$
$$\varsigma_j(t) \;=\; 0, \text{ for } t \ge t_j \qquad (10.137)$$

where t_j is the time instant when the jth actuator is failed. Notice that $\varsigma_j(t)$ changes abruptly at each failure time instant (if the corresponding jth actuator is still alive, κ_j changes its value to satisfy the matching equations (10.106) in the new situation, and if jth actuator is failed at that time, $\varsigma_j(t)$ becomes 0 because the failed actuator has no effect on the system anymore). Thus the disturbance caused by actuator failures only exists in $\varsigma_j(t)$, $j = 1, 2, \ldots, m$, as some jumping with finite values at the failure time instants, which results in some signal discontinuity.

The effect of the disturbance caused by the actuator failures is to drive the signals $\varsigma_j(t)$, $j = 1, 2, \ldots, m$, to some new values as the new initial conditions of the system at the corresponding failure time instants, which is considered in the stability proof. By choosing the Lyapunov function $V = V_{\sigma_{i-1}}$ for each time interval (t_{i-1}, t_i) as defined in (10.111), we prove that the signals $z(t)$, $\hat{\theta}(t)$, and $\hat{\kappa}_j(t)$ are bounded on each time interval (t_{i-1}, t_i). It should be mentioned that from the Lyapunov function, we first prove that $\hat{\kappa}_j(t)$, $j \ne j_1, j_2, \ldots, j_p$, are bounded and then conclude $\hat{\kappa}_j(t)$, $j = j_1, j_2, \ldots, j_p$, are also bounded from the choice of adaptive laws. With Assumption (A10.5a) to ensure the minimum phase property of the zero dynamics, it follows that $\eta(t)$ is also bounded. Therefore, the proof of closed-loop stability is completed without any parameter drifting in the system.

The same conclusion holds for the adaptive design in Section 10.2.2.

10.2.4 Longitudinal Control of a Twin Otter Aircraft

In this section, the nonlinear longitudinal dynamics of a twin otter aircraft is used for our actuator failure compensation control study, with system modeling, control design, stability analysis, and performance simulation results.

Longitudinal Model. The longitudinal motion dynamics of the twin otter aircraft [86] can be described as

$$\dot{V} \;=\; \frac{F_x \cos(\alpha) + F_z \sin(\alpha)}{m}$$
$$\dot{\alpha} \;=\; q + \frac{-F_x \sin(\alpha) + F_z \cos(\alpha)}{mV}$$
$$\dot{\theta} \;=\; q$$
$$\dot{q} \;=\; \frac{M}{I_y} \qquad (10.138)$$

where V is the velocity, α is the angle of attack, θ is the pitch angle, and q is the pitch rate, m is the mass, I_y is the moment of inertia, and

$$
\begin{aligned}
F_x &= \bar{q}SC_x(\alpha, q, u_1, u_2) + T_x - mg\sin(\theta) \\
F_z &= \bar{q}SC_z(\alpha, q, u_1, u_2) + T_z + mg\cos(\theta) \\
M &= \bar{q}cSC_m(\alpha, q, u_1 u_2)
\end{aligned}
\tag{10.139}
$$

for which $\bar{q} = \frac{1}{2}\rho V^2$ is the dynamic pressure, ρ is the air density, S is the wing area, c is the mean chord, and T_x and T_z are the components of thrust along the body x and z. The nonlinear functions C_x, C_z, and C_m are

$$
\begin{aligned}
C_x &= C_{x1}\alpha + C_{x2}\alpha^2 + C_{x3} + C_{x4}(d_1 u_1 + d_2 u_2) \\
C_z &= C_{z1}\alpha + C_{z2}\alpha^2 + C_{z3} + C_{z4}(d_1 u_1 + d_2 u_2) + C_{z5}q \\
C_m &= C_{m1}\alpha + C_{m2}\alpha^2 + C_{m3} + C_{m4}(d_1 u_1 + d_2 u_2) + C_{m5}q
\end{aligned}
\tag{10.140}
$$

where u_1 and u_2 are the elevator angles of an augmented two-piece elevator used as two actuators for our failure compensation study.

Remark 10.2.6. We note that the elevator force in this twin otter aircraft model is linear in the elevator angle u although nonlinear in the velocity V. Since V is almost the same for all elevator segments, the elevator segments act on the system through the same nonlinear function with different parameters, i.e., they are proportional to each other.   □

Choosing V, α, θ, and q as the states x_1, x_2, x_3, and x_4, and considering u_1 and u_2 as the inputs, we rewrite the nonlinear aircraft plant as

$$
\begin{aligned}
\dot{x}_1 &= (c_1^T \varphi_0(x_2)x_1^2 + \varphi_1(x))\cos(x_2) \\
&\quad + (c_2^T \varphi_0(x_2)x_1^2 + \varphi_2(x))\sin(x_2) + d_1 g_1(x)u_1 + d_2 g_1(x)u_2 \\
\dot{x}_2 &= x_4 - \left(c_1^T \varphi_0(x_2)x_1 + \varphi_1(x)\frac{1}{x_1}\right)\sin(x_2) \\
&\quad + \left(c_2^T \varphi_0(x_2)x_1 + \varphi_2(x)\frac{1}{x_1}\right)\cos(x_2) + d_1 g_2(x)u_1 + d_2 g_2(x)u_2 \\
\dot{x}_3 &= x_4 \\
\dot{x}_4 &= \theta^T \phi(x) + b_1 x_1^2 u_1 + b_2 x_1^2 u_2
\end{aligned}
\tag{10.141}
$$

where

$$
\begin{aligned}
\varphi_0(x_2) &= [x_2, x_2^2, 1]^T \\
\varphi_1(x) &= p_{11} + p_{12}x_4 x_1^2 - p_0 \sin(x_3) \\
\varphi_2(x) &= p_{21} + p_{22}x_4 x_1^2 + p_0 \cos(x_3)
\end{aligned}
$$

$$g_1(x) = a_1 x_1^2 \cos(x_2) + a_2 x_1^2 \sin(x_2)$$
$$g_2(x) = -a_1 x_1 \sin(x_2) + a_2 x_1 \cos(x_2)$$
$$\phi(x) = [x_1^2 x_2, x_1^2 x_2^2, x_1^2, x_1^2 x_4]^T. \tag{10.142}$$

The control objective is to design an adaptive scheme to control the elevator angles such that the pitch angle x_3 tracks a reference signal generated from a reference system even if one piece of the elevator is stuck at an unexpected angle and fails taking action anymore.

Canonical Structure. For aircraft control, x_3 is chosen as the output y. The $[x_3, x_4]^T$ system is already in the parametric-strict-feedback form

$$\dot{x}_3 = x_4$$
$$\dot{x}_4 = \theta^T \phi(x) + b^T x_1^2 u \tag{10.143}$$

with some unknown constant vectors $\theta \in R^4$ and $b = [b_1, b_2]^T$.

The parametric-strict-feedback form of $[x_3, x_4]^T$ indicates that the relative degree condition (A10.4b) is satisfied. Furthermore, it can be seen that the nonlinear actuation functions for u_1 and u_2, i.e., $[d_1 g_1(x), d_1 g_2(x), 0, b_1 x_1^2]^T$ and $[d_2 g_1(x), d_2 g_2(x), 0, b_2 x_1^2]^T$, are parallel (as $b_1/b_2 = d_1/d_2$ follows from (10.140)), which implies that the more restrictive structure condition (A10.4a) is also satisfied so that the corresponding control design may be used.

Zero Dynamics. Since the inputs u_1, u_2 enter the zero dynamics (10.141), we use a change of coordinates $\eta = [\eta_1, \eta_2]^T = [T_1(x), T_2(x)]^T$ such that

$$\frac{\partial T_1}{x_1} g_1(x) + \frac{\partial T_1}{x_2} g_2(x) + \frac{\partial T_1}{x_4} k x_1^2 = 0$$
$$\frac{\partial T_2}{x_1} g_1(x) + \frac{\partial T_2}{x_2} g_2(x) + \frac{\partial T_2}{x_4} k x_1^2 = 0 \tag{10.144}$$

where $k = b_1/d_1 = b_2/d_2$, to transform the zero dynamics into the form of $\dot{\eta} = \psi(\eta, \xi) + \Psi(\eta, \xi)\theta$ independent of input, where $\xi = [x_3, x_4]^T$.

Solving the partial differential equations (10.144), we obtain

$$\eta_1 = x_1 \cos(x_2) - q_1 x_4$$
$$\eta_2 = x_1 \sin(x_2) - q_2 x_4 \tag{10.145}$$

where $q_1 = a_1/k$ and $q_2 = a_2/k$. The Jacobian matrix

$$\frac{\partial T(x)}{\partial x} = \begin{bmatrix} \cos(x_2) & -x_1 \sin(x_2) & 0 & q_1 \\ \sin(x_2) & x_1 \cos(x_2) & 0 & q_2 \\ 0 & 0 & 1 & 0 \\ 0 & 0 & 0 & 1 \end{bmatrix} \tag{10.146}$$

of $T(x) = [T_1(x), T_2(x), x_3, x_4]^T$ is nonsingular for $x_1 \in (0, \infty)$ (as $x_1 > 0$ is the aircraft velocity), which implies that $[\eta, \xi]^T = T(x)$ is a diffeomorphism.

Within the new coordinates, the zero dynamics can be expressed as

$$
\begin{aligned}
\dot{\eta}_1 &= \psi_1(\eta, \xi) + \Psi_1^T(\eta, \xi)\theta = f_1(x, \theta) \\
&= c_1^T \varphi_0(x_2) x_1^2 + \varphi_1(x) - x_1 \sin(x_2) x_4 - q_1 \theta^T \phi(x) \\
\dot{\eta}_2 &= \psi_2(\eta, \xi) + \Psi_2^T(\eta, \xi)\theta = f_2(x, \theta) \\
&= c_2^T \varphi_0(x_2) x_1^2 + \varphi_2(x) + x_1 \cos(x_2) x_4 - q_2 \theta^T \phi(x). \quad (10.147)
\end{aligned}
$$

Input-to-State Stability. For a nonlinear system $\dot{x} = f(t, x, u)$, where the function $f(t, x, u)$ is continuously differentiable and the Jacobian matrices $[\frac{\partial f}{\partial x}]$ and $[\frac{\partial f}{\partial u}]$ are bounded in some neighborhood of the equilibrium point $(x = x_e, u = 0)$, if the unforced system $\dot{x} = f(t, x, 0)$ has a uniformly asymptotically stable equilibrium point $x = x_e$, then the system is locally input-to-state stable (ISS) [67] (see its definition in Section 9.2).

It is now shown that the zero dynamics (10.147) are locally ISS with the real aircraft constants and the parameters obtained in a certain operation condition [86]: $S = 310.02\text{m}^2$, $c = 1.98\text{m}$, and $T_x = 4864\text{N}$, $T_z = 212\text{N}$ with the power maintained at a constant thrust, $\rho = 0.7377\text{kg/m}^3$ at the altitude of 5000m, and for the $0°$ flap setting,

$$
\begin{aligned}
C_{x1} &= 0.39, \ C_{x2} = 2.9099, \ C_{x3} = -0.0758, \ C_{x4} = 0.0961 \\
C_{z1} &= -7.0186, \ C_{z2} = 4.1109, \ C_{z3} = -0.3112 \\
C_{z4} &= -0.2340, \ C_{z5} = -0.1023 \\
C_{m1} &= -0.8789, \ C_{m2} = -3.8520, \ C_{m3} = -0.0108 \\
C_{m4} &= -1.8987, \ C_{m5} = -0.6266. \quad (10.148)
\end{aligned}
$$

Analyzing the unforced zero dynamics

$$
\begin{aligned}
\dot{\eta} &= \psi(\eta, 0) + \Psi(\eta, 0)\theta \\
&= \begin{bmatrix} (\eta_1^2 + \eta_2^2)(c_1^T - q_1 \bar{\theta}^T)\varphi_0(\arctan(\frac{\eta_2}{\eta_1})) + p_{11} \\ (\eta_1^2 + \eta_2^2)(c_2^T - q_2 \bar{\theta}^T)\varphi_0(\arctan(\frac{\eta_2}{\eta_1})) + p_{21} + p_0 \end{bmatrix} \quad (10.149)
\end{aligned}
$$

where $\bar{\theta} = [\theta_1, \theta_2, \theta_3]^T$, we get the equilibrium point of $\dot{\eta} = \psi(\eta, 0) + \Psi(\eta, 0)\theta$: $\eta_e = [71.2401, 1.70089]^T$. Investigate the stability of the unforced zero dynamics at the equilibrium point by checking $(\partial(\psi(\eta, 0) + \Psi(\eta, 0)\theta))/\partial \eta$ at η_e:

$$
\left[\frac{\partial(\psi(\eta, 0) + \Psi(\eta, 0)\theta)}{\partial \eta} \right]\Bigg|_{\eta_e} - \begin{bmatrix} -0.0423159 & 0.138259 \\ -0.229867 & -1.96684 \end{bmatrix}. \quad (10.150)
$$

The eigenvalues of the matrix are -1.95018 and -0.0589739, which indicates that the zero dynamics are locally asymptotically stable. It in turn implies that the zero dynamics are locally input-to-state stable in a neighborhood of η_e with ξ in a neighborhood of $[0, 0]^T$ as the input.

Remark 10.2.7. In this example, the system zero dynamics is only locally input-to-state stable. For a nonadaptive design with known system parameters and actuator failures, a local initial condition results in a local system trajectory to meet the condition for local input-to-state stability (ISS) of the zero dynamics so that the local stability of the closed-loop system can be ensured. However, in the adaptive control case, a local initial condition may not result in a local system trajectory as an adaptive control design may not ensure asymptotic stability of the closed-loop system. In this case, the condition for local ISS of the zero dynamics may not be satisfied. For the twin otter aircraft model, the boundedness of the states x_3 and x_4 is ensured by the backstepping design, while in theory the boundedness of the states x_1 and x_2 depends on the ISS of the zero dynamics, which may not be guaranteed because the system state may go beyond the local ISS region even if the initial state starts from that region. We note that in our simulation results, the boundedness of x_1 and x_2 is observed.   □

Adaptive Compensation Design. An adaptive compensation scheme for the twin otter nonlinear longitudinal dynamic model is now derived from the design in Section 10.2.1, for output tracking of the pitch angle.

The adaptive controller is

$$v_i(t) = k_{1,i}(t) \frac{\alpha(x, \hat{\theta}, y_m, \dot{y}_m, \ddot{y}_m)}{x_1^2} + k_{2,i}(t), \ i = 1, 2 \tag{10.151}$$

along with the adaptive laws for $\hat{\theta}$, $k_{1,i}(t)$ and $k_{2,i}(t)$

$$\dot{\hat{\theta}} = z_2 \Gamma \phi(x)$$
$$\dot{k}_{1i} = -\text{sign}[b_i] \gamma_{1i} z_2 \alpha, \ i = 1, 2$$
$$\dot{k}_{2i} = -\text{sign}[b_i] \gamma_{2i} z_2 x_1^2, \ i = 1, 2 \tag{10.152}$$

with z_2 and α given by

$$z_1 = x_3 - y_m$$
$$z_2 = x_4 + c_1 z_1 - \dot{y}_m$$
$$\alpha = -c_2 z_2 - z_1 - \phi^T \hat{\theta} - c_1 x_4 + c_1 \dot{y}_m + \ddot{y}_m \tag{10.153}$$

where $\Gamma = \Gamma^T > 0$, $\gamma_{1i} > 0$, $\gamma_{2i} > 0$, and $c_i > 0$, $i = 1, 2$.

Simulation Results. Applying the controller (10.151) with the adaptive laws (10.152) to the nonlinear longitudinal model (10.141), we present the simulation results to illustrate the effectiveness of the adaptive actuator failure compensation design, for both regulation and output tracking (of a signal from a reference system) to demonstrate the desired system performance.

Regulation. For the simulation of regulation, the initials of state and estimates are $x(0) = [85\ 0\ 0.05\ 0]^T$, $\hat{\theta}(0) = [0\ 0\ -0.01\ 0]^T$, and $k_{1,1}(0) = -1.2$, $k_{1,2}(0) = -0.8$, $k_{2,1}(0) = 0$, $k_{2,2}(0) = -0.05$. The gains in the adaptive laws are chosen as $c_1 = 1.05$, $c_2 = 1.05$, $\Gamma = 0.001I$, $\gamma_{1,i} = 1$, $\gamma_{2,i} = 0.005$, $i = 1, 2$. Figure 10.3 shows the regulation results when one actuator (u_1) fails at 150 sec with an unknown failure value 0.04 (rad).

Output tracking. For the simulation of output tracking, the reference signal $y_m(t)$ to be tracked is generated from the reference system $W(s) = 1/(s^2 + 5s + 6)$ with $r(t) = 0.1\sin(0.05t)$ as the reference input, that is, $y_m(t) = W_m(s)[r](t)$. The initial values and adaptation gains are the same as above. Figure 10.4 shows the tracking results when one actuator (u_1) also fails at 150 sec with an unknown failure value 0.04 (rad).

The results indicate that the closed-loop system is stable in the sense that all signals are bounded and both regulation and asymptotic output tracking are ensured even though one actuator fails during operation.

In summary, in this chapter we have formulated and solved the actuator failure compensation problem for nonlinear parametric-strict-feedback systems with unknown actuator failures without or with zero dynamics. We characterize the system structures with actuator failures and derive matching conditions for actuator failure compensation. Several adaptive compensation schemes have been developed based on desired controllers and parameter update laws, which ensure global stability and asymptotic output tracking. The desired performance of the developed adaptive control schemes applied to an aircraft wing model and a twin otter aircraft model with actuator failures is verified by simulation results.

While in theory, the desired system performance, that is, closed-loop signal boundedness and asymptotic tracking, is achieved, despite the uncertain plant parameters and actuator failures, modifications to the developed adaptive parameter update laws are needed, and they can be derived to ensure system performance robustness with respect to system modeling errors such as unmodeled dynamics and bounded disturbances [55], [95].

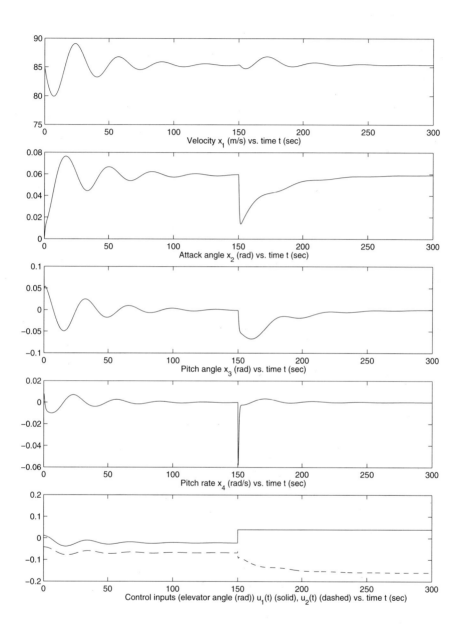

Fig. 10.3. System response of regulation $(r(t) = 0)$.

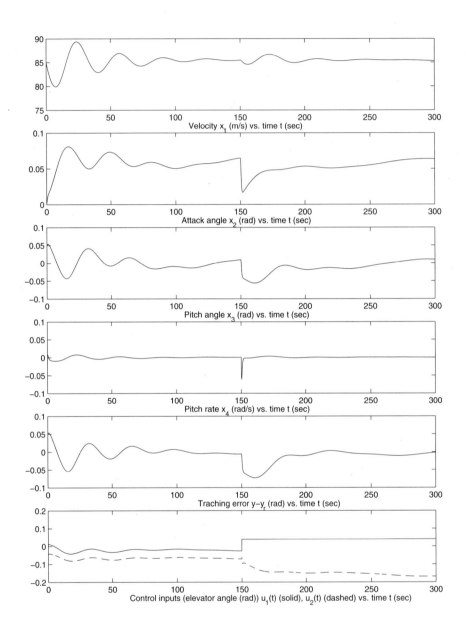

Fig. 10.4. System response for $r(t) = 0.1 \sin(0.05t)$.

Chapter 11

Nonlinear Output Feedback Designs

In this chapter, we address adaptive *output feedback* actuator failure compensation control for nonlinear systems with uncertain actuator failures. In Section 11.1, the actuator failure compensation control problem is formulated for for the class of output-feedback systems. An adaptive state observer for output feedback control is proposed in Section 11.2. Combined with the state observer, an adaptive backstepping controller is developed in Section 11.3, which ensures closed-loop signal boundedness and asymptotic output tracking despite the presence of the uncertain system parameters and actuator failure parameters. In Section 11.4, an extension to systems whose dynamics are state-dependent (beyond the class of output-feedback systems) is demonstrated by a second-order example based on either a reduced-order observer or a full-order observer. Such an observer-based adaptive actuator failure compensation design is illustrated in Section 11.4.4 by simulation results of an application to controlling the angle of attack of a nonlinear hypersonic aircraft model in the presence of elevator segment failures. The developed nonlinear output feedback designs can also be applied to linear systems, which may lead to improved system performance.

11.1 Problem Statement

Consider a nonlinear plant described by

$$\dot{\chi} = f(\chi) + \sum_{j=1}^{m} \left(g_{j,0}(\chi) + \sum_{l=1}^{q} \theta_{j,l} g_{j,l}(\chi) \right) u_j$$

$$y = h(\chi) \tag{11.1}$$

where $\chi \in R^n$ is the unmeasured state vector, $u_j \in R$, $j = 1, 2, \ldots, m$, are input signals subject to actuator failures, $y \in R$ is the output, $f(\chi) \in R^n$, $g_{j,0}(\chi) \in R^n$, $g_{j,l}(\chi) \in R^n$, $l = 1, 2, \ldots, q$, $j = 1, 2, \ldots, m$, and $h(\chi) \in R$ are smooth functions, and $\theta_{j,l} \in R$, $l = 1, 2, \ldots, q$, $j = 1, 2, \ldots, m$, are some unknown constant parameters.

Under certain geometric differential conditions on the system dynamics [71], the plant (11.1) can be transformed into the output-feedback form

$$\dot{x}_1 = x_2 + \varphi_1(y)$$
$$\dot{x}_2 = x_3 + \varphi_2(y)$$
$$\vdots$$
$$\dot{x}_{\rho-1} = x_\rho + \varphi_{\rho-1}(y)$$
$$\dot{x}_\rho = x_{\rho+1} + \varphi_\rho(y) + \sum_{j=1}^{m} b_{n^*,j}\beta_j(y)u_j$$
$$\vdots$$
$$\dot{x}_{n-1} = x_n + \varphi_{n-1}(y) + \sum_{j=1}^{m} b_{1,j}\beta_j(y)u_j$$
$$\dot{x}_n = \varphi_n(y) + \sum_{j=1}^{m} b_{0,j}\beta_j(y)u_j$$
$$y = x_1 \qquad\qquad (11.2)$$

where $b_{r,j}$, $r = 0, 1, \ldots, n^* = n - \rho$, $j = 1, 2, \ldots, m$, are unknown constant parameters, $\varphi_i(y)$, $i = 1, 2, \ldots, n$, and $\beta_j(y)$, $j = 1, 2, \ldots, m$, are known smooth nonlinear functions, via a parameter-independent diffeomorphism $x = T(\chi) = [T_1(\chi), T_2(\chi), \ldots, T_n(\chi)]^T$ that satisfies

$$T_1(\chi) = h(\chi)$$

$$\begin{cases} \frac{\partial T_i(\chi)}{\partial \chi} f(\chi) = T_{i+1}(\chi) + \varphi_i(h(\chi)), \\ \frac{\partial T_i(\chi)}{\partial \chi} g_{j,0}(\chi) = 0, \quad \frac{\partial T_i(\chi)}{\partial \chi} g_{j,l}(\chi) = 0, \end{cases} \quad i = 1, \ldots, \rho - 1$$

$$\begin{cases} \frac{\partial T_i(\chi)}{\partial \chi} f(\chi) = T_{i+1}(\chi) + \varphi_i(h(\chi)), \\ \frac{\partial T_i(\chi)}{\partial \chi} \left(g_{j,0}(\chi) + \sum_{l=1}^{p} \theta_{j,l}g_{j,l}(\chi)\right) = b_{n-i,j}\beta_j(h(\chi)), \end{cases} \quad i = \rho, \ldots, n - 1$$

$$\frac{\partial T_n(\chi)}{\partial \chi} f(\chi) = \varphi_n(h(\chi))$$

$$\frac{\partial T_n(\chi)}{\partial \chi} \left(g_{j,0}(\chi) + \sum_{l=1}^{p} \theta_{j,l}g_{j,l}(\chi)\right) = b_{0,j}\beta_j(h(\chi)) \qquad (11.3)$$

for $l = 1, 2, \ldots, q,$, $j = 1, 2, \ldots, m$. We should note that the more general case where $\phi_i(y)$ contains unknown parameters [71] can be treated by a similar design to that developed in this chapter. Such a case also includes the class of linear time-invariant systems. Thus, nonlinear output feedback designs can also be applied to linear systems, which may lead to improved system performance, as in the case of no actuator failures [70], [71].

For a simple presentation, we consider the actuator failure model (1.2):

$$u_j(t) = \bar{u}_j, \ t \geq t_j, \ j \in \{1, 2, \ldots, m\} \tag{11.4}$$

where the failure value $\bar{u}_j$ and failure time instant t_j are unknown, and so is the failure index j. The more general failure model (1.6) or (1.8) can be dealt with in a similar way. The basic assumption for the actuator failure compensation problem may be restated as: The plant (11.2) is so constructed that for any up to $m - 1$ actuator failures, the remaining actuators can still achieve a desired control objective, when implemented with the knowledge of the plant parameters and failure parameters.

Suppose that p_k actuators fail at a time instant t_k, $k = 1, 2, \ldots, q$, and $t_0 < t_1 < t_2 < \ldots < t_q < \infty$. Obviously, it follows from this basic assumption that $\sum_{k=1}^{q} p_k \leq m - 1$. In other words, at time $t \in (t_k, t_{k+1})$, $k = 0, 1, \ldots, q$, with $t_{q+1} = \infty$, there are $p = \sum_{i=1}^{k} p_i$ failed actuators, that is, $u_j(t) = \bar{u}_j$, $j = j_1, j_2, \ldots, j_p$, $0 \leq p \leq m - 1$, and $u_j(t) = v_j$, $j \neq j_1, j_2, \ldots, j_p$, where $v_j(t)$, $j = 1, 2, \ldots, m$, are applied control inputs to be designed. Then the output-feedback plant (11.2) can be rewritten as

$$\begin{aligned}
\dot{x}_1 &= x_2 + \varphi_1(y) \\
\dot{x}_2 &= x_3 + \varphi_2(y) \\
&\vdots \\
\dot{x}_{\rho-1} &= x_\rho + \varphi_{\rho-1}(y) \\
\dot{x}_\rho &= x_{\rho+1} + \varphi_\rho(y) + \sum_{j=j_1,\ldots,j_p} b_{n^*,j}\beta_j(y)\bar{u}_j + \sum_{j \neq j_1,\ldots,j_p} b_{n^*,j}\beta_j(y)v_j \\
&\vdots \\
\dot{x}_{n-1} &= x_n + \varphi_{n-1}(y) + \sum_{j=j_1,\ldots,j_p} b_{1,j}\beta_j(y)\bar{u}_j + \sum_{j \neq j_1,\ldots,j_p} b_{1,j}\beta_j(y)v_j \\
\dot{x}_n &= \varphi_n(y) + \sum_{j=j_1,\ldots,j_p} b_{0,j}\beta_j(y)\bar{u}_j + \sum_{j \neq j_1,\ldots,j_p} b_{0,j}\beta_j(y)v_j \\
y &= x_1.
\end{aligned} \tag{11.5}$$

The control objective is to design an output feedback control scheme for the plant (11.5) with p, the number of failed actuators, changing at time instants t_k, $k = 1, 2, \ldots, q$, such that the plant output $y(t)$ asymptotically tracks a prescribed reference signal $y_m(t)$ (the ρth derivative of $y_m(t)$, $y_m^{(\rho)}(t)$, is piecewise continuous) and the closed-loop signals are all bounded, despite the presence of unknown actuator failures and unknown plant parameters. It is clear that the key task is to design adaptive feedback control laws for $v_j(t)$, $j = 1, 2, \ldots, m$, without knowing which of these $v_j(t)$ will have action on the plant dynamics, as there are arbitrary p failed actuators: $u_j(t) = \bar{u}_j$, $j = j_1, \ldots, j_p$, $u_j(t) = v_j$, $j \neq j_1, \ldots, j_p$, $0 \leq p \leq m - 1$, that is, for any $\{j_1, \ldots, j_p\} \subset \{1, 2, \ldots, m\}$, with any $p \in \{0, 1, 2, \ldots, m - 1\}$.

11.2 Adaptive Compensation Control

For the plant (11.5) in the output-feedback form with p failed actuators, at each time interval (t_k, t_{k+1}), $k = 0, 1, \ldots, q$, an adaptive control scheme will be developed in this section and the control scheme will be proved to achieve the control objective proposed above for $\forall t > t_0$ in the next section.

Since the failure pattern $u_j(t) = \bar{u}_j, j = j_1, \ldots, j_p, 0 \leq p \leq m - 1$, is assumed to be unknown in this problem, a desirable adaptive control design is expected to achieve the control objective for any possible failure pattern. For a fixed failure pattern, there is a set of failed actuators, while there is another set of working actuators. For this set of working actuators, there is a resulting zero dynamics pattern for the plant (11.5). For closed-loop stabilization and tracking, the zero dynamics system of the plant (11.5) needs to be stable. As such zero dynamics depend on the pattern of working actuators whose number may range from 1 to m, which leads to many possible characterizations of required stable zero dynamics conditions related to different possible control designs, we choose to specify the following one with which a stable adaptive control scheme can be developed to achieve the control objective.

For the plant (11.2), we assume that

(A11.1) The polynomials $\sum_{j \neq j_1, \ldots, j_p} \text{sign}[b_{n^*,j}] B_j(s)$ are stable, $\forall$ $\{j_1, \ldots, j_p\} \subset \{1, 2, \ldots, m\}$, $\forall p \in \{0, 1, 2, \ldots, m - 1\}$, where

$$B_j(s) = b_{n^*,j} s^{n^*} + b_{n^*-1,j} s^{n^*-1} + \cdots$$
$$+ b_{1,j} s + b_{0,j}, \ j = 1, 2, \ldots, m. \tag{11.6}$$

Remark 11.2.1. Assumption (A11.1) implies that the plant (11.2) is minimum phase for all possible actuator failure cases under an actuation scheme (see (11.8) ahead). The plant (11.5) has relative degree ρ, and under the actuation scheme (11.8), it has a linear zero dynamics system

$$\dot{\eta} = A_z \eta + \phi(y) \tag{11.7}$$

where the eigenvalues of $A_z \in R^{n^* \times n^*}$ are the roots of the polynomial $\sum_{j \neq j_1, \ldots, j_p} \text{sign}[b_{n^*,j}] B_j(s)$, which is stable from Assumption (A11.1). Note that the zero dynamics depend on the actuator failures, that is, A_z is determined by the actuator failure pattern, while $\phi(y)$ depends on both the failure pattern and failure values.   □

For an adaptive control design, the following assumption is needed:

(A11.2) The sign of $b_{n^*,j}$ is known, $j = 1, 2, \ldots, m$.

To develop a solution to this adaptive actuator failure compensation problem, we apply the proportional-actuation scheme

$$v_j = \text{sign}[b_{n^*,j}] \frac{1}{\beta_j(y)} v_0, \quad j = 1, 2, \ldots, m \tag{11.8}$$

where v_0 is a control signal from a backstepping design in Section 11.2.2.

To express the plant (11.5) with the actuation scheme (11.8), we define

$$x = [x_1, x_2, \ldots, x_n]^T$$
$$\varphi(y) = [\varphi_1(y), \varphi_2(y), \ldots, \varphi_n(y)]^T \tag{11.9}$$
$$k_{1,r} = \sum_{j \neq j_1, \ldots, j_p} \text{sign}[b_{n^*,j}] b_{r,j}, \quad r = 0, 1, \ldots, n^*$$
$$k_{2,rj} = b_{r,j} \bar{u}_j, \quad r = 0, 1, \ldots, n^*, \ j = j_1, \ldots, j_p$$
$$k_{2,rj} = 0, \quad r = 0, 1, \ldots, n^*, \ j \neq j_1, \ldots, j_p \tag{11.10}$$

and rewrite the plant (11.5) in a compact form as

$$\dot{x} = Ax + \varphi(y) + \sum_{r=0}^{n^*} e_{n-r} \sum_{j=1}^{m} k_{2,rj} \beta_j(y) + \sum_{r=0}^{n^*} e_{n-r} k_{1,r} v_0$$
$$y = c^T x \tag{11.11}$$

where e_i is the ith coordinate vector in R^n, and

$$A = \begin{bmatrix} 0 & 1 & 0 & \cdots & 0 & 0 \\ 0 & 0 & 1 & \cdots & 0 & 0 \\ & & \cdots & \cdots & & \\ 0 & 0 & \cdots & 0 & 0 & 1 \\ 0 & 0 & \cdots & 0 & 0 & 0 \end{bmatrix} \in R^{n \times n}, \quad c = \begin{bmatrix} 1 \\ 0 \\ \vdots \\ 0 \\ 0 \end{bmatrix} \in R^{n \times 1}. \tag{11.12}$$

It follows from Assumption (A11.1) that $k_{1,n^*} s^{n^*} + k_{1,n^*-1} s^{n^*-1} + \cdots + k_{1,1} s + k_{1,0}$ is a stable polynomial and, in addition, from (11.8), that $k_{1,n^*} > 0$.

11.2.1 State Observer

Since the states of the plant (11.11) are not available for feedback control, an observer is needed to provide the estimates of the unavailable state variables.

To develop a state observer, similar to that in [71], we choose a vector $l \in R^n$ such that $A_o = A - lc^T$ is stable, and define the filters

$$\dot{\xi} = A_o \xi + ly + \varphi(y)$$

$$\dot{\zeta}_{rj} = A_o \zeta_{rj} + e_{n-r} \beta_j(y), \ 0 \le r \le n^*, \ 1 \le j \le m$$

$$\dot{\mu}_r = A_o \mu_r + e_{n-r} v_0, \ 0 \le r \le n^*. \tag{11.13}$$

With the knowledge of $k_{1,r}$ and $k_{2,rj}$, $j = 1, 2, \ldots, m$, $r = 0, 1, \ldots, n^*$, the nominal state estimate vector is

$$\bar{x} = \xi + \sum_{r=0}^{n^*} \sum_{j=1}^{m} k_{2,rj} \zeta_{rj} + \sum_{r=0}^{n^*} k_{1,r} \mu_{rj}. \tag{11.14}$$

With $\epsilon = x - \bar{x}$, it follows from (11.11) and (11.13) that

$$\dot{\epsilon} = A_o \epsilon, \ \lim_{t \to \infty} \epsilon(t) = 0 \text{ exponentially.} \tag{11.15}$$

Denoting $\hat{k}_{1,r}$ and $\hat{k}_{2,rj}$ as the adaptive estimates of $k_{1,r}$ and $k_{2,rj}$, $j = 1, 2, \ldots, m$, $r = 0, 1, \ldots, n^*$, we construct an adaptive estimate of x as

$$\hat{x} = \xi + \sum_{r=0}^{n^*} \sum_{j=1}^{m} \hat{k}_{2,rj} \zeta_{rj} + \sum_{r=0}^{n^*} \hat{k}_{1,r} \mu_{rj} \tag{11.16}$$

for which the adaptive update laws for $\hat{k}_{1,r}$ and $\hat{k}_{2,rj}$ are to be developed together with a failure compensation control law for $v_0(t)$ in (11.8) to ensure closed-loop stability and output tracking.

11.2.2 Backstepping Design

The backstepping technique [71] is now applied to derive a stable adaptive control scheme for the plant (11.11), with a design procedure of ρ steps.

Define the unknown constant vectors

$$k_1 = [k_{1,0}, k_{1,1}, \ldots, k_{1,n^*-1}]^T$$

$$k_2 = [k_{2,01}, \ldots, k_{2,0m}, k_{2,11}, \ldots, k_{2,1m}, \ldots, k_{2,n^*m}]^T \tag{11.17}$$

and the measured vector signals

$$\omega_i = [\mu_{0,i}, \mu_{1,i}, \ldots, \mu_{n^*-1,i}]^T, \ i = 1, 2, \ldots, n$$

$$\varepsilon_i = [\zeta_{01,i}, \ldots, \zeta_{0m,i}, \zeta_{11,i}, \ldots, \zeta_{1m,i}, \ldots, \zeta_{n^*m,i}]^T, \ i = 1, 2, \ldots, n \quad (11.18)$$

where $\mu_{r,i}$ and $\zeta_{rj,i}$, $i = 1, 2, \ldots, n$, are the ith variable of μ_r and ζ_{rj}, $r = 0, 1, \ldots, n^* - 1$, $j = 1, 2, \ldots, m$.

A backstepping based adaptive actuator failure compensation control design for the plant (11.11) consists of the following ρ steps:

Step 1: Define the output tracking error

$$z_1 = y - y_m \quad (11.19)$$

where y_m is a reference signal with the piecewise continuous ρth derivative $y_m^{(\rho)}$. It follows from (11.11)–(11.14) that

$$\begin{aligned}
\dot{z}_1 &= \epsilon_2 + \bar{x}_2 + \varphi_1(y) - \dot{y}_m \\
&= \epsilon_2 + k_{1,n^*}\mu_{n^*,2} + \omega_2^T k_1 + \varepsilon_2^T k_2 + \xi_2 + \varphi_1(y) - \dot{y}_m. \quad (11.20)
\end{aligned}$$

Choose the auxiliary error signal

$$z_2 = \mu_{n^*,2} - \hat{\kappa}\dot{y}_m - \alpha_1 \quad (11.21)$$

where $\hat{\kappa}$ is an estimate of $\kappa = 1/k_{1,n^*}$, and

$$\begin{aligned}
\alpha_1 &= \hat{\kappa}\bar{\alpha}_1 \\
\bar{\alpha}_1 &= -c_1 z_1 - d_1 z_1 - \omega_2^T \hat{k}_1 - \varepsilon_2^T \hat{k}_2 - \xi_2 - \varphi_1(y). \quad (11.22)
\end{aligned}$$

Substituting (11.21) into (11.20) results in

$$\begin{aligned}
\dot{z}_1 &= -c_1 z_1 - d_1 z_1 + \epsilon_2 + \hat{k}_{1,n^*} z_2 + \hat{k}_{1,n^*}(\mu_{n^*,2} - \hat{\kappa}\dot{y}_m - \alpha_1) \\
&\quad + \omega_2^T \tilde{k}_1 + \varepsilon_2^T \tilde{k}_2 - k_{1,n^*}\tilde{\kappa}(\dot{y}_m + \bar{\alpha}_1) \quad (11.23)
\end{aligned}$$

where $\tilde{k}_{1,n^*} = k_{1,n^*} - \hat{k}_{1,n^*}$, $\tilde{k}_1 = k_1 - \hat{k}_1$, $\tilde{k}_2 = k_2 - \hat{k}_2$, $\tilde{\kappa} = \kappa - \hat{\kappa}$, and c_1, d_1 are some positive constant design parameters.

The time derivative of the partial Lyapunov candidate function

$$V_1 = \frac{1}{2}z_1^2 + \frac{k_{1,n^*}}{2\lambda_1}\tilde{\kappa}^2 \quad (11.24)$$

where $\lambda_1 > 0$ is a chosen constant gain, and k_{1,n^*} is positive due to Assumption (A11.1), is derived as

$$\begin{aligned}
\dot{V}_1 &= -c_1 z_1^2 - d_1 z_1^2 + z_1 \epsilon_2 + \hat{k}_{1,n^*} z_1 z_2 + \upsilon_1 \tilde{k}_{1,n^*} + \tau_1^T \tilde{k}_1 + \upsilon_1^T \tilde{k}_2 \\
&\quad - z_1 k_{1,n^*}(\dot{y}_m + \alpha_1)\tilde{\kappa} - \frac{k_{1,n^*}}{\lambda_1}\tilde{\kappa}\dot{\hat{\kappa}} \\
&= -c_1 z_1^2 - d_1 z_1^2 + z_1 \epsilon_2 + \hat{k}_{1,n^*} z_1 z_2 + \upsilon_1 \tilde{k}_{1,n^*} + \tau_1^T \tilde{k}_1 + \upsilon_1^T \tilde{k}_2 \quad (11.25)
\end{aligned}$$

with the choice of the adaptive law for $\hat{\kappa}$:

$$\dot{\hat{\kappa}} = -\lambda_1 z_1 (\dot{y}_m + \bar{\alpha}_1) \qquad (11.26)$$

where υ_1, τ_1 and ν_1, which are the first tuning functions for $\hat{k}_{1,n^*}$, $\hat{k}_1$, and $\hat{k}_2$, respectively, are given as

$$\upsilon_1 = z_1 (\mu_{n^*,2} - \hat{\kappa}\dot{y}_m - \alpha_1)$$
$$\tau_1 = z_1 \omega_2$$
$$\nu_1 = z_1 \varepsilon_2. \qquad (11.27)$$

Step $i = 2$: With z_2 defined in (11.21), the time derivative of z_2 is given by

$$
\begin{aligned}
\dot{z}_2 &= \dot{\mu}_{n^*,2} - \dot{\hat{\kappa}}\ddot{y}_m - \hat{\kappa}\dot{\ddot{y}}_m - \dot{\alpha}_{i-1} \\
&= \mu_{n^*,3} - l_2 \mu_{n^*,1} - \hat{\kappa}\ddot{y}_m - \dot{\hat{\kappa}}\dot{y}_m - \frac{\partial \alpha_1}{\partial \hat{\kappa}}\dot{\hat{\kappa}} \\
&\quad - \frac{\partial \alpha_1}{\partial y}(\epsilon_2 + k_{1,n^*}\mu_{n^*,2} + \omega_2^T k_1 + \varepsilon_2^T k_2 + \xi_2 + \varphi_1(y)) \\
&\quad - \frac{\partial \alpha_1}{\partial \xi_2}(\xi_3 - l_2 \xi_1 + l_2 y + \varphi_2(y)) - \frac{\partial \alpha_1}{\partial \varepsilon_2}(\varepsilon_3 - l_2 \varepsilon_1) \\
&\quad - \frac{\partial \alpha_1}{\partial \omega_2}(\omega_3 - l_2 \omega_1) - \frac{\partial \alpha_1}{\partial \hat{k}_1}\dot{\hat{k}}_1 - \frac{\partial \alpha_1}{\partial \hat{k}_2}\dot{\hat{k}}_2. \qquad (11.28)
\end{aligned}
$$

Choose the stabilizing function α_2 as

$$
\begin{aligned}
\alpha_2 &= -\hat{k}_{1,n^*} z_1 - c_2 z_2 - d_2 \left(\frac{\partial \alpha_1}{\partial y}\right)^2 z_2 + l_2 \mu_{n^*,1} + \left(\dot{y}_m + \frac{\partial \alpha_1}{\partial \hat{\kappa}}\right)\dot{\hat{\kappa}} \\
&\quad + \frac{\partial \alpha_1}{\partial y}(\hat{k}_{1,n^*}\mu_{n^*,2} + \omega_2^T \hat{k}_1 + \varepsilon_2^T \hat{k}_2 + \xi_2 + \varphi_1(y)) \\
&\quad + \frac{\partial \alpha_1}{\partial \xi_2}(\xi_3 - l_2 \xi_1 + l_2 y + \varphi_2(y)) + \frac{\partial \alpha_1}{\partial \varepsilon_2}(\varepsilon_3 - l_2 \varepsilon_1) \\
&\quad + \frac{\partial \alpha_1}{\partial \omega_2}(\omega_3 - l_2 \omega_1) + \frac{\partial \alpha_1}{\partial \hat{k}_{1,n^*}}\lambda_2 \upsilon_2 + \frac{\partial \alpha_1}{\partial \hat{k}_1}\Gamma_1 \tau_2 + \frac{\partial \alpha_1}{\partial \hat{k}_2}\Gamma_2 \nu_2 \qquad (11.29)
\end{aligned}
$$

where c_2 and d_2 are positive constants to be chosen, and the tuning functions, υ_2, τ_2, and ν_2, are designed as

$$\upsilon_2 = \upsilon_1 - \frac{\partial \alpha_1}{\partial y}\mu_{n^*,2} z_2$$
$$\tau_2 = \tau_1 - \frac{\partial \alpha_1}{\partial y}\omega_2 z_2$$
$$\nu_2 = \nu_1 - \frac{\partial \alpha_1}{\partial y}\varepsilon_2 z_2. \qquad (11.30)$$

With the definition

$$z_3 = \mu_{n^*,3} - \hat{\kappa} y_m^{(3)} - \alpha_2 \tag{11.31}$$

we can rewrite (11.28) as

$$
\begin{aligned}
\dot{z}_2 = {} & -\hat{k}_{1,n^*} z_1 - c_2 z_2 - d_2 \left(\frac{\partial \alpha_1}{\partial y}\right)^2 z_2 + z_3 - \frac{\partial \alpha_1}{\partial y} \epsilon_2 \\
& -\frac{\partial \alpha_1}{\partial y} \mu_{n^*,2} \tilde{k}_{1,n^*} - \frac{\partial \alpha_1}{\partial y} \omega_2 \tilde{k}_1 - \frac{\partial \alpha_1}{\partial y} \varepsilon_2 \tilde{k}_2 - \frac{\partial \alpha_1}{\partial \hat{k}_1} \dot{\hat{k}}_1 - \frac{\partial \alpha_1}{\partial \hat{k}_2} \dot{\hat{k}}_2 \\
& + \frac{\partial \alpha_1}{\partial \hat{k}_{1,n^*}} \lambda_2 v_2 + \frac{\partial \alpha_1}{\partial \hat{k}_1} \Gamma_1 \tau_2 + \frac{\partial \alpha_1}{\partial \hat{k}_2} \Gamma_2 v_2.
\end{aligned} \tag{11.32}
$$

Introducing the partial Lyapunov candidate function

$$V_2 = V_1 + \frac{1}{2} z_2^2 \tag{11.33}$$

we derive the time derivative of V_2 from (11.32) as

$$
\begin{aligned}
\dot{V}_2 = {} & -\sum_{q=1}^{2} c_q z_q^2 - d_1 z_1^2 - d_2 \left(\frac{\partial \alpha_1}{\partial y}\right)^2 z_2^2 + z_2 z_3 + z_1 \epsilon_2 - z_2 \frac{\partial \alpha_1}{\partial y} \epsilon_2 \\
& + v_2 \tilde{k}_{1,n^*} + \tau_2^T \tilde{k}_1 + v_2^T \tilde{k}_2 + z_2 \frac{\partial \alpha_1}{\partial \hat{k}_{1,n^*}} \lambda_2 v_2 \\
& + z_2 \frac{\partial \alpha_1}{\partial \hat{k}_1} (\Gamma_1 \tau_2 - \dot{\hat{k}}_1) + z_2 \frac{\partial \alpha_1}{\partial \hat{k}_2} (\Gamma_2 v_2 - \dot{\hat{k}}_2).
\end{aligned} \tag{11.34}
$$

Step $i = 3, 4, \ldots, \rho - 1$: According to (11.31), define z_i, $i = 2, 3, \ldots, \rho - 1$, in a similar way as

$$z_i = \mu_{n^*,i} - \hat{\kappa} y_m^{(i-1)} - \alpha_{i-1}. \tag{11.35}$$

Differentiating (11.35) with respect to t, we obtain

$$
\begin{aligned}
\dot{z}_i = {} & \dot{\mu}_{n^*,i} - \hat{\kappa} y_m^{(i)} - \dot{\hat{\kappa}} y_m^{(i-1)} - \dot{\alpha}_{i-1} \\
= {} & \mu_{n^*,i+1} - l_i \mu_{n^*,1} - \hat{\kappa} y_m^{(i)} - \dot{\hat{\kappa}} y_m^{(i-1)} - \frac{\partial \alpha_{i-1}}{\partial \hat{\kappa}} \dot{\hat{\kappa}} \\
& -\frac{\partial \alpha_{i-1}}{\partial y} (\epsilon_2 + k_{1,n^*} \mu_{n^*,2} + \omega_2^T k_1 + \varepsilon_2^T k_2 + \xi_2 + \varphi_1(y)) \\
& -\sum_{q=1}^{i-1} \frac{\partial \alpha_{i-1}}{\partial \mu_{n^*,q}} (\mu_{n^*,q+1} - l_q \mu_{n^*,1}) - \sum_{q=1}^{i-1} \frac{\partial \alpha_{i-1}}{\partial y_m^{(q-1)}} y_m^{(q)} \\
& -\sum_{q=1}^{i} \frac{\partial \alpha_{i-1}}{\partial \xi_q} (\xi_{q+1} - l_q \xi_1 + l_q y + \varphi_q(y))
\end{aligned}
$$

$$-\sum_{q=1}^{i}\frac{\partial\alpha_{i-1}}{\partial\varepsilon_q}(\varepsilon_{q+1}-l_q\varepsilon_1)-\sum_{q=1}^{i}\frac{\partial\alpha_{i-1}}{\partial\omega_q}(\omega_{q+1}-l_q\omega_1)$$

$$-\frac{\partial\alpha_{i-1}}{\partial\hat{k}_{1,n^*}}\dot{\hat{k}}_{1,n^*}-\frac{\partial\alpha_{i-1}}{\partial\hat{k}_1}\dot{\hat{k}}_1-\frac{\partial\alpha_{i-1}}{\partial\hat{k}_2}\dot{\hat{k}}_2. \tag{11.36}$$

Choose the stabilizing function α_i as

$$\alpha_i = -z_{i-1}-c_iz_i-d_i\left(\frac{\partial\alpha_{i-1}}{\partial y}\right)^2 z_i + l_i\mu_{n^*,1}+\left(y_m^{(i-1)}+\frac{\partial\alpha_{i-1}}{\partial\hat{\kappa}}\right)\dot{\hat{\kappa}}$$

$$+\frac{\partial\alpha_{i-1}}{\partial y}(\hat{k}_{1,n^*}\mu_{n^*,2}+\omega_2^T\hat{k}_1+\varepsilon_2^T\hat{k}_2+\xi_2+\varphi_1(y))$$

$$+\sum_{q=1}^{i-1}\frac{\partial\alpha_{i-1}}{\partial\mu_{n^*,q}}(\mu_{n^*,q+1}-l_q\mu_{n^*,1})+\sum_{q=1}^{i-1}\frac{\partial\alpha_{i-1}}{\partial y_m^{(q-1)}}y_m^{(q)}$$

$$+\sum_{q=1}^{i}\frac{\partial\alpha_{i-1}}{\partial\xi_q}(\xi_{q+1}-l_q\xi_1+l_qy+\varphi_q(y))$$

$$+\sum_{q=1}^{i}\frac{\partial\alpha_{i-1}}{\partial\varepsilon_q}(\varepsilon_{q+1}-l_q\varepsilon_1)+\sum_{q=1}^{i}\frac{\partial\alpha_{i-1}}{\partial\omega_q}(\omega_{q+1}-l_q\omega_1)$$

$$+\frac{\partial\alpha_{i-1}}{\partial\hat{k}_{1,n^*}}\lambda_2v_i-\sum_{q=2}^{i-1}\frac{\partial\alpha_{q-1}}{\partial\hat{k}_{1,n^*}}\lambda_2\frac{\partial\alpha_{i-1}}{\partial y}\mu_{n^*,2}z_q$$

$$+\frac{\partial\alpha_{i-1}}{\partial\hat{k}_1}\Gamma_1\tau_i-\sum_{q=2}^{i-1}\frac{\partial\alpha_{q-1}}{\partial\hat{k}_1}\Gamma_1\frac{\partial\alpha_{i-1}}{\partial y}\omega_2 z_q$$

$$+\frac{\partial\alpha_{i-1}}{\partial\hat{k}_2}\Gamma_2\nu_i-\sum_{q=2}^{i-1}\frac{\partial\alpha_{q-1}}{\partial\hat{k}_2}\Gamma_2\frac{\partial\alpha_{i-1}}{\partial y}\varepsilon_2 z_q \tag{11.37}$$

where c_i and d_i, $i=3,4,\ldots,\rho-1$, are some positive constants to be chosen, and v_i, τ_i, and ν_i, $i=3,4,\ldots,\rho-1$, are the tuning functions given as

$$v_i = v_{i-1}-\frac{\partial\alpha_{i-1}}{\partial y}\mu_{n^*,2}z_i$$

$$\tau_i = \tau_{i-1}-\frac{\partial\alpha_{i-1}}{\partial y}\omega_2 z_i$$

$$\nu_i = \nu_{i-1}-\frac{\partial\alpha_{i-1}}{\partial y}\varepsilon_2 z_i. \tag{11.38}$$

With the definition

$$z_{i+1}=\mu_{n^*,i+1}-\hat{\kappa}y_m^{(i)}-\alpha_i \tag{11.39}$$

we rewrite (11.36) as

$$\dot{z}_i = -z_{i-1} - c_i z_i - d_i \left(\frac{\partial \alpha_{i-1}}{\partial y} \right)^2 z_i + z_{i+1} - \frac{\partial \alpha_{i-1}}{\partial y} \epsilon_2$$

$$- \frac{\partial \alpha_{i-1}}{\partial y} \mu_{n^*,2} \tilde{k}_{1,n^*} - \frac{\partial \alpha_{i-1}}{\partial y} \omega_2 \tilde{k}_1 - \frac{\partial \alpha_{i-1}}{\partial y} \epsilon_2 \tilde{k}_2$$

$$+ \frac{\partial \alpha_{i-1}}{\partial \hat{k}_{1,n^*}} \lambda_2 \upsilon_i - \sum_{q=2}^{i-1} \frac{\partial \alpha_{q-1}}{\partial \hat{k}_{1,n^*}} \lambda_2 \frac{\partial \alpha_{i-1}}{\partial y} \mu_{n^*,2} z_q - \frac{\partial \alpha_{i-1}}{\partial \hat{k}_{1,n^*}} \dot{\hat{k}}_{1,n^*}$$

$$+ \frac{\partial \alpha_{i-1}}{\partial \hat{k}_1} \Gamma_1 \tau_i - \sum_{q=2}^{i-1} \frac{\partial \alpha_{q-1}}{\partial \hat{k}_1} \Gamma_1 \frac{\partial \alpha_{i-1}}{\partial y} \omega_2 z_q - \frac{\partial \alpha_{i-1}}{\partial \hat{k}_1} \dot{\hat{k}}_1$$

$$+ \frac{\partial \alpha_{i-1}}{\partial \hat{k}_2} \Gamma_2 \upsilon_i - \sum_{q=2}^{i-1} \frac{\partial \alpha_{q-1}}{\partial \hat{k}_2} \Gamma_2 \frac{\partial \alpha_{i-1}}{\partial y} \epsilon_2 z_q - \frac{\partial \alpha_{i-1}}{\partial \hat{k}_2} \dot{\hat{k}}_2. \tag{11.40}$$

Based on the partial Lyapunov candidate function

$$V_i = V_{i-1} + \frac{1}{2} z_i^2 \tag{11.41}$$

it follows from (11.40) that the time derivative of V_i is

$$\dot{V}_i = -\sum_{q=1}^{i} c_q z_q^2 - d_1 z_1^2 - \sum_{q=2}^{i} d_q \left(\frac{\partial \alpha_{q-1}}{\partial y} \right)^2 z_q^2$$

$$+ z_i z_{i+1} + z_1 \epsilon_2 - \sum_{q=2}^{i} z_q \frac{\partial \alpha_{q-1}}{\partial y} \epsilon_2$$

$$+ \upsilon_i \tilde{k}_{1,n^*} + \tau_i^T \tilde{k}_1 + \upsilon_i^T \tilde{k}_2 + \sum_{q=2}^{i} z_q \frac{\partial \alpha_{q-1}}{\partial \hat{k}_{1,n^*}} (\lambda_2 \upsilon_i - \dot{\hat{k}}_{1,n^*})$$

$$+ \sum_{q=2}^{i} z_q \frac{\partial \alpha_{q-1}}{\partial \hat{k}_1} (\Gamma_1 \tau_i - \dot{\hat{k}}_1) + \sum_{q=2}^{i} z_q \frac{\partial \alpha_{q-1}}{\partial \hat{k}_2} (\Gamma_2 \upsilon_i - \dot{\hat{k}}_2). \tag{11.42}$$

Step ρ: Considering

$$z_\rho = \mu_{n^*,\rho} - \hat{\kappa} y_m^{(\rho-1)} - \alpha_{\rho-1} \tag{11.43}$$

we obtain its time derivative as

$$\dot{z}_\rho = \dot{\mu}_{n^*,\rho} - \hat{\kappa} y_m^{(\rho)} - \dot{\hat{\kappa}} y_m^{(\rho-1)} - \dot{\alpha}_{\rho-1}$$

$$= \mu_{n^*,\rho+1} - l_\rho \mu_{n^*,1} + v_0 - \hat{\kappa} y_m^{(\rho)} - \dot{\hat{\kappa}} y_m^{(\rho-1)} - \frac{\partial \alpha_{\rho-1}}{\partial \hat{\kappa}} \dot{\hat{\kappa}}$$

$$- \frac{\partial \alpha_{\rho-1}}{\partial y} (\epsilon_2 + k_{1,n^*} \mu_{n^*,2} + \omega_2^T k_1 + \varepsilon_2^T k_2 + \xi_2 + \varphi_1(y))$$

$$-\sum_{q=1}^{\rho-1}\frac{\partial\alpha_{\rho-1}}{\partial\mu_{n^*,q}}(\mu_{n^*,q+1}-l_q\mu_{n^*,1})-\sum_{q=1}^{\rho-1}\frac{\partial\alpha_{\rho-1}}{\partial y_m^{(q-1)}}y_m^{(q)}$$

$$-\sum_{q=1}^{\rho}\frac{\partial\alpha_{\rho-1}}{\partial\xi_q}(\xi_{q+1}-l_q\xi_1+l_qy+\varphi_q(y))-\sum_{q=1}^{\rho}\frac{\partial\alpha_{\rho-1}}{\partial\omega_q}(\omega_{q+1}-l_q\omega_1)$$

$$-\sum_{q=1}^{\rho}\frac{\partial\alpha_{\rho-1}}{\partial\varepsilon_q}(\varepsilon_{q+1}-l_q\varepsilon_1)-\sum_{j=1}^{m}\frac{\partial\alpha_{\rho-1}}{\partial\zeta_{n^*j,\rho}}\beta_j(y)$$

$$-\frac{\partial\alpha_{\rho-1}}{\partial\hat{k}_{1,n^*}}\dot{\hat{k}}_{1,n^*}-\frac{\partial\alpha_{\rho-1}}{\partial\hat{k}_1}\dot{\hat{k}}_1-\frac{\partial\alpha_{\rho-1}}{\partial\hat{k}_2}\dot{\hat{k}}_2. \tag{11.44}$$

Design the control signal $v_0(t)$ for the control law '(11.8) as

$$v_0=\alpha_\rho+\hat{\kappa}y_m^{(\rho)} \tag{11.45}$$

where α_ρ is constructed as

$$\alpha_\rho=-z_{\rho-1}-c_\rho z_\rho-d_\rho\left(\frac{\partial\alpha_{\rho-1}}{\partial y}\right)^2 z_\rho+l_\rho\mu_{n^*,1}$$

$$-\mu_{n^*,\rho+1}+\left(y_m^{(i-1)}+\frac{\partial\alpha_{\rho-1}}{\partial\hat{\kappa}}\right)\dot{\hat{\kappa}}$$

$$+\frac{\partial\alpha_{\rho-1}}{\partial y}(\epsilon_2+\hat{k}_{1,n^*}\mu_{n^*,2}+\omega_2^T\hat{k}_1+\varepsilon_2^T\hat{k}_2+\xi_2+\varphi_1(y))$$

$$+\sum_{q=1}^{\rho-1}\frac{\partial\alpha_{\rho-1}}{\partial\mu_{n^*,q}}(\mu_{n^*,q+1}-l_q\mu_{n^*,1})+\sum_{q=1}^{\rho-1}\frac{\partial\alpha_{\rho-1}}{\partial y_m^{(q-1)}}y_m^{(q)}$$

$$+\sum_{q=1}^{\rho}\frac{\partial\alpha_{\rho-1}}{\partial\xi_q}(\xi_{q+1}-l_q\xi_1+l_qy+\varphi_q(y))+\sum_{q=1}^{\rho}\frac{\partial\alpha_{\rho-1}}{\partial\omega_q}(\omega_{q+1}-l_q\omega_1)$$

$$+\sum_{q=1}^{\rho}\frac{\partial\alpha_{\rho-1}}{\partial\varepsilon_q}(\varepsilon_{q+1}-l_q\varepsilon_1)+\sum_{j=1}^{m}\frac{\partial\alpha_{\rho-1}}{\partial\zeta_{n^*j,\rho}}\beta_j(y)$$

$$+\frac{\partial\alpha_{\rho-1}}{\partial\hat{k}_{1,n^*}}\lambda_2 v_\rho-\sum_{q=2}^{\rho-1}\frac{\partial\alpha_{q-1}}{\partial\hat{k}_{1,n^*}}\lambda_2\frac{\partial\alpha_{\rho-1}}{\partial y}\mu_{n^*,2}z_q$$

$$+\frac{\partial\alpha_{\rho-1}}{\partial\hat{k}_1}\Gamma_1\tau_\rho-\sum_{q=2}^{\rho-1}\frac{\partial\alpha_{q-1}}{\partial\hat{k}_1}\Gamma_1\frac{\partial\alpha_{\rho-1}}{\partial y}\omega_2 z_q$$

$$+\frac{\partial\alpha_{\rho-1}}{\partial\hat{k}_2}\Gamma_2 v_\rho-\sum_{q=2}^{\rho-1}\frac{\partial\alpha_{q-1}}{\partial\hat{k}_2}\Gamma_2\frac{\partial\alpha_{\rho-1}}{\partial y}\varepsilon_2 z_q \tag{11.46}$$

in which $c_\rho>0$ and $d_\rho>0$ are designing constants.

Consider the Lyapunov function candidate

$$V = V_\rho = \frac{1}{2}z^T z + \frac{k_{1,n^*}}{2\lambda_1}\tilde{\kappa}^2 + \frac{1}{2\lambda_2}\tilde{k}_{1,n^*}^2$$

$$+ \frac{1}{2}\tilde{k}_1^T \Gamma_1^{-1}\tilde{k}_1 + \frac{1}{2}\tilde{k}_2^T \Gamma_2^{-1}\tilde{k}_2 + \sum_{i=1}^{\rho}\frac{1}{2d_i}\epsilon^T P \epsilon \qquad (11.47)$$

where P, a positive definite matrix, satisfies the Lyapunov equation

$$P A_o + A_o^T P = -I \qquad (11.48)$$

and $\lambda_2 > 0$, $\Gamma_1 = \Gamma_1^T > 0$, $\Gamma_2 = \Gamma_2^T > 0$. The time derivative of V is

$$\dot{V} = -\sum_{i=1}^{\rho} c_i z_i^2 - d_1 z_1^2 - \sum_{i=2}^{\rho} d_i \left(\frac{\partial \alpha_{i-1}}{\partial y}\right)^2 z_i^2 + z_1 \epsilon_2$$

$$-\sum_{i=2}^{\rho} z_i \frac{\partial \alpha_{i-1}}{\partial y}\epsilon_2 - \sum_{i=1}^{\rho}\frac{1}{2d_i}\epsilon^T \epsilon + \left(v_\rho - \frac{1}{\lambda_2}\dot{\hat{k}}_{1,n^*}\right)\tilde{k}_{1,n^*}$$

$$+(\tau_\rho^T - \dot{\hat{k}}_1^T \Gamma_1^{-1})\tilde{k}_1 + (v_\rho^T - \dot{\hat{k}}_2^T \Gamma_2^{-1})\tilde{k}_2$$

$$+\sum_{i=2}^{\rho} z_i \frac{\partial \alpha_{i-1}}{\partial \hat{k}_{1,n^*}}(\lambda_2 v_\rho - \dot{\hat{k}}_{1,n^*}) + \sum_{i=2}^{\rho} z_i \frac{\partial \alpha_{i-1}}{\partial \hat{k}_1}(\Gamma_1 \tau_\rho - \dot{\hat{k}}_1)$$

$$+\sum_{i=2}^{\rho} z_i \frac{\partial \alpha_{i-1}}{\partial \hat{k}_2}(\Gamma_2 v_\rho - \dot{\hat{k}}_2). \qquad (11.49)$$

With the choice of the update laws

$$\dot{\hat{k}}_{1,n^*} = \lambda_2 v_\rho$$

$$\dot{\hat{k}}_1 = \Gamma_1 \tau_\rho$$

$$\dot{\hat{k}}_2 = \Gamma_2 v_\rho \qquad (11.50)$$

we derive the time derivative of V as

$$\dot{V} = -\sum_{i=1}^{\rho} c_i z_i^2 - d_1 z_1^2 - \sum_{i=2}^{\rho} d_i \left(\frac{\partial \alpha_{i-1}}{\partial y}\right)^2 z_i^2 + z_1 \epsilon_2$$

$$-\sum_{i=2}^{\rho} z_i \frac{\partial \alpha_{i-1}}{\partial y}\epsilon_2 - \sum_{i=1}^{\rho}\frac{1}{2d_i}\epsilon^T \epsilon$$

$$= -\sum_{i=1}^{\rho} c_i z_i^2 - d_1 \left(z_1 - \frac{1}{2d_1}\epsilon_2\right)^2 - \sum_{i=2}^{\rho} d_i \left(\frac{\partial \alpha_{i-1}}{\partial y}z_i + \frac{1}{2d_i}\epsilon_2\right)^2$$

$$-\sum_{i=1}^{\rho}\frac{1}{4d_i}\epsilon_2^2 - \sum_{i=1}^{\rho}\frac{1}{2d_i}(\epsilon_1^2 + \epsilon_3^2 + \cdots + \epsilon_n^2)$$

$$\leq -\sum_{i=1}^{\rho} c_i z_i^2 - \sum_{i=1}^{\rho}\frac{1}{4d_i}\epsilon_2^2 - \sum_{i=1}^{\rho}\frac{1}{2d_i}(\epsilon_1^2 + \epsilon_3^2 + \cdots + \epsilon_n^2). \qquad (11.51)$$

In summary, we have developed an adaptive actuator failure compensation scheme for the system (11.11), which consists of the controller law

$$v_j = \text{sign}[b_{n^*,j}]\frac{1}{\beta_j(y)}v_0, \; j = 1, 2, \ldots, m$$

$$v_0 = \alpha_\rho + \hat{\kappa}y_m^{(\rho)} \qquad (11.52)$$

and the adaptive laws for updating the controller parameters,

$$\dot{\hat{\kappa}} = -\lambda_1 z_1(\dot{y}_m + \bar{\alpha}_1)$$

$$\dot{\hat{k}}_{1,n^*} = \lambda_2 v_\rho$$

$$\dot{\hat{k}}_1 = \Gamma_1 \tau_\rho$$

$$\dot{\hat{k}}_2 = \Gamma_2 \nu_\rho \qquad (11.53)$$

where α_ρ, v_ρ, τ_ρ, and ν_ρ are derived from the backstepping procedure.

11.3 Stability Analysis

Now we prove that when the adaptive controller developed in Section 11.2 is applied to the plant (11.11), the closed-loop signal boundedness and asymptotic output tracking are guaranteed, so that the stated adaptive actuator failure compensation control objective is achieved.

Theorem 11.3.1. *The adaptive output feedback control scheme consisting of the controller (11.52) and the filters (11.13) along with the parameter update laws (11.53) applied to the system (11.11), based on Assumptions (A11.1) and (A11.2), ensures global boundedness of all closed-loop signals and global asymptotic output tracking:* $\lim_{t\to\infty}(y(t) - y_m(t)) = 0.$

Proof: For each time interval (t_k, t_{k+1}), $k = 0, 1, \ldots, q$, the Lyapunov function candidate V in (11.47) can be evaluated. From the smoothness of the nonlinear functions, the solution of the closed-loop system exists and is unique. In view of (11.51), the time derivative of V has the property

$$\dot{V} \leq -\sum_{i=1}^{\rho} c_i z_i^2 - \sum_{i=1}^{\rho} \frac{1}{4d_i}\epsilon_2^2 - \sum_{i=1}^{\rho}\frac{1}{2d_i}(\epsilon_1^2 + \epsilon_3^2 + \cdots + \epsilon_n^2) \leq 0. \qquad (11.54)$$

Starting from the first time interval, it can be seen that $V(t) \leq V(t_0)$ and $\dot{V} \leq 0$ for $\forall t \in [t_0, t_1)$. Hence we conclude that z, $\hat{\kappa}$, $\hat{k}_{1,n^*}$, $\hat{k}_1$, $\hat{k}_2$, and ϵ are bounded for $t \in [t_0, t_1)$. From the boundedness of y_m and z_1, y is bounded.

It follows from (11.13) that ξ and ζ_{rj}, $r = 0, 1, \ldots, n^*$, $j = 1, 2, \ldots, m$, are bounded. Also from (11.13), it can be obtained that

$$\mu_{r,i} = e_i^T (sI - A_o)^{-1} e_{n-r} v_0, \ 0 \le r \le n^*, \ 1 \le i \le n \qquad (11.55)$$

where e_i is the ith coordinate vector in R^n. Express the plant (11.11) in the differential equation form

$$y^{(n)} = \sum_{i=1}^{n} \varphi_i^{(n-i)}(y) + \sum_{r=0}^{n^*} \sum_{j=1}^{m} k_{2,rj} \beta_j^{(r)}(y) + \sum_{r=0}^{n^*} k_{1,r} v_0^{(r)}. \qquad (11.56)$$

Express system (11.56) in the input–output form with transfer function $1/(k_{1,n^*} s^{n^*} + \cdots + k_{1,1}s + k_{1,0})$, input $y^{(n)} - \sum_{i=1}^{n} \varphi_i^{(n-i)}(y) - \sum_{r=0}^{n^*} \sum_{j=1}^{m} k_{2,rj}$ $\cdot \beta_j^{(r)}(y)$, and output v_0:

$$\begin{aligned} v_0 &= \frac{1}{k_{1,n^*} s^{n^*} + \cdots + k_{1,1}s + k_{1,0}} [y^{(n)} \\ &\quad - \sum_{i=1}^{n} \varphi_i^{(n-i)}(y) - \sum_{r=0}^{n^*} \sum_{j=1}^{m} k_{2,rj} \beta_j^{(r)}(y)]. \end{aligned} \qquad (11.57)$$

Substituting (11.57) into (11.55), we have

$$\begin{aligned} \mu_{r,i} &= e_i^T (sI - A_o)^{-1} e_{n-r} \frac{1}{k_{1,n^*} s^{n^*} + \cdots + k_{1,1}s + k_{1,0}} \\ &\quad \cdot [y^{(n)} - \sum_{i=1}^{n} \varphi_i^{(n-i)}(y) - \sum_{r=0}^{n^*} \sum_{j=1}^{m} k_{2,rj} \beta_j^{(r)}(y)] \end{aligned} \qquad (11.58)$$

which results in the boundedness of $\mu_{r,i}$, $r = 0, 1, \ldots, n^*$, $i = 1, 2, \ldots, n$, because y is bounded and $\varphi_i(\cdot)$, $i = 1, 2, \ldots, n$, and $\beta_j(\cdot)$, $j = 1, 2, \ldots, m$, are smooth, and the matrix A_o and the polynomial $k_{1,n^*} s^{n^*} + \cdots + k_{1,1}s + k_{1,0}$ are stable. It is in turn implied from (11.14) and the boundedness of ϵ that x is bounded. Based on (11.57), v_0 is a bounded signal. Since $\beta_j(y) \ne 0$ for $\forall y \in R$, the boundedness of v_j is guaranteed, too, $j = 1, 2, \ldots, m$. Therefore, all closed-loop signals are bounded for $t \in [t_0, t_1)$.

At time $t = t_1$, there occur p_1 actuator failures, which result in the abrupt change of κ, k_{1,n^*}, k_1, and k_2. Since the change of values of these parameters are finite and z, $\hat{\kappa}$, $\hat{k}_{1,n^*}$, $\hat{k}_1$, and $\hat{k}_2$ are continuous, it can be concluded from $\dot{V} \le 0$ that $V(t) \le V(t_1^+) = V(t_1^-) + \bar{V}_1 \le V(t_0) + \bar{V}_1$ for $t \in (t_1, t_2)$ with a constant $\bar{V}_1$. By repeating the argument above, the boundedness of all the signals are proved for the time interval (t_1, t_2). Continuing in the same way, finally we have that $V(t) \le V(t_q^+) = V(t_q^-) + \bar{V}_q \le V(t_0) + \sum_{k=1}^{q} \bar{V}_k$ for $t \in (t_q, \infty)$ with some constants $\bar{V}_k$, $k = 1, 2, \ldots, q$. Since there is only a finite number of actuator failures, it can be concluded that $V(t)$ is bounded for $\forall t > t_0$, and so are all the closed-loop signals.

To prove output tracking, considering the last time interval (t_q, ∞) with a positive initial $V(t_q^+)$, we see that it follows from (11.54) that $z \in L^2$ and $\epsilon \in L^2$. On the other hand, we can conclude that $\dot{z} \in L^\infty$ and $\dot{\epsilon} \in L^\infty$ from the boundedness of the closed-loop signals. In turn it follows that over the time interval (t_q, ∞), $\lim_{t \to \infty} z_1(t) = 0$, which means that the output tracking error $y(t) - y_m(t)$ is such that $\lim_{t \to \infty} (y(t) - y_m(t)) = 0$. ∇

11.4 Design for State-Dependent Nonlinearities

So far the adaptive actuator failure problem has been addressed for output feedback systems whose nonlinearities depend on the system output only. In this section, we extend our investigation to a class of systems whose nonlinearities also depend on system states for adaptive actuator failure compensation with output feedback design. To solve the failure compensation problem for systems with state-dependent nonlinearities, an observer whose state estimates can exponentially converge to the real states, when implemented with known system and failure parameters, is crucial for the feedback control design. Due to the nature of nonlinearity, such an observer can only be developed for certain classes of nonlinear systems.

For a nonlinear system

$$\dot{\chi} = f(\chi) + g(\chi)u, \; y = h(\chi) \tag{11.59}$$

where $\chi \in R^n$, $u \in R^m$, and only y is measured, there are still open issues in designing an observer to obtain the estimate $\hat{\chi}$ that converges to ξ exponentially. In our problem, even if there exists an observer design whose nominal observer with known system parameters and actuator failures achieves the convergence of the actual state by the state estimate, major issues still remain to be solved for designing an adaptive actuator failure compensation scheme with adaptive laws for parameter estimates which ensures the stability of the adaptive observer and the closed-loop system.

To address this problem, we consider a class of systems in the form

$$\dot{x} = A(y)x + B(y)u, \; y = c^T x \tag{11.60}$$

where $A(y) \in R^{n \times n}$ and $B(y) \in R^{n \times m}$ denote the matrices of functions of y, $c \in R^n$ is a constant vector. Without loss of generality, we define $c = [1, 0, 0, \ldots, 0]^T$, because for an arbitrary c, we can always find a linear transformation $T = [T_1^T, T_2^T, \ldots, T_n^T]^T$ with $T_1 = c$ to transform the system into the form (5.2). In fact, under a diffeomorphism $x = T(\chi) = [T_1(\chi), T_2(\chi), \ldots, T_n(\chi)]^T$, with $T_1(\chi) = h(\chi)$, such that

$$\frac{\partial T(\chi)}{\partial \chi} f(\chi) = A(T_1(\chi))T(\chi)$$

$$\frac{\partial T(\chi)}{\partial \chi} g(\chi) = B(T_1(\chi)) \qquad (11.61)$$

the system (11.59) can be transformed into (11.60) with $c = [1, 0, 0, \ldots, 0]^T$.

A nominal observer may be proposed for the system (11.60):

$$\dot{\hat{x}} = [A(y) - l(y)c^T]\hat{x} + l(y)y + B(y)u \qquad (11.62)$$

where $l(y) \in R^n$ is to be chosen for an exponentially stable error system, that is, $\dot{\epsilon} = [A(y) - l(y)c^T]\epsilon$ is exponentially stable, where $\epsilon = x - \hat{x}$. The construction of such an $l(y)$ is also essential for us to solve the failure compensation problem. It is clear that additional conditions are needed on $A(y)$ such that there exists an $l(y)$ to ensure an exponentially stable error system.

While the general case is still subject to further investigation, next for second-order systems in the form of (11.60), we design two observers: one reduced order and one full order, whose state exponentially converges to the unmeasured system state. Then, based on the exponential stability of a nominal observer, the actuator failure compensation problem is solved by using the adaptive method to handle the uncertainties. For the output tracking objective, the assumption on the system minimum phase property is needed, that is, the zero dynamics should be input-to-state stable.

11.4.1 Reduced-Order Observer

Express the second-order system (11.60) as

$$\dot{x}_1 = a_{11}(y)x_1 + a_{12}(y)x_2 + b_1(y)u$$
$$\dot{x}_2 = a_{21}(y)x_1 + a_{22}(y)x_2 + b_2(y)u$$
$$y = x_1 \qquad (11.63)$$

with x_2 to be estimated, for which we construct a reduced-order observer

$$\hat{x}_2 = w + l(y)y - \varrho(y) \qquad (11.64)$$

where w is generated from the dynamic system

$$\dot{w} = (a_{22}(y) - l(y)a_{12}(y)) w + (a_{22}(y) - l(y)a_{12}(y)) (l(y)y - \varrho(y))$$
$$+ a_{21}(y)y - l(y)a_{11}(y)y - l(y)b_1u + b_2(y)u \qquad (11.65)$$

the nonlinear gain $l(y)$ is chosen such that $a_{22}(y) - l(y)a_{12}(y) \leq -\lambda_0 < 0$ for a constant $\lambda_0 > 0$, and

$$\varrho(y) = \int_{y_0}^{y} \frac{dl(y)}{dy} y \, dy \qquad (11.66)$$

in which y_0 is the initial value of y. It follows from (11.64) and (11.65) that

$$
\begin{aligned}
\dot{\hat{x}}_2 &= (a_{22}(y) - l(y)a_{12}(y))\, \hat{x}_2 + a_{21}(y)y \\
&\quad -l(y)a_{11}(y)y - l(y)b_1 u + b_2(y)u + l(y)\dot{y}.
\end{aligned}
\qquad (11.67)
$$

Defining the partial state estimation error $\tilde{x}_2 = x_2 - \hat{x}_2$, we obtain the observation error equation from (11.63) and (11.67) as

$$\dot{\tilde{x}}_2 = (a_{22}(y) - l(y)a_{12}(y))\, \tilde{x}_2 \qquad (11.68)$$

which results in exponential convergence of $\tilde{x}_2$ to zero.

11.4.2 Full-Order Observer

To illustrate a full-order observer design, we consider the system (11.63) with $a_{12}(y) = \bar{a}_{12}$ and $a_{22}(y) \leq \bar{a}_{22}$ for some constant $\bar{a}_{12} \neq 0$ and $\bar{a}_{22}$. In this case, a full-order observer can be constructed as

$$
\begin{aligned}
\dot{\hat{x}}_1 &= \left(a_{11}(y) - \bar{l}_1(y) - l_1\right) \hat{x}_1 + \bar{a}_{12}\hat{x}_2 \\
&\quad + \left(\bar{l}_1(y) + l_1\right) y + b_1(y)u \\
\dot{\hat{x}}_2 &= \left(a_{21}(y) - \bar{l}_2(y) - l_2\right) \hat{x}_1 + a_{22}(y)\hat{x}_2 \\
&\quad + \left(\bar{l}_2(y) + l_2\right) y + b_2(y)u
\end{aligned}
\qquad (11.69)
$$

where $l_1(y) = a_{11}(y) - a_{22}(y) + \bar{a}_{22}$, $\bar{l}_2(y) = a_{21}(y)$, and $l = [l_1, l_2]^T$ is chosen such that $A_o = \bar{A} - lc^T$ is stable, where $\bar{A} = \begin{bmatrix} 0 & \bar{a}_{12} \\ 0 & \bar{a}_{22} \end{bmatrix}$ so that $(\bar{A}, c)$ is observable. For $\hat{x} = [\hat{x}_1, \hat{x}_2]^T$, we rewrite the full-order observer (11.69) as

$$
\begin{aligned}
\dot{\hat{x}} &= \begin{bmatrix} \dot{\hat{x}}_1 \\ \dot{\hat{x}}_1 \end{bmatrix} \\
&= A_o\hat{x} + \begin{bmatrix} a_{11}(y) - \bar{l}_1(y) & 0 \\ a_{21}(y) - \bar{l}_2(y) & a_{22}(y) - \bar{a}_{22} \end{bmatrix} \hat{x} \\
&\quad + \begin{bmatrix} \bar{l}_1(y) + l_1 \\ \bar{l}_2(y) + l_2 \end{bmatrix} y + \begin{bmatrix} b_1(y) \\ b_2(y) \end{bmatrix} u.
\end{aligned}
\qquad (11.70)
$$

With $\tilde{x} = x - \hat{x}$, the error equation is derived from (11.63) and (11.70) as

$$\dot{\tilde{x}} = A_o\tilde{x} + (a_{22}(y) - \bar{a}_{22})\, \tilde{x} \qquad (11.71)$$

from which it follows that $\tilde{x}_2$ goes to zero exponentially, because $a_{22}(y) - \bar{a}_{22} \leq 0$ and the constant matrix A_o is stable.

Remark 11.4.1. In the presence of actuator failures, uncertainties are introduced into the system. With the choice of an appropriate actuation scheme such as the proportional-actuation scheme (11.8), unknown actuator failures can be parametrized, together with unknown system parameters. An adaptive observer will be employed for which the unknown parameters are updated by the adaptive laws designed with the adaptive compensation controller, under the condition that the state estimate from the nominal observer (which needs to be redesigned to take into account actuator failures) converges to the actual system state exponentially. □

Remark 11.4.2. System zero dynamics play a crucial role in feedback control system performance. For the asymptotic tracking objective, a feedback control design works based on the minimum phase property of the zero dynamics, because the boundedness of the zero dynamics states that are unobservable due to the input–output cancellation by the feedback control input is essential for the stability of the closed-loop system.

On the other hand, for the observer design, the system output is used through a dynamic process to generate the estimates of the unmeasured states, that is, the observer estimates all the states in the system, including the zero dynamics states. Hence it is not needed to specify the zero dynamics of the system for an observer design. □

11.4.3 Design Procedure

The design procedure for a stable adaptive output feedback actuator failure compensation control scheme, based on a state observer, for the nonlinear system (11.60) with uncertain actuator failures can be given as follows.

Assume that an exponentially stable nominal state observer of the form

$$\dot{\bar{x}} = [A(y) - l(y)c^T]\bar{x} + l(y)y + B(y)u \tag{11.72}$$

is available for a nonlinear system without actuator failures such that there is a Lyapunov function $V_o(\tilde{x})$ whose time derivative $\dot{V}_o(\tilde{x}) \leq -W_1(\tilde{x})$, where $\tilde{x} = x - \bar{x}$ and $W_1(\tilde{x})$ is a continuous positive definite function.

Step 1: A suitable actuation scheme for handling actuator failure uncertainty is chosen. For example, the equal-actuation scheme is

$$v_1 = v_2 = \cdots = v_m = v_0 \tag{11.73}$$

where v_j, $j = 1, 2, \ldots, m$, are the applied control inputs, and v_0 is a control signal to be designed.

Step 2: Based on the chosen actuation scheme such as (11.73), a nominal observer for actuator failures is reparametrized as

$$\dot{\hat{x}} = [A(y) - l(y)c^T]\hat{x} + l(y)y + B(y)k_2 + B(y)k_1 v_0 \qquad (11.74)$$

where $k_1 = [k_{11}, k_{12}, \ldots, k_{1m}]^T \in R^m$ and $k_2 = [k_{21}, k_{22}, \ldots, k_{2m}]^T \in R^m$ with $k_{1j} = 0$, $k_{2j} = \bar{u}_j$ if the jth actuator is failed at an unknown value $\bar{u}_j$; otherwise, $k_{1j} = 1$, $k_{2j} = 0$, for $j \in \{1, 2, \ldots, m\}$.

Step 3: An adaptive observer for actuator failures is designed with estimates of k_1 and k_2 as

$$\dot{\hat{x}} = [A(y) - l(y)c^T]\hat{x} + l(y)y + B(y)\hat{k}_2 + B(y)\hat{k}_1 v_0 \qquad (11.75)$$

where $\hat{k}_1$ and $\hat{k}_2$ are the estimates of k_1 and k_2 whose adaptive update laws are developed in combination with the adaptive controller.

Step 4: An adaptive feedback control law is

$$v_1 = v_2 = \cdots = v_m = v_0(\hat{x}, y, \hat{k}_1, \hat{k}_2, y_m, \dot{y}_m, \ldots, y_m^{(\rho)}) \qquad (11.76)$$

along with the update laws for $\hat{k}_1$ and $\hat{k}_2$:

$$\begin{aligned}
\dot{\hat{k}}_1 &= \psi_1(\hat{x}, y, y_m, \dot{y}_m, \ldots, y_m^{(\rho)}) \\
\dot{\hat{k}}_2 &= \psi_2(\hat{x}, y, y_m, \dot{y}_m, \ldots, y_m^{(\rho)})
\end{aligned} \qquad (11.77)$$

where y_m is the prescribed reference signal to be tracked.

Step 5: Closed-loop stability analysis can be made based on the Lyapunov function:

$$V = V_c(z) + \frac{1}{2}\tilde{k}_1^T \Gamma_1 \frac{1}{2}\tilde{k}_1 + \frac{1}{2}\tilde{k}_2^T \Gamma_2 \frac{1}{2}\tilde{k}_2 + V_o(\tilde{x}) \qquad (11.78)$$

where Γ_1 and Γ_2 are the gain matrices, and $V_c(z)$ is a continuous positive definite function of $z(\hat{x}, \tilde{x}, y, \tilde{k}_1, \tilde{k}_2, y_m, \dot{y}_m, \ldots, y_m^{(\rho)}) = [z_1, z_2, \ldots, z_\rho]^T \in R^\rho$ with z_i, $i = 1, 2, \ldots, \rho$, the ρ independent state variables. The exponential stability of a nominal observer is used for handling the additional variable $\tilde{x}$ in $V_c(z)$ caused by the state estimate error from the adaptive observer via $V_o(\tilde{x})$ such that

$$\dot{V} \leq -W_2(e) - W_3(\tilde{x}) \qquad (11.79)$$

where $e = y - y_m$ is the tracking error, and $W_2(e)$ and $W_3(\tilde{x})$ are some continuous positive definite functions. From $V_c(z)$, it can be concluded that ρ state variables represented in a partial system equation are bounded, while, in addition, the other $n^* = n - \rho$ state variables that construct the system zero dynamics are also ensured to be bounded by the assumed input-to-state stability of the zero dynamics with respect to z as the input.

Next we will apply this adaptive compensation scheme for a dynamic model of an aircraft, based on the two observers given above, with simulation results to demonstrate the effectiveness of the observer-based adaptive actuator failure compensation schemes.

11.4.4 Longitudinal Control of a Hypersonic Aircraft

In this section, we consider the dynamic system of the attack angle from a rigid-body longitudinal model of a hypersonic aircraft cruising at a speed of 15 Mach and at an altitude of 110,000 ft, described by

$$
\begin{aligned}
\dot{x}_1 &= x_2 + \psi_1(x_3, x_4, y) \\
\dot{x}_2 &= \psi_2(x_2, x_4, y) + b_1 x_4^2 u_1 + b_2 x_4^2 u_2 \\
\dot{x}_3 &= \psi_3(x_3, x_4, y) \\
\dot{x}_4 &= \psi_4(x_3, x_4, y) \\
y &= x_1
\end{aligned}
\tag{11.80}
$$

where x_1 is the angle of attack, x_2 is the pitch rate, which is unmeasured, x_3 is the flight-path angle, x_4 is the velocity, $u_1(t)$ and $u_2(t)$ are the elevator segment deflection angles (which are introduced as redundant actuators for our adaptive actuator failure compensation study), b_1 and b_2 are unknown constants with known signs, and $\psi_1(x_3, y)$, $\psi_2(x_2, y)$, $\psi_3(x_3, y)$, and $\psi_4(x_3, y)$ are known nonlinear functions

$$
\begin{aligned}
\psi_1(x_3, x_4, y) &= \theta_1 x_4 y + \theta_2 x_4 \sin(y) + \theta_3 x_4 y^2 \sin(y) \\
&\quad + \theta_4 \frac{\cos(x_3)}{x_4} + \theta_5 x_4 \cos(x_3) \\
\psi_2(x_2, x_4, y) &= \theta_6 x_4^2 y^2 + \theta_7 x_4^2 y + \theta_8 x_4 y^2 x_2 + \theta_9 x_4 y x_2 + \theta_{10} x_4 x_2 \\
\psi_3(x_3, x_4, y) &= -\theta_1 x_4 y - \theta_2 x_4 \sin(y) - \theta_3 x_4 y^2 \sin(y) \\
&\quad - \theta_4 \frac{\cos(x_3)}{x_4} - \theta_5 x_4 \cos(x_3) \\
\psi_4(x_3, x_4, y) &= \theta_{11} x_4^2 \cos(y) + \theta_{12} x_4^2 y^2 \cos(y) + \theta_{13} x_4^2 y^2 \\
&\quad + \theta_{14} x_4^2 y + \theta_{15} x_4^2 + \theta_{16} \sin(x_3)
\end{aligned}
\tag{11.81}
$$

with known constants θ_i, $i = 1, 2, \ldots, 16$, which are determined by the gravitational constant, aircraft mass, moment of inertia, air density, and aircraft altitude. The equations $\dot{x}_3 = \psi_3(x_3, x_4, y)$ and $\dot{x}_4 = \psi_4(x_3, x_4, y)$ represent the zero dynamics. Note that the term $\theta_4(\cos(x_3)/x_4)$ will go to infinity when x_4 goes to zero so that the zero dynamics of the aircraft model are not globally input-to-state stable. However, since x_4 is the velocity of the aircraft, it cannot be zero during operation (in fact, the aircraft model is established based on a cruising speed of 15 Mach). In a neighborhood of the equilibrium point $(x_1 = 0, x_2 = 0, x_3 = -\pi/2, x_4 = V_0)$, where $V_0 = 15$ Mach (15,060 ft/s), it can be shown that the zero dynamics are locally input-to-state stable.

Input-to-state stability. A nonlinear system $\dot{x} = f(t, x, u)$ with an equilibrium point $x = x_e$ is locally input-to-state stable if in a neighborhood of $[x^T, u] = [x_e, 0]^T$ its solution trajectory $\tilde{x}(t) = x(t) - x_e$ satisfies $\|\tilde{x}(t)\| \leq \beta(\|x(t_0) - x_e\|, t - t_0) + \gamma(\sup_{t_0 \leq \tau \leq t} \|u(\tau)\|)$ for some class-$\mathcal{K}$ functions γ (i.e., $\gamma(0) = 0$, $\gamma(r) > 0$ for $r > 0$, and $\gamma(\cdot)$ is strictly increasing) and β (with respect to $\|x(t_0) - x_e\|$), while β is decreasing with respect to $t - t_0$ and $\lim_{t \to \infty} \beta(\|x(t_0) - x_e\|, t - t_0) = 0$ [67].

For a nonlinear system $\dot{x} = f(t, x, u)$, where the function $f(t, x, u)$ is continuously differentiable and the Jacobian matrices $[\frac{\partial f}{\partial x}]$ and $[\frac{\partial f}{\partial u}]$ are bounded in some neighborhood of the equilibrium point $(x = x_e, u = 0)$, if the unforced system $\dot{x} = f(t, x, 0)$ has a uniformly asymptotically stable equilibrium point $x = x_e$, then the system is locally input-to-state stable [67].

Consider the zero dynamics system consisting of $\dot{x}_3 = \psi_3(x_3, x_4, y)$, $\dot{x}_4 = \psi_4(x_3, x_4, y)$, of the aircraft model and study the unforced system with $y = 0$ around the equilibrium point $(x_3 = -\pi/2, x_4 = V_0)$. It can be verified that $\psi_3(x_3, x_4, y)$ and $\psi_4(x_3, x_4, y)$ are continuously differentiable and the

Jacobian matrices $\begin{bmatrix} \frac{\partial \psi_3}{\partial x_3} & \frac{\partial \psi_3}{\partial x_4} \\ \frac{\partial \psi_4}{\partial x_3} & \frac{\partial \psi_4}{\partial x_4} \end{bmatrix}$ and $\begin{bmatrix} \frac{\partial \psi_3}{\partial y} \\ \frac{\partial \psi_4}{\partial y} \end{bmatrix}$ are bounded in some neigh-

borhood of $(x_3 = -\pi/2, x_4 = V_0, y = 0)$. The unforced system is uniformly asymptotically stable at the equilibrium point $(x_3 = -\pi/2, x_4 = V_0)$, because the linearized system matrix is

$$
\begin{aligned}
A_z &= \left[\begin{array}{cc} \frac{\partial \psi_3(x_3,x_4,0)}{\partial x_3} & \frac{\partial \psi_3(x_3,x_4,0)}{\partial x_4} \\ \frac{\partial \psi_4(x_3,x_4,0)}{\partial x_3} & \frac{\partial \psi_4(x_3,x_4,0)}{\partial x_4} \end{array} \right]_{x_3=-\frac{\pi}{2},\, x_4=V_0} \\
&= \left[\begin{array}{cc} -0.0012 & 0 \\ 0 & -2.017 \times 10^{-4} \end{array} \right]
\end{aligned}
\tag{11.82}
$$

which is asymptotically stable. Therefore, the zero dynamics are locally input-to-state stable with respect to y as the input (to be around $y = 0$).

The control objective for actuator failure compensation of this aircraft model is that the angle of attack y asymptotically converges to zero, while the closed-loop system is stabilized in the presence of an actuator failure in either one of the elevator segments.

Remark 11.4.3. For a nonadaptive design with known system parameters and actuator failures, due to the resulting closed-loop asymptotic stability, a local initial condition results in a local system trajectory so that the local input-to-state stability condition can always be met. In this case the local stability of the closed-loop system can be ensured under the locally input-to-state stability of the zero dynamics system.

However, in the adaptive control case, asymptotic stability of the closed-loop system is not guaranteed in general. A feedback control law may be able to ensure the boundedness of the first two state variables x_1 and x_2, while the boundedness of zero dynamics states depend on the stability of the zero dynamics system, which is only a local property in our example system. Without asymptotic closed-loop stability, such a local property may not be retained in the adaptive control case. This is also the case for adaptive actuator failure compensation control, where the local stability may not be guaranteed because the system state may go beyond the local input-to-state stability region even if the initial conditions start from that region.

Given that adaptive stabilization of systems with locally stable zero dynamics systems is an important open problem that is beyond the scope of our research, from a theory point of view, we assume that the closed-loop system under the global output feedback control law (which ensures the boundedness of the state variables x_1 and x_2, as shown in the next subsections) is such that the conditions for locally input-to-state stable zero dynamics are always satisfied, with the initial conditions of the system chosen to be "local" enough, so that the states of the zero dynamics system are also bounded as implied by the boundedness of the states x_1 and x_2.   □

To proceed, we rewrite the system (11.80) as

$$
\begin{aligned}
\dot{x}_1 &= x_2 + \varphi_1(x_3, x_4, y) \\
\dot{x}_2 &= \phi(x_4, y)x_2 + \varphi_2(x_4, y) + b_1 x_4^2 u_1 + b_2 x_4^2 u_2 \\
\dot{x}_3 &= \psi_3(x_3, x_4, y) \\
\dot{x}_4 &= \psi_4(x_3, x_4, y) \\
y &= x_1
\end{aligned}
\tag{11.83}
$$

where $\varphi_1(x_3, x_4, y) = \psi_1(x_3, x_4, y)$, $\varphi_2(x_4, y) = \theta_6 x_4^2 y^2 + \theta_7 x_4^2 y$, and $\phi(x_4, y) = \theta_8 x_4 y^2 + \theta_9 x_4 y + \theta_{10} x_4$. Next we present two control designs based on the two observers developed in Section 11.4 with simulation results.

Reduced-Order Observer Based Design (I). Applying the proportional-actuation scheme that is designed for the aircraft system (11.83) as

$$v_j = \text{sign}[b_j]\frac{v_0}{x_4^2}, \quad j = 1, 2 \tag{11.84}$$

where v_0 is the nominal control signal to be designed, the adaptive reduced-order observer for estimating x_2 is given by

$$\hat{x}_2 = w + l(x_4, y)y - \varrho(x_4, y) - \frac{\partial l(x_4, y)}{\partial x_4}\psi_4(x_3, x_4, y)$$

$$w = \xi + \hat{k}_2\zeta + \hat{k}_1\mu \tag{11.85}$$

where $\hat{x}_2$ is the estimate of the state x_2, and $\hat{k}_1$ and $\hat{k}_2$ are the estimates of the parameters k_1 and k_2. As defined in (11.9) and (11.10), $k_1 = \sum_{j=1}^{2}\text{sign}[b_j]b_j > 0$, $k_2 = 0$ if there is no actuator failure, and $k_1 = \sum_{j \neq j_0}\text{sign}[b_j]b_j > 0$, $k_2 = \sum_{j=j_0}b_j\bar{u}_j$ if the j_0th actuator fails with $j_0 = 1$ or 2. The signals ξ, ζ, and μ are from the filters

$$\dot{\xi} = (\phi(x_4, y) - l(x_4, y))\xi + (\phi(x_4, y) - l(x_4, y))(l(x_4, y)y - \varrho(x_4, y)$$
$$- \frac{\partial l(x_4, y)}{\partial x_4}\psi_4(x_3, x_4, y)) + \varphi_2(x_4, y) - l(x_4, y)\varphi_1(x_3, x_4, y)$$

$$\dot{\zeta} = (\phi(x_4, y) - l(x_4, y))\zeta + x_4^2$$

$$\dot{\mu} = (\phi(x_4, y) - l(x_4, y))\mu + v_0 \tag{11.86}$$

where $l(x_4, y)$ is chosen such that $\phi(x_4, y) - l(x_4, y) \leq -\lambda_0 < 0$ for a constant $\lambda_0 > 0$, and $\varrho(x_4, y) = \int_{y_0}^{y}(\partial l(x_4, y)/\partial y)y\,dy$ in which y_0 is the initial value of y. For the function $\phi(x_4, y)$, we have

$$\phi(x_4, y) = \theta_8\left(y + \frac{\theta_9}{2\theta_8}\right)^2 x_4^2 + \left(\theta_{10} - \frac{\theta_9^2}{2\theta_8}\right)x_4^2 \tag{11.87}$$

with $\theta_8 < 0$. Hence, we choose $l(x_4, y) = \bar{\theta}x_4^2 - \lambda_0$ for some constant $\bar{\theta} \geq \theta_{10} - (\theta_9^2/2\theta_8)$, which in turn implies that $\varrho(x_4, y) = 0$. Thus, the adaptive observer (11.85) is rewritten as

$$\hat{x}_2 = w + l(x_4)y - 2\bar{\theta}x_4\psi_4(x_3, x_4, y)$$

$$w = \xi + \hat{k}_2\zeta + \hat{k}_1\mu \tag{11.88}$$

with the filters defined in (11.86) while $l(x_4, y) = \bar{\theta}x_4^2 - \lambda_0$.

With the knowledge of k_1 and k_2, the nominal state error from the nominal (nonadaptive) reduced-order observer is $\epsilon = x_2 - \bar{x}_2$, where $\bar{x}_2 = \bar{w} + l(x_4)y - 2\bar{\theta}x_4\psi_4(x_3, x_4, y)$ and $\bar{w} = \xi + k_2\zeta + k_1\mu$. From (11.83), (11.88), and (11.86), it follows that $\dot{\epsilon} = (\phi(x_4, y) - l(x_4, y))\,\epsilon$, which indicates that ϵ converges to zero as fast as $e^{-\lambda_0 t}$ with the choice of $l(x_4, y) = \bar{\theta}x_4^2 + \lambda_0$.

The adaptive control law based on the adaptive observer (11.88) is

$$v_j = \operatorname{sign}[b_j]\frac{v_0}{x_4^2}, \; j = 1, 2$$

$$v_0 = \alpha_2 \tag{11.89}$$

where α_2 is derived from the backstepping procedure. The adaptive laws for $\hat{\kappa}$, the estimate of $\kappa = 1/k_1$, and $\hat{k}_1$, $\hat{k}_2$, are given by

$$\dot{\hat{\kappa}} = -\lambda z_1 \bar{\alpha}_1$$
$$\dot{\hat{k}}_1 = \lambda_1 v_2$$
$$\dot{\hat{k}}_2 = \lambda_2 v_2 \tag{11.90}$$

where z_1, α_1, v_2, and ν_2 are derived from the backstepping procedure:

$$z_1 = y$$
$$\bar{\alpha}_1 = -c_1 z_1 - d_1 z_1 - \hat{k}_2\zeta - \xi - \varphi_1(x_3, x_4, y)$$
$$\alpha_1 = \hat{\kappa}\bar{\alpha}_1$$
$$v_1 = z_1(\mu - \alpha_1)$$
$$\nu_1 = z_1\zeta_2$$
$$z_2 = \mu - \alpha_1$$
$$
\begin{aligned}
\alpha_2 = {} & -\hat{\kappa}z_1 - c_2 z_2 - d_2\left(\frac{\partial\alpha_1}{\partial y}\right)^2 z_2 - (\phi(x_4, y) - l(x_4, y))\,\mu \\
& + \frac{\partial\alpha_1}{\partial y}(\hat{k}_1\mu + \hat{k}_2\zeta + \xi + l(x_4)y - 2\bar{\theta}x_4\psi_4(x_3, x_4, y) + \varphi_1(x_3, x_4, y)) \\
& + \frac{\partial\alpha_1}{\partial \xi}\left((\phi(x_4, y) - l(x_4))(\xi + l(x_4) - 2\bar{\theta}x_4\psi_4(x_3, x_4, y))\right) \\
& + \varphi_2(x_4, y) - l(x_4)\varphi_1(x_3, x_4, y)) + \frac{\partial\alpha_1}{\partial \zeta}\left((\phi(x_4, y) - l(x_4, y))\zeta + x_4^2\right) \\
& + \frac{\partial\alpha_1}{\partial x_3}\psi_3(x_3, x_4, y) + \frac{\partial\alpha_1}{\partial x_4}\psi_4(x_3, x_4, y) + \frac{\partial\alpha_1}{\partial \hat{\kappa}}\dot{\hat{\kappa}} + \frac{\partial\alpha_1}{\partial \hat{k}_2}\lambda_2 v_2
\end{aligned}
$$
$$v_2 = v_1 - \frac{\partial\alpha_1}{\partial y}\mu_2 z_2$$
$$\nu_2 = \nu_1 - \frac{\partial\alpha_1}{\partial y}\zeta_2 z_2 \tag{11.91}$$

where the design constants are chosen as $c_i = 0.01$, $d_i = 1.1$, $i = 1, 2$, $\lambda = 1.05$, $\lambda_1 = 1.05$, and $\lambda_2 = 50.5$.

Similar to the stability analysis in Section 11.3, we consider the following Lyapunov function

$$V = \frac{1}{2}(z_1^2 + z_2^2) + \frac{k_1}{2\lambda}\tilde{\kappa}^2 + \frac{1}{2\lambda_1}\tilde{k}_1^2 + \frac{1}{2\lambda_2}\tilde{k}_2^2 + \left(\frac{1}{2d_1} + \frac{1}{2d_2}\right)\frac{1}{\lambda_0}\epsilon^2 \quad (11.92)$$

which is piecewise continuous in t with at most one jumping at some time t_1 when one of the actuators is failed. Except for the time instant t_1,

$$\dot{V} \leq -c_1 z_1^2 - c_2 z_2^2 - \left(\frac{1}{4d_1} + \frac{1}{4d_2}\right)\epsilon^2 \leq 0 \quad (11.93)$$

which implies that $V(t) \leq V(0)$ for $t \in [0, t_1)$ and $V(t) \leq V(t_1^+)$ for $t \in (t_1, \infty)$, where $V(t_1^+)$ is finite because of $V(t_1^-) \leq V(0)$ and a finite jumping from the abrupt changes of κ, k_1, and k_2. Therefore, the signals z_1, z_2, $\hat{\kappa}$, $\hat{k}_1$, $\hat{k}_2$, ϵ, ξ, ζ, and μ are bounded for $\forall t \geq 0$. It follows that x_2 is bounded because $x_2 = \epsilon + \bar{x}_2$, where $\bar{x}_2$ is also bounded.

Given that the zero dynamics are locally input-to-state stable with respect to y as a bounded input, under the assumption that the conditions for such a stability property are satisfied (see Remark 11.5.1), the states x_3 and x_4 are bounded in the local sense that the initials of x_3 and x_4 are close to the equilibrium point (in this case, $y(t) = x_1(t)$ is assumed to be close to 0). From (11.83), we see that $\dot{x}_1$ is bounded, which implies that $\dot{y} \in L^\infty$. In addition, it follows from (11.93) that $z_1 \in L^2$. Thus we may conclude that the system output y converges to zero as time goes to infinity.

The simulation is based on the following flight conditions: The altitude is 110,000 ft, the aircraft velocity is 15 Mach, and the throttle setting is 0.2. The aircraft parameters used in the simulation are the mass of the aircraft, 9,375 slugs, the reference area, 3,603 ft^2, the reference length, 80 ft, and the moment of inertia, 7,000,000 slug-ft^2. We consider a failure happening in u_2 after 50 sec with an unknown failure value 0.1 rad. Figure 11.1 shows the transient output, and Figure 11.2 shows the control inputs with u_2 failing after 50 sec. The simulation results show that the angle of attack $y = x_1(t)$ converges to small values in the presence of unknown actuator failures and system parameters. This simulation result indicates that the condition for local input-to-state stability of the zero dynamics system may be satisfied.

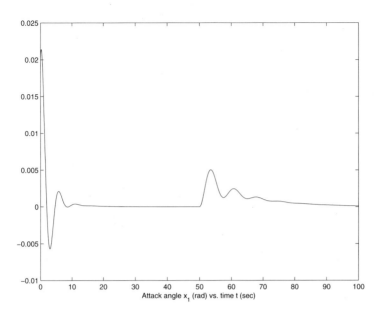

Fig. 11.1. System output: angle of attack $y = x_1$ (rad) (Design I).

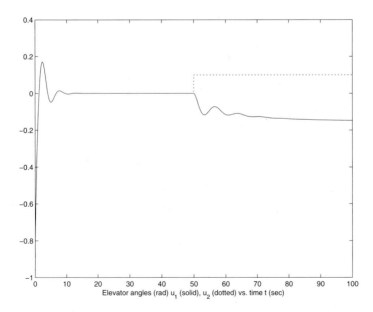

Fig. 11.2. Control inputs: elevator deflections u_1 and u_2 (rad) (Design I).

Full-Order Observer Based Design (II). Recall the system dynamics (11.83) and the function $\phi(x_4, y)$ in the form of (11.87). We notice that the condition for designing a full-order observer given in Section 11.4.1 is satisfied with $\bar{a}_{12} = 1$ and $\bar{a}_{22} = \bar{\theta} \geq \theta_{10} - (\theta_9^2/2\theta_8)$. Using the same proportional-actuation scheme (11.84), we construct the adaptive observer for $x = [x_1, x_2]^T$ based on the full-order design

$$\hat{x} = \xi + \hat{k}_2 \zeta + \hat{k}_1 \mu \tag{11.94}$$

where $\hat{x}$ is the estimate of x, and $\hat{k}_1$ and $\hat{k}_2$ are the estimates of k_1 and k_2, which are defined in the reduced-order observer based design. The signals ξ, ζ, and μ are from the filters

$$
\begin{aligned}
\dot{\xi} &= A_o \xi + \begin{bmatrix} -\bar{l}_1(y) & 0 \\ 0 & \phi(x_4, y) - \bar{\theta} \end{bmatrix} \xi \\
&\quad + \begin{bmatrix} \bar{l}_1(y) + l_1 \\ \bar{l}_2(y) + l_2 \end{bmatrix} y + \begin{bmatrix} \varphi_1(x_3, x_4, y) \\ \varphi_2(x_4, y) \end{bmatrix} \\
\dot{\zeta} &= A_o \zeta + \begin{bmatrix} -\bar{l}_1(y) & 0 \\ 0 & \phi(x_4, y) - \bar{\theta} \end{bmatrix} \zeta + \begin{bmatrix} 0 \\ x_4^2 \end{bmatrix} \\
\dot{\mu} &= A_o \mu + \begin{bmatrix} -\bar{l}_1(y) & 0 \\ 0 & \phi(x_4, y) - \bar{\theta} \end{bmatrix} \mu + \begin{bmatrix} 0 \\ v_0 \end{bmatrix} \tag{11.95}
\end{aligned}
$$

where $A_o = \bar{A} - lc^T = \begin{bmatrix} -l_1 & 1 \\ -l_2 & \bar{\theta} \end{bmatrix}$ is stable by choosing appropriate $l = [l_1, l_2]^T$, $\bar{l}_1(y) = \bar{\theta} - \phi(x_4, y)$, and $\bar{l}_1(y) = 0$. Based on the knowledge of k_1 and k_2, the nominal state error from the nonadaptive full-order observer is $\epsilon = x_2 - \bar{x}_2$, where $\bar{x}_2 = \xi + k_2 \zeta + k_1 \mu$, which satisfies $\dot{\epsilon} = A_o \epsilon + (\phi(x_4, y) - \bar{\theta})\epsilon$, so that ϵ converges to zero exponentially because from (11.87) we have that $\phi(x_4, y) - \bar{\theta} \leq 0$.

The control law based on the adaptive observer (11.94) and (11.95) is

$$
\begin{aligned}
v_j &= \text{sign}[b_j] \frac{v_0}{x_4^2}, \quad j = 1, 2 \\
v_0 &= \alpha_2 \tag{11.96}
\end{aligned}
$$

where α_2 is derived from the backstepping procedure. The adaptive laws for $\hat{\kappa}$, the estimate of $\kappa = 1/k_1$, and $\hat{k}_1$, $\hat{k}_2$, are given by

$$
\begin{aligned}
\dot{\hat{\kappa}} &= -\lambda z_1 \bar{\alpha}_1 \\
\dot{\hat{k}}_1 &= \lambda_1 v_2 \\
\dot{\hat{k}}_2 &= \lambda_2 v_2 \tag{11.97}
\end{aligned}
$$

where z_1, α_1, υ_2, and ν_2 are derived from the backstepping procedure:

$$z_1 = y$$
$$\bar{\alpha}_1 = -c_1 z_1 - d_1 z_1 - \hat{k}_2 \zeta_2 - \xi_2 - \varphi_1(x_3, x_4, y)$$
$$\alpha_1 = \hat{\kappa} \bar{\alpha}_1$$
$$\upsilon_1 = z_1(\mu_2 - \alpha_1)$$
$$\nu_1 = z_1 \zeta_2$$
$$z_2 = \mu_2 - \alpha_1$$
$$\alpha_2 = -\hat{\kappa} z_1 - c_2 z_2 - d_2 \left(\frac{\partial \alpha_1}{\partial y}\right)^2 z_2 + l_2 \mu_1 - \phi(x_4, y)\mu_2$$
$$+ \frac{\partial \alpha_1}{\partial y}(\hat{k}_1 \mu_2 + \hat{k}_2 \zeta_2 + \xi_2 + \varphi_1(x_3, x_4, y))$$
$$+ \frac{\partial \alpha_1}{\partial \xi_2}(-l_2 \xi_1 + l_2 y + \phi(x_4, y)\xi_2 + \varphi_2(x_4, y))$$
$$+ \frac{\partial \alpha_1}{\partial \zeta_2}(-l_2 \zeta_1 + \phi(x_4, y)\zeta_2 + x_4^2) + \frac{\partial \alpha_1}{\partial x_3}\psi_3(x_3, x_4, y)$$
$$+ \frac{\partial \alpha_1}{\partial x_4}\psi_4(x_3, x_4, y) + \frac{\partial \alpha_1}{\partial \hat{\kappa}}\dot{\hat{\kappa}} + \frac{\partial \alpha_1}{\partial \hat{k}_2}\lambda_2 \nu_2$$
$$\upsilon_2 = \upsilon_1 - \frac{\partial \alpha_1}{\partial y}\mu_2 z_2$$
$$\nu_2 = \nu_1 - \frac{\partial \alpha_1}{\partial y}\zeta_2 z_2 \tag{11.98}$$

where the design constants are chosen as $c_i = 0.01$, $d_i = 1.1$, $i = 1, 2$, $\lambda = 1.05$, $\lambda_1 = 1.05$, and $\lambda_2 = 50.5$.

Consider the Lyapunov function

$$V = \frac{1}{2}(z_1^2 + z_2^2) + \frac{k_1}{2\lambda}\tilde{\kappa}^2 + \frac{1}{2\lambda_1}\tilde{k}_1^2 + \frac{1}{2\lambda_2}\tilde{k}_2^2 + \left(\frac{1}{2d_1} + \frac{1}{2d_2}\right)\epsilon^T P \epsilon \tag{11.99}$$

where P satisfies the Lyapunov equation $PA_o + A_o^T P = -I$. Except for the time instant t_1 when one of the actuators is failed,

$$\dot{V} \leq -c_1 z_1^2 - c_2 z_2^2 - \left(\frac{1}{4d_1} + \frac{1}{4d_2}\right)\epsilon^T \epsilon \leq 0. \tag{11.100}$$

Since the changes in the values of κ, k_1, and k_2 are finite if one actuator is failed, we conclude that the signals z_1, z_2, $\hat{\kappa}$, $\hat{k}_1$, $\hat{k}_2$, ϵ, ξ, ζ, and μ are bounded for $\forall t \geq 0$. It follows that x_2 is bounded because $x_2 = \epsilon_2 + \bar{x}_2$, where $\bar{x}_2 = \xi_2 + k_2 \zeta_2 + k_1 \mu_2$ is bounded. With the locally input-to-state stable zero dynamics with respect to y as a bounded input and under the assumption that its stability conditions are satisfied (see Remark 11.5.1), the states x_3 and x_4 are bounded in the local sense that the initials of x_3 and x_4 are close to the equilibrium point. With bounded x_2, x_3, x_4, and y, we have that $\dot{x}_1$ is bounded from (11.83), which implies that $\dot{y} \in L^\infty$. In addition, it follows from (11.100) that $z_1 \in L^2$. Thus we may conclude that y converges to zero as time goes to infinity and establish the local stability of the closed-loop system in the presence of actuator failures.

The simulation is based on the same flight conditions given in the reduced-order observer based design and we consider the same failure case, that is, a failure happens in u_2 after 50 sec with an unknown failure value 0.1 rad. The output of the system is shown in Figure 11.3, and the control inputs with u_2 failing after 50 sec are shown in Figure 11.4. It can be seen that the angle of attack $x_1(t)$ converges to small values in the presence of unknown actuator failures and system parameters.

The simulation results we presented so far are for the cases when some of the actuators had a jumping failure value. Our other simulation results indicate that much smaller transient responses result from a failed actuator whose failure value is its value at the time when a failure occurs.

11.5 Concluding Remarks

This chapter, together with Chapters 9 and 10, presents an introductory study of adaptive actuator failure compensation control of nonlinear dynamic systems with uncertain actuator failures. Like the linear case, there are different control designs for different performance requirements, based on different conditions, such as state feedback for state tracking of feedback linearizable systems (Chapter 9), state feedback for output tracking of parametric-strict-feedback systems (Chapter 10), and output feedback for output tracking of output-feedback systems (Chapter 11). Unlike the linear case, adaptive control of nonlinear systems with uncertainties still has many open issues, and the presence of uncertain actuator failures makes the control problems more challenging. Key issues include specification of design conditions for both feedback control and actuator failure compensation, construction of failure compensation control schemes, characterization of system zero dynamics in the presence of actuator failures, and stability analysis of adaptive compensation control systems. We develop several solutions to these issues, which ensure desired asymptotic tracking despite uncertain actuator failures that caused structural and environmental modeling errors.

In particular, in this chapter, an adaptive output feedback actuator failure compensation control scheme for output-feedback nonlinear systems is devel-

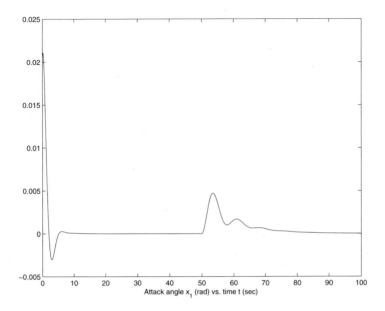

Fig. 11.3. System output: angle of attack $y = x_1$ (rad) (Design II).

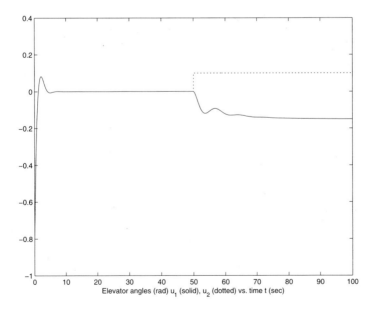

Fig. 11.4. Control inputs: elevator deflections u_1 and u_2 (rad) (Design II).

oped. In order to estimate the unmeasured states of the controlled plant, an adaptive observer for the uncertain plant dynamics in the presence of actuator failures is constructed and used for an adaptive control design. With the state observer, the backstepping method is applied to develop an adaptive compensation scheme to ensure that the plant output tracks a reference output asymptotically and that all closed-loop signals are bounded in spite of the unknown actuator failures in addition to unknown plant parameters. This adaptive compensation approach is extended to a class of nonlinear systems whose nonlinearities depend on both system output and state. An application to controlling the angle of attack of an aircraft is presented on detailed compensating designs based on two observers to handle state-dependent nonlinearities: a reduced-order one and a full-order one. Simulation results demonstrated their effectiveness in compensating uncertain actuator failures.

The developed adaptive failure compensation approach has many potential extensions. First, adaptive compensation of actuator failures of varying values as modeled in (1.6) or (1.8) can be similarly handled, based on augmented parametrizations of failure-related terms, plus additional damping functions in the control law (similar to the linear case considered in Section 2.3 for those unparametrizable failures).

More parameter uncertainties can be treated for systems with actuator failures. For the feedback linearizable systems in Chapter 9, parametric-strict-feedback systems in Chapter 10, and output-feedback systems in Chapter 11, unknown parameters are allowed to be present in the nonlinear dynamics, which can be handled by adaptation in estimating them [60], [71].

Adaptive compensation of actuator failures in multivariable nonlinear systems with multi-output and multi-input (i.e., multigroup actuators) is an important extension of the current results. The significance of this problem is that it can deal with the case when the actuators have different physical effects on the system dynamics. A key issue is to specify the system zero structure at infinity (similar to that of an interactor matrix; see Chapter 5) for nonlinear systems, that is, the system relative degree. A main design task is to decouple the system dynamics. Such adaptive designs can be developed based on the failure compensation techniques of this book.

Adaptive control of nonlinear systems is an active area of research. The presence of actuator failures leads to more and greater challenges for control designs. Solutions to such challenges will be of major significance in both control systems theory and practice.

Chapter 12

Conclusions and Research Topics

This research is aimed at developing an effective approach for control of uncertain systems with uncertain actuator failures: *adaptive actuator failure compensation*. The technical goal is to achieve both closed-loop stabilization and asymptotic tracking in the presence of uncertain failures of redundant actuators, with unknown failure values, failure time instants, and failure patterns, in addition to system parameter uncertainties.

The work reported in this book has been focused on the development of a theoretical framework for adaptive actuator failure compensation, which provides guidelines for designing control systems with guaranteed stability and tracking performance in the presence of system parameter and failure uncertainties. It has solved some key related issues such as controller structures, matching conditions, error models, adaptive laws, and stability and performance analysis, for uncertain linear, multivariable, nonminimum phase, or nonlinear dynamics with uncertain actuator failures.

Compensation Designs

This book has presented a collection of adaptive actuator failures compensation designs for control of uncertain dynamics systems with uncertain actuator failures, based on different design conditions and performance requirements. Unlike some other design methods applied to actuator failure compensation, adaptive designs adjust controller parameters to continually reduce tracking errors over time, in the presence of uncertain actuator failures. The developed adaptive compensation control designs may be classified as

- state feedback designs for state tracking (Chapters 2, 7, 9)
- state feedback designs for output tracking (Chapters 3, 7, 10)
- output feedback designs for output tracking (Chapters 4–6, 8, 11).

They can also be classified as

- model reference designs for minimum phase systems (Chapters 2–5, 7–11)

- pole placement designs for nonminimum phase systems (Chapter 6)
- designs for linear time-invariant systems (Chapters 2–8)
- designs for nonlinear time-invariant systems (Chapters 9–11)
- designs for multivariable linear time-invariant systems (Chapters 2, 5).

System Performance

Extensive simulation results are presented to verify the effectiveness of our developed adaptive compensation control designs in handling uncertain actuator failures, for various aircraft control system models:

- Boeing 737 longitudinal dynamics model (elevator/stabilizer failure)
- Boeing 737 lateral-directional dynamics model (rudder/aileron failure)
- Boeing 747 lateral-directional dynamics model (rudder failure)
- DC-8 lateral-directional dynamics model (aileron failure)
- F-18 wing dynamics model (aileron failure)
- Twin Otter longitudinal nonlinear dynamics model (elevator failure)
- a hypersonic aircraft longitudinal nonlinear model (elevator failure).

All simulation results show that system responses are as desired and expected: Signal boundedness and asymptotic tracking are achieved, and at the time instant when an actuator failure occurs there is a transient response in tracking errors, which are reduced to small value by controller adaptation.

Robustness Issue

The robustness issue is important for control systems. In Section 2.3.3, we address this issue for the unparametrizable failure case when a bounded error term is present in the tracking error equation and is handled by modifying the controller parameter update law with a projection or switching-σ modification. This issue can be systematically addressed for other adaptive actuator failure compensation designs developed in other chapters of this book. In the presence of modeling errors such as system structural changes, parameter variations, external disturbance, and unmodeled dynamics, a standard adaptive law $\dot{\theta}(t) = -\Gamma\phi(t)$ needs to be modified to ensure robust stability (parameter boundedness and mean error smallness). The modification is typically done by introducing a design signal $f(t)$ in the adaptive law: $\dot{\theta}(t) = -\Gamma\phi(t) + f(t)$, where $f(t)$ can be a dead-zone modification [95], σ-modification [53], switching-σ modification [55], [56], ϵ-modification [95], and parameter projection [44], [125]. A main function of such a modification is to prevent the parameter estimate from drifting unboundedly.

In the case of state tracking control, the plant may have a bounded disturbance $w(t)$ in its dynamic equation, that is,

$$\dot{x} = Ax + Bu + w(t). \tag{12.1}$$

The adaptive controller structure (2.25) leads to the error equation

$$\dot{e}(t) = A_M e(t) + b_M \sum_{j \neq j_1, \dots, j_p} \frac{1}{k_{s2j}^*} \left(\tilde{k}_{1j}^T x(t) + \tilde{k}_{2j} r(t) + \tilde{k}_{3j} \right) + w(t). \tag{12.2}$$

The adaptive laws (2.33)–(2.35) lead the time derivative of V_p in (2.31) to

$$\dot{V}_p = -e^T(t) Q e(t) + e^T(t) P w(t). \tag{12.3}$$

Since $w(t)$ is bounded, this equation implies that $e(t)$ cannot go unbounded. However, the parameters $\tilde{k}_{1j}(t)$, $\tilde{k}_{2j}(t)$, and $\tilde{k}_{3j}(t)$ are not guaranteed to be bounded due to possible "parameter drift" instability [54] caused by the disturbance $w(t)$. With the modified adaptive laws

$$\dot{k}_{1j}(t) = -\text{sign}[k_{s2j}^*] \Gamma_{1j} x(t) e^T(t) P b_M + f_{1j}(t) \tag{12.4}$$

$$\dot{k}_{2j}(t) = -\text{sign}[k_{s2j}^*] \gamma_{2j} r(t) e^T(t) P b_M + f_{2j}(t) \tag{12.5}$$

$$\dot{k}_{3j}(t) = -\text{sign}[k_{s2j}^*] \gamma_j e^T(t) P b_M + f_{3j}(t), \tag{12.6}$$

the time derivative of V_p becomes

$$\dot{V}_p = -e^T(t) Q e(t) + e^T(t) P w(t)$$
$$+ \sum_{j \neq j_1, \dots, j_p} \frac{1}{|k_{s2j}^*|} \left(\tilde{k}_{1j}^T \Gamma_{1j}^{-1} f_{1j} + \tilde{k}_{2j} \gamma_{2j}^{-1} f_{2j} + \tilde{k}_j \gamma_j^{-1} f_{3j} \right). \tag{12.7}$$

A projection design of $f_{ij}(t)$ ensures that $k_{ij}(t)$ stays in the bounded interval and that $\tilde{k}_{ij} \Gamma_{ij}^{-1} f_{ij} \leq 0$, while a switching-$\sigma$ design of $f_{ij}(t)$ ensures that $\tilde{k}_{ij} \Gamma_{ij}^{-1} f_{ij} \leq 0$ and that $\lim_{\|\tilde{k}_{ij}\|_2 \to \infty} \tilde{k}_{ij} \Gamma_{ij}^{-1} f_{ij} = -\infty$ (see Section 2.3.3). In both cases, $\tilde{k}_{ij}(t)$ will remain bounded despite $w(t)$.

Similar robust parameter adaptation techniques can be used for other adaptive compensation control designs developed in this book to ensure robust stability in the presence of bounded disturbances and small stable unmodeled dynamics. The typical error model for these cases is

$$\epsilon(t) = \rho^* \tilde{\theta}^T(t) \zeta(t) + \tilde{\rho}(t) \xi(t) + w(t) \tag{12.8}$$

for some disturbance or unmolded dynamics dependent term $w(t)$. For this error equation, robust adaptive laws can also be developed. For more details of robust adaptive control theory, see [55], [95], [118], and [125].

Open Problems

There are still many open and challenging theoretical problems in this area of adaptive failure compensation research, including

– characterization and improvement of adaptive control system transient behavior, especially at the time instants when actuator failures occur
– adaptive control designs for nonminimum phase systems, especially the issue of control singularity for systems with actuator failures
– adaptive control designs for nonlinear dynamic systems beyond those common canonical forms, especially output feedback designs
– adaptive control designs for systems with other failures such as dynamics failures (caused by system damages) and sensor failures
– failure compensability, which is a new theoretic issue for feedback control systems: characterization of an inherent property of a feedback control system, with which a set of failure patterns can be compensated, and
– other adaptive or intelligent feedback control designs such as neural networks based designs for nonlinear systems, backstepping designs for time-varying systems, and variable structure designs for systems with uncertainties, for handling uncertain actuator, dynamics, or sensor failures.

Potential Applications

Actuator failures in control systems may cause severe system performance deterioration and even lead to catastrophic instability. Such failures are often uncertain in time, value, and pattern. Adaptive actuator failure compensation designs developed in this book, whose effectiveness of accommodating uncertain actuator failures has been verified by extensive simulation results, should have the potential for improving many applications, including

– aircraft flight control (elevator and rudder may fail)
– temperature control (heating devices may fail)
– power system control (failure in power generation/distribution)
– liquid level control (valve servo mechanisms or pumps may fail)
– magnetic bearing control (magnetic actuators may fail)
– active vibration control (acoustic or mechanical actuators may fail)
– bio-agent control (failure in chemical distribution)
– ship motion control (stabilizer and rudder may fail)
– robotic system control (electric/mechanical actuators may fail).

Appendix

This appendix presents the continuous-time designs of model reference adaptive control using state feedback for state tracking, state feedback for output tracking, or output feedback for output tracking (and its discrete-time version), and multivariable design, as well as that of adaptive pole placement control. Key issues such as *a prior* system knowledge, controller structure, plant-model matching, adaptive laws, and stability are addressed.

A.1 Model Reference Adaptive Control

There are three types of model reference adaptive control (MRAC) designs: state feedback for state tracking; state feedback for output tracking; and output feedback for output tracking, all described in this section.

A.1.1 MRAC: State Feedback for State Tracking

Consider a linear time-invariant plant described by

$$\dot{x}(t) = Ax(t) + bu(t), \ x(t) \in R^n, \ u(t) \in R \qquad (A.1)$$

where $A \in R^{n \times n}$, $b \in R^n$ are unknown constant parameter matrices, and the state vector $x(t)$ is available for measurement.

The control objective is to design a *state feedback* control signal $u(t)$ that ensures that all signals in the closed-loop system are bounded and the plant state vector $x(t)$ asymptotically tracks a given reference state vector $x_m(t)$ generated from the reference model system

$$\dot{x}_m(t) = A_m x_m(t) + b_m r(t), \ x_m(t) \in R^n, \ r(t) \in R \qquad (A.2)$$

where $A_m \in R^{n \times n}$, $b_m \in R^n$ are known constant matrices such that all eigenvalues of A_m are in the open left-half complex plane, and $r(t)$ is bounded.

To meet this control objective, we assume

(A.1-1) There exist a constant vector $k_1^* \in R^n$ and a constant scalar $k_2^* \in R$ such that

$$A + bk_1^{*T} = A_m, \ bk_2^* = b_m \qquad (A.3)$$

(A.1-2) sign$[k_2^*]$, the sign of the parameter k_2^*, is known.

While Assumption (A.1-2) is needed for implementing an adaptive law, Assumption (A.1-1) is the so-called matching condition such that if the parameters of A and b are known and (A.3) is satisfied, then the control law

$$u(t) = k_1^{*T} x(t) + k_2^* r(t) \qquad (A.4)$$

achieves the desired control objective that the closed-loop system becomes

$$\dot{x}(t) = Ax(t) + b(k_1^{*T} x(t) + k_2^* r(t)) = A_m x(t) + b_m r(t) \qquad (A.5)$$

whose state vector $x(t)$ is bounded, and so is the control $u(t)$ in (A.4), and that the tracking error $e(t) = x(t) - x_m(t)$ satisfies

$$\dot{e}(t) = A_m e(t), \ e(0) = x(0) - x_m(0) \qquad (A.6)$$

which indicates that $\lim_{t\to\infty} e(t) = 0$ exponentially. It is clear that Condition (A.3) is also necessary for the control law (A.3) to achieve the stated control objective, even if the plant parameters A and b are known.

In the adaptive control problem, the parameters of A and b are unknown so that the control law (A.4) cannot be used for control. In this case, the following adaptive controller structure is used:

$$u(t) = k_1^T(t)x(t) + k_2(t)r(t) \qquad (A.7)$$

where $k_1(t)$ and $k_2(t)$ are the estimates of k_1^* and k_2^*, respectively. The adaptive control design task now is to choose adaptive laws to update these estimates so that the stated control objective is still achievable.

Defining the parameter errors as

$$\tilde{k}_1(t) = k_1(t) - k_1^*, \ \tilde{k}_2(t) = k_2(t) - k_2^* \qquad (A.8)$$

and using (A.1), (A.3), and (A.7), we obtain

$$\begin{aligned} \dot{x}(t) &= Ax(t) + b\left(k_1^T(t)x(t) + k_2(t)r(t)\right) \\ &= A_m x(t) + b_m r(t) + b_m \left(\frac{1}{k_2^*}\tilde{k}_1^T(t)x(t) + \frac{1}{k_2^*}\tilde{k}_2(t)r(t)\right). \end{aligned} \qquad (A.9)$$

Substituting (A.2) in (A.9), we have the tracking error equation

$$\dot{e}(t) = A_m e(t) + b_m \left(\frac{1}{k_2^*} \tilde{k}_1^T(t) x(t) + \frac{1}{k_2^*} \tilde{k}_2(t) r(t) \right). \tag{A.10}$$

Consider the positive definite function

$$V(e, \tilde{k}_1, \tilde{k}_2) = e^T P e + \frac{1}{|k_2^*|} \tilde{k}_1^T \Gamma^{-1} \tilde{k}_1 + \frac{1}{|k_2^*|} \tilde{k}_2^2 \gamma^{-1} \tag{A.11}$$

as a measure of the system error signals $e(t)$, $\tilde{k}_1(t)$, and $\tilde{k}_2(t)$, where $P \in R^{n \times n}$ is constant, $P = P^T > 0$ and satisfies

$$P A_m + A_m^T P = -Q \tag{A.12}$$

for some constant $Q \in R^{n \times n}$ such that $Q = Q^T > 0$, $\Gamma \in R^{n \times n}$ is constant and $\Gamma = \Gamma^T > 0$, and $\gamma > 0$ is a constant scalar.

The time derivative of $V(e, \tilde{k}_1, \tilde{k}_2)$ is

$$\begin{aligned}
\dot{V} &= \frac{d}{dt} V = \left(\frac{\partial V}{\partial e} \right)^T \dot{e}(t) + \left(\frac{\partial V}{\partial \tilde{k}_1} \right)^T \dot{\tilde{k}}_1(t) + \frac{\partial V}{\partial \tilde{k}_2} \dot{\tilde{k}}_2(t) \\
&= 2 e^T(t) P \dot{e}(t) + \frac{2}{|k_2^*|} \tilde{k}_1^T(t) \Gamma^{-1} \dot{\tilde{k}}_1(t) + \frac{2}{|k_2^*|} \tilde{k}_2(t) \gamma^{-1} \dot{\tilde{k}}_2(t). \tag{A.13}
\end{aligned}$$

Substituting (A.10) and (A.12) in (A.13), we have

$$\begin{aligned}
\dot{V} &= -e^T(t) Q e(t) + e^T(t) P b_m \frac{2}{k_2^*} \tilde{k}_1^T(t) x(t) + e^T(t) P b_m \\
&\quad + \frac{2}{k_2^*} \tilde{k}_2(t) r(t) + \frac{2}{|k_2^*|} \tilde{k}_1^T(t) \Gamma^{-1} \dot{\tilde{k}}_1(t) + \frac{2}{|k_2^*|} \tilde{k}_2(t) \gamma^{-1} \dot{\tilde{k}}_2(t). \tag{A.14}
\end{aligned}$$

To make $\dot{V} \le 0$, we choose the adaptive laws for $k_1(t)$ and $k_2(t)$ as

$$\dot{k}_1(t) = -\text{sign}[k_2^*] \Gamma x(t) e^T(t) P b_m \tag{A.15}$$

$$\dot{k}_2(t) = -\text{sign}[k_2^*] \gamma r(t) e^T(t) P b_m \tag{A.16}$$

with $\Gamma = \Gamma^T > 0$, $\gamma > 0$, $k_1(0)$ and $k_2(0)$ being arbitrary.

Indeed, with this choice of $\dot{k}_1(t)$ and $\dot{k}_2(t)$, (A.14) becomes

$$\dot{V} = -e^T(t) Q e(t) \le 0. \tag{A.17}$$

From (A.17), we conclude that $x(t)$, $k_1(t)$, and $k_2(t)$ are all bounded, and so are $u(t)$ in (A.7) and $\dot{e}(t)$ in (A.10), and that $e(t) \in L^2$.[1]

---

[1] A *continuous-time* vector signal $x(t) \in R^n$ belongs to L^2 if

$$\sqrt{\int_0^\infty (x_1^2(t) + \cdots + x_n^2(t)) \, dt} < \infty$$

To show that $\lim_{t\to\infty} e(t) = 0$, we see that for $e = [e_1, \cdots, e_n]^T$, $e_i(t) \in L^2$ and $\dot{e}_i(t) \in L^\infty$, $i = 1, 2, \ldots, n$. It follows that

$$\int_0^t e_i^2(\tau)|\dot{e}_i(\tau)|d\tau \leq \sup_{t\geq 0} |\dot{e}_i(t)| \int_0^\infty e_i^2(\tau)d\tau < \infty \qquad (A.18)$$

for any $t \geq 0$. This implies that $\lim_{t\to\infty} \int_0^t e_i^2(\tau)|\dot{e}_i(\tau)|d\tau$ exists and is finite, and, therefore, $\lim_{t\to\infty} \int_0^t e_i^2(\tau)\dot{e}_i(\tau)d\tau$ exists and is finite. From the identity

$$e_i^2(t) = |e_i^3(t)|^{\frac{2}{3}} = |3\int_0^t e_i^2(\tau)\dot{e}_i(\tau)d\tau + e_i^3(0)|^{\frac{2}{3}} \qquad (A.19)$$

we have that $\lim_{t\to\infty} e_i^2(t)$ exists and is zero as $e_i(t) \in L^2$. This proves that $\lim_{t\to\infty} e_i(t) = 0$, $i = 1, 2, \ldots, n$, so that $\lim_{t\to\infty} e(t) = 0$.

In summary, we have the following result.

Theorem A.1. *The adaptive controller (A.7), with the adaptive laws (A.15) and (A.16), and applied to the plant (A.1), ensures that all closed-loop signals are bounded and the tracking error $e(t) = x(t) - x_m(t)$ goes to zero as $t \to \infty$.*

A.1.2   MRAC: State Feedback for Output Tracking

Consider a linear time-invariant plant in the state variable form

$$\begin{aligned} \dot{x}(t) &= Ax(t) + bu(t) \\ y(t) &= Cx(t) \end{aligned} \qquad (A.20)$$

for some unknown constant matrices $A \in R^{n\times n}$, $b \in R^{n\times 1}$, and $C \in R^{1\times n}$, with $n > 0$. The input–output description of this plant is

$$y(s) = C(sI - A)^{-1}bu(s) = \frac{Z(s)}{P(s)}u(s) \qquad (A.21)$$

where $P(s) = \det(sI - A)$ and

$$Z(s) = z_m s^m + \cdots + z_1 s + z_0, \quad z_m \neq 0, \quad m < n \qquad (A.22)$$

with s being the Laplace transform variable or the time differentiation operator: $s[x](t) = \dot{x}(t)$, $t \in [0, \infty)$, as the case may be.

---

and it belongs to L^∞ if

$$\sup_{t\geq 0} \|x(t)\|_\infty = \sup_{t\geq 0} \max_{1\leq i\leq n} |x_i(t)| < \infty.$$

A signal $x(t)$ is bounded if it belongs to L^∞.

To design an adaptive *state feedback* model reference controller for generating the plant input $u(t)$, which ensures closed-loop signal boundedness and asymptotic tracking of a given reference signal $y_m(t)$ by the plant output $y(t)$, we need the following assumptions:

(A.2-1) (A, b, C) is stabilizable and detectable,
(A.2-2) $Z(s)$ is a stable polynomial,
(A.2-3) The degree m of $Z(s)$ is known, and
(A.2-4) sign$[z_m]$, the sign of z_m, is known.

Assumption (A.2-1) is needed for output matching, Assumption (A.2-2) is needed for model reference control, and Assumption (A.2-3) is needed for constructing a reference model system, while Assumption (A.2-4) is used for designing an adaptive parameter update law.

The reference model, independent of the dynamics of (A.20), is chosen as

$$y_m(t) = W_m(s)[r](t), \quad W_m(s) = \frac{1}{P_m(s)} \tag{A.23}$$

where $P_m(s)$ is a desired stable polynomial of degree $n - m$, and $r(t)$ is a bounded reference input signal.

The adaptive model reference controller structure is

$$u(t) = k_1^T(t)x(t) + k_2(t)r(t) \tag{A.24}$$

where $k_1(t) = [k_{11}(t), k_{12}(t), \ldots, k_{1n}(t)]^T \in R^n$ and $k_2(t) \in R$ are the adaptive estimates of the unknown parameters $k_1^* = [k_{11}^*, k_{12}^*, \ldots, k_{1n}^*]^T \in R^n$ and k_2^*, which satisfy the matching eqaution

$$\det(sI - A - bk_1^{*T}) = P_m(s)Z(s)\frac{1}{z_m}, \quad k_2^* = \frac{1}{z_m} \tag{A.25}$$

that is, all zeros of $\det(sI - A - bk_1^{*T})$ are stable [118].

With this definition of k_1^* and k_2^*, the fixed version of (A.24) is

$$u(t) = k_1^{*T}x(t) + k_2^*r(t) \tag{A.26}$$

which would lead to the desired closed-loop system: $y(s) = W_m(s)r(s)$.

In the adaptive control problem when A, b, and z_m are all unknown parameters, we need to develop adaptive laws to update the parameter estimates $k_1(t)$ and $k_2(t)$. With the controller (A.24), the closed-loop system is

$$\begin{aligned} \dot{x}(t) &= (A + bk_1^{*T})x(t) + bk_2^*r(t) + b((k_1(t) - k_1^*)^T x(t) + (k_2(t) - k_2^*)r(t)) \\ y(t) &= Cx(t). \end{aligned} \tag{A.27}$$

From (A.25) it follows that

$$C(sI - A - bk_1^{*T})^{-1}bk_2^* = \frac{Z(s)k_2^*}{\det(sI - A - bk_1^{*T})} = \frac{1}{P_m(s)} = W_m(s). \quad \text{(A.28)}$$

In view of (A.23), (A.27), and (A.28), the tracking error equation is

$$
\begin{aligned}
e(t) &= y(t) - y_m(t) \\
&= \rho^* W_m(s)[(k_1 - k_1^*)^T x + (k_2 - k_2^*)r](t) + Ce^{(A + bk_1^{*T})t}x(0) \quad \text{(A.29)}
\end{aligned}
$$

where $\rho^* = z_m$, and $\lim_{t \to \infty} Ce^{(A + bk_1^{*T})t}x(0) = 0$ exponentially.

To derive an estimation error equation, we define

$$\theta(t) = \left[k_1^T(t), k_2(t)\right]^T \quad \text{(A.30)}$$

$$\theta^* = \left[k_1^{*T}, k_2^*\right]^T \quad \text{(A.31)}$$

$$\omega(t) = \left[x^T(t), r(t)\right]^T \quad \text{(A.32)}$$

$$\zeta(t) = W_m(s)[\omega](t) \quad \text{(A.33)}$$

$$\xi(t) = \theta^T(t)\zeta(t) - W_m(s)[\theta^T\omega](t) \quad \text{(A.34)}$$

$$\epsilon(t) = e(t) + \rho(t)\xi(t) \quad \text{(A.35)}$$

where $\rho(t)$ is an estimate of $\rho^* = z_m$. Then, from (A.29)–(A.35), ignoring the exponentially decaying term $Ce^{(A + bk_1^{*T})t}x(0)$, we have

$$\epsilon(t) = \rho^*(\theta(t) - \theta^*)^T\zeta(t) + (\rho(t) - \rho^*)\xi(t) \quad \text{(A.36)}$$

which is linear in the parameters errors $\theta(t) - \theta^*$ and $\rho(t) - \rho^*$.

We then choose the adaptive laws for $\theta(t)$ and $\rho(t)$ as

$$\dot{\theta}(t) = -\frac{\Gamma \text{sign}[z_m]\zeta(t)\epsilon(t)}{1 + \zeta^T(t)\zeta(t) + \xi^2(t)} \quad \text{(A.37)}$$

$$\dot{\rho}(t) = -\frac{\gamma\xi(t)\epsilon(t)}{1 + \zeta^T(t)\zeta(t) + \xi^2(t)} \quad \text{(A.38)}$$

where $\Gamma = \Gamma^T > 0$ and $\gamma > 0$ are adaptation gains.

This adaptive control scheme has the following properties [55], [95], [118].

Lemma A.1. *The adaptive laws (A.37) and (A.38) ensures that* $\theta(t) \in L^\infty$, $\rho(t) \in L^\infty$, $\epsilon(t)/m(t) \in L^2 \cap L^\infty$, $\dot{\theta}(t) \in L^2 \cap L^\infty$, *and* $\dot{\rho}(t) \in L^2 \cap L^\infty$, *where*

$$m(t) = \sqrt{1 + \zeta^T(t)\zeta(t) + \xi^2(t)}. \quad \text{(A.39)}$$

Proof: Consider the positive definite function

$$V(\tilde{\theta}, \tilde{\rho}) = \frac{1}{2}(|\rho^*|\tilde{\theta}^T \Gamma^{-1}\tilde{\theta} + \gamma^{-1}\tilde{\rho}^2) \tag{A.40}$$

where

$$\tilde{\theta}(t) = \theta(t) - \theta^*, \quad \tilde{\rho}(t) = \rho(t) - \rho^*. \tag{A.41}$$

The time derivative of $V(\tilde{\theta}, \tilde{\rho})$, along the trajectories of (A.37) and (A.38), is

$$\dot{V} = -\frac{\epsilon^2(t)}{m^2(t)}. \tag{A.42}$$

This, together with (A.36)–(A.40), leads to the results of the lemma. ∇

Theorem A.2. *All signals in the closed-loop control system, with the plant (A.20), the reference model (A.23), and the controller (A.24) updated by the adaptive law (A.37) and (A.38), are bounded, and the tracking error $e(t) = y(t) - y_m(t)$ satisfies*

$$\lim_{t \to \infty} (y(t) - y_m(t)) = 0 \tag{A.43}$$

$$\int_0^\infty (y(t) - y_m(t))^2 dt < \infty. \tag{A.44}$$

The proof of Theorem A.2 can be found in [118].

A.1.3 MRAC: Output Feedback for Output Tracking

Consider the linear time-invariant plant

$$y(t) = G(s)[u](t) \tag{A.45}$$

where $G(s) = k_p(Z(s)/P(s))$, k_p is a constant high frequency gain, and $Z(s)$ and $P(s)$ are monic polynomials of degrees n and m, respectively.

Given a reference model system

$$y_m(t) = W_m(s)[r](t) \tag{A.46}$$

where $W_m(s)$ is stable and $r(t)$ is bounded, the control objective is to find an *output feedback* control $u(t)$ such that all closed-loop signals are bounded and the plant output y tracks the reference output y_m asymptotically.

In this case, the plant state variables are not needed for control.

To meet this objective, we make the following assumptions:

(A.3-1) $Z(s)$ is a stable polynomial,

(A.3-2) The degree n of $P(s)$ is known,

(A.3-3) The relative degree $n^* = n - m$ of $G(s)$ is known,

(A.3-4) The sign of k_p is known, and

(A.3-5) $W_m(s) = 1/P_m(s)$ for a stable polynomial $P_m(s)$ of degree n^*.

The desired model reference controller structure is

$$u(t) = \theta_1^T \omega_1(t) + \theta_2^T \omega_2(t) + \theta_{20} y(t) + \theta_3 r(t) \qquad (A.47)$$

where

$$\omega_1(t) = \frac{a(s)}{\Lambda(s)}[u](t), \quad \omega_2(t) = \frac{a(s)}{\Lambda(s)}[y](t) \qquad (A.48)$$

with $a(s) = [1, s, \cdots, s^{n-2}]^T$, $\theta_1, \theta_2 \in R^{n-1}$, θ_{20}, and $\theta_3 \in R$, and $\Lambda(s)$ being a monic stable polynomial of degree $n - 1$.

It can be shown [55], [118] that with $\theta_3^* = k_p^{-1}$, there exist constant parameters $\theta_1^*, \theta_2^* \in R^{n-1}$, and $\theta_{20}^* \in R$ such that

$$\theta_1^{*T} a(s) P(s) + (\theta_2^{*T} a(s) + \theta_{20}^* \Lambda(s)) k_p Z(s)$$
$$= \Lambda(s)(P(s) - k_p \theta_3^* Z(s) P_m(s)) \qquad (A.49)$$

and that the controller (A.47) implemented with $\theta_1^*, \theta_2^*, \theta_{20}^*$, and θ_3^*, that is,

$$u(t) = \theta_1^{*T} \omega_1(t) + \theta_2^{*T} \omega_2(t) + \theta_{20}^* y(t) + \theta_3^* r(t) \qquad (A.50)$$

ensures that all the closed-loop signals are bounded and the output tracking is achieved: $y(t) = y_m(t) + \eta_0(t)$, for some exponentially decaying $\epsilon_0(t)$ that depends on system initial conditions.

For the adaptive control problem when $G(s)$ is unknown, an adaptive version of the controller (A.47) is implemented with $\theta_1 = \theta_1(t)$, $\theta_2 = \theta_2(t)$, $\theta_{20} = \theta_{20}(t)$, and $\theta_3 = \theta_3(t)$, where $\theta_1(t), \theta_2(t), \theta_{20}(t)$, and $\theta_3(t)$ are the estimates of $\theta_1^*, \theta_2^*, \theta_{20}^*$, and θ_3^*, to be updated from an adaptive law.

To derive an error equation, operating both sides of (A.49) on $y(t)$ and using (A.45): $P(s)[y](t) = k_p Z(s)[u](t)$, we have

$$\theta_1^{*T} a(s) k_p Z(s) [u](t) + (\theta_2^{*T} a(s) + \theta_{20}^* \Lambda(s)) k_p Z(s)[y](t)$$
$$= \Lambda(s) k_p Z(s)[u](t) - \Lambda(s) Z(s) P_m(s)[y](t). \qquad (A.51)$$

This equality, with $Z(s)$ and $\Lambda(s)$ stable and the effect of initial conditions ignored, leads to the parametrized plant model

$$u(t) = \theta_1^{*T} \frac{a(s)}{\Lambda(s)}[u](t) + \theta_2^{*T} \frac{a(s)}{\Lambda(s)}[y](t) + \theta_{20}^* y(t) + \theta_3^* W_m^{-1}(s)[y](t). \quad (A.52)$$

Introducing $\rho^* = 1/\theta_3^* = k_p$ and

$$\omega(t) = \left[\omega_1^T(t), \omega_2^T(t), y(t), r(t)\right]^T \tag{A.53}$$

$$e(t) = y(t) - y_m(t), \ \tilde{\theta}(t) = \theta(t) - \theta^* \tag{A.54}$$

where

$$\theta(t) = \left[\theta_1^T(t), \theta_2^T(t), \theta_{20}(t), \theta_3(t)\right]^T \tag{A.55}$$

$$\theta^* = \left[\theta_1^{*T}, \theta_2^{*T}, \theta_{20}^*, \theta_3^*\right]^T \tag{A.56}$$

substituting (A.52) in (A.47) with adaptive parameter estimates, ignoring the effect of the initial conditions, and using (A.55) and (A.56), we obtain the tracking error equation

$$\begin{aligned} e(t) &= \rho^* W_m(s)[\tilde{\theta}^T \omega](t) \\ &= -\rho^* \left(\theta^{*T} W_m(s)[\omega](t) - W_m(s)[\theta^T \omega](t)\right). \end{aligned} \tag{A.57}$$

Since both θ^* and ρ^* are unknown, the second equality of (A.57) suggests that we define the estimation error

$$\epsilon(t) = e(t) + \rho(t)\xi(t) \tag{A.58}$$

where $\rho(t)$ is the estimate of ρ^*, and

$$\xi(t) = \theta^T(t)\zeta(t) - W_m(s)[\theta^T \omega](t) \tag{A.59}$$

$$\zeta(t) = W_m(s)[\omega](t). \tag{A.60}$$

For $\tilde{\rho}(t) = \rho(t) - \rho^*$, using (A.57), (A.59), and (A.60), we express $\epsilon(t)$ as

$$\epsilon(t) = \rho^* \tilde{\theta}^T(t)\zeta(t) + \tilde{\rho}(t)\xi(t) \tag{A.61}$$

which is linear in the parameter errors $\tilde{\theta}(t)$ and $\tilde{\rho}(t)$.

Finally, we choose the adaptive laws for $\theta(t)$ and $\rho(t)$:

$$\dot{\theta}(t) = \frac{-\text{sign}[k_p]\Gamma \epsilon(t)\zeta(t)}{m^2(t)} \tag{A.62}$$

$$\dot{\rho}(t) = \frac{-\gamma\epsilon(t)\xi(t)}{m^2(t)} \tag{A.63}$$

where $\Gamma = \Gamma^T > 0$ and $\gamma > 0$ are constant adaptation gains, and

$$m(t) = \sqrt{1 + \zeta^T(t)\zeta(t) + \xi^2(t)}. \tag{A.64}$$

This adaptive controller has the following desired properties [55], [95], [118].

Lemma A.2. *The adaptive update law (A.62) and (A.63) ensures that* $\theta(t) \in L^\infty$, $\rho(t) \in L^\infty$, $\epsilon(t)/m(t) \in L^2 \cap L^\infty$, *and* $\dot\theta(t) \in L^2 \cap L^\infty$.

Theorem A.3. *All signals in the closed-loop control system with the plant (A.45), reference model (A.46), and controller (A.47) updated by the adaptive laws (A.62) and (A.63) are bounded, and*

$$y(t) - y_m(t) \in L^2, \quad \lim_{t \to \infty} (y(t) - y_m(t)) = 0. \tag{A.65}$$

The proof of Lemma A.2 is similar to that of Lemma A.1, and the proof of Theorem A.2 can be found in [118], [125].

Discrete-Time Design. While both state feedback for output tracking and output feedback for output tracking designs can be developed in discrete time, here we only illustrate how to develop a discrete-time adaptive output feedback control design for output tracking. Such a design can be obtained in a way similar to that for the continuous-time case: the same controller structure (A.47), the same parametrization (A.49)–(A.57), and the same estimation error (A.58), but in (A.45)–(A.61), we need to replace s by z which, for the discrete-time case, is the z-transform variable or the time advance operator: $z[x](k) = x(k+1)$, $k \in \{0, 1, 2, 3, \ldots\}$. In particular, a stable polynomial of degree n_p can be chosen as z^{n_p}, for example, $\Lambda(z) = z^{n-1}$ for (A.47) and $W_m(z) = z^{-n^*}$ for (A.46). However, the adaptive law for updating the controller parameter vector $\theta(k)$ in (A.47) is different:

$$\theta(k+1) = \theta(k) - \frac{\text{sign}[k_p]\Gamma\epsilon(k)\omega(k - n^*)}{m^2(k)} \tag{A.66}$$

$$\rho(k+1) = \rho(k) - \frac{\gamma\epsilon(k)\xi(k)}{m^2(k)} \tag{A.67}$$

where the adaptation gains Γ and γ satisfy

$$0 < \Gamma = \Gamma^T < \frac{2}{k_p^0} I_{2n}, \ 0 < \gamma < 2 \tag{A.68}$$

for a known constant $k_p^0 \geq |k_p|$.

The adaptive laws (A.66) and (A.67) ensures that $\theta(k) \in L^\infty$, $\rho(k) \in L^\infty$, $\epsilon(k)/m(k) \in L^2 \cap L^\infty$, and $\theta(k + i_0) - \theta(k) \in L^2$ for any finite integer $i_0 > 0$.[2]

---

[2] A *discrete-time* vector signal $x(k) \in R^n$ belongs to L^2 if

$$\sqrt{\sum_{k=0}^{\infty} (x_1^2(k) + \cdots + x_n^2(k))} < \infty$$

The proof of this result is similar to that for the continuous-time case. The time increment of the positive definite function

$$V(\tilde{\theta}, \tilde{\rho}) = |\rho^*| \tilde{\theta}^T \Gamma^{-1} \tilde{\theta} + \gamma^{-1} \tilde{\rho}^2 \tag{A.69}$$

along the trajectories of (A.66) and (A.67) is

$$
\begin{aligned}
& V(\tilde{\theta}(k+1), \tilde{\rho}(k+1)) - V(\tilde{\theta}(k), \tilde{\rho}(k)) \\
& = -\left(2 - \frac{|k_p| \omega^T(k-n^*) \Gamma \omega(k-n^*) + \gamma \xi^2(k)}{m^2(k)}\right) \frac{\epsilon^2(k)}{m^2(k)} \\
& \leq -\alpha_1 \frac{\epsilon^2(k)}{m^2(k)} \tag{A.70}
\end{aligned}
$$

for some constant $\alpha_1 > 0$. This implies that $\theta(k) \in L^\infty$, $\rho(k) \in L^\infty$, and $\epsilon(k)/m(k) \in L^2$. From (A.61), we have $\epsilon(k)/m(k) \in L^\infty$, and from (A.66), we have $\theta(k+1) - \theta(k) \in L^2$. Finally, using the inequality

$$\|\theta(k+i_0) - \theta(k)\|_2 \leq \sum_{i=0}^{i_0-1} \|\theta(k+i+1) - \theta(k+i)\|_2 \tag{A.71}$$

we have that $\theta(k+i_0) - \theta(k) \in L^2$ for any finite integer i_0.

Based on this result, we can also establish the result of Theorem A.3 for the discrete-time adaptive control scheme [118].

A.2   Multivariable MRAC

Consider the linear time-invariant plant

$$y(t) = G(s)[u](t) \tag{A.72}$$

where $y(t) \in R^M$, $u(t) \in R^M$, $G(s) = Z(s)P^{-1}(s)$ is strictly proper and has full rank, and $Z(s)$ and $P(s)$ are $M \times M$ right co-prime polynomial matrices with $P(s)$ column proper [95]. The symbol s is used as the Laplace transform variable or the time differentiation operator: $s[x](t) = \dot{x}(t)$, $t \in [0, \infty)$. The following development of multivariable MRAC is applicable to the discrete-time case, with different system stability conditions (which lead to the special choice of adaptation gains in parameter adaptive laws).

---

and it belongs to L^∞ (i.e., it is bounded) if

$$\sup_{k \geq 0} \|x(k)\|_\infty = \sup_{k \geq 0} \max_{1 \leq i \leq n} |x_i(k)| < \infty.$$

The control objective is to find a feedback control $u(t)$ for the plant (A.72) with unknown $G(s)$ such that all signals in the closed-loop system are bounded and $y(t)$ asymptotically tracks the reference output

$$y_m(t) = W_m(s)[r](t), \ W_m(s) = \xi_m^{-1}(s) \tag{A.73}$$

where $\xi_m(s)$ is a modified interactor matrix of $G(s)$ [32], [95], [123], and $r(t)$ is a bounded signal. The modified interactor matrix $\xi_m(s)$ has a stable inverse and represents the zero structure at infinity of $G(s)$. The high frequency gain matrix K_p of $G(s)$ is defined as $K_p = \lim_{D \to \infty} \xi_m(s)G(s)$, finite and nonsingular. To design MRAC schemes, we assume

(A.4-1) All zeros of $G(s)$ (that is, zeros of $\det[Z(s)]$) are stable,
(A.4-2) $\bar{\nu} \geq \nu$, the observability index of $G(s)$, is known,
(A.4-3) For some known $S_p \in R^{M \times M}$, $\Gamma_p = K_p^T S_p^{-1} = \Gamma_p^T > 0$, and
(A.4-4) A modified interactor matrix $\xi_m(s)$ of $G(s)$ is known.

Note that Assumption (A.4-3) may be relaxed, as all leading principal minors of K_p are nonzero and their signs are known (see Section 5.3.4).

Controller Structure. The MRAC controller structure for (A.72) is

$$u(t) = \Theta_1^T \omega_1(t) + \Theta_2^T \omega_2(t) + \Theta_{20} y(t) + \Theta_3 r(t) \tag{A.74}$$

where $\Theta_1 = [\Theta_{11}, \ldots, \Theta_{1\bar{\nu}-1}]^T$, $\Theta_2 = [\Theta_{21}, \ldots, \Theta_{2\bar{\nu}-1}]^T$, Θ_{20}, Θ_3, $\Theta_{ij} \in R^{M \times M}$, $i = 1, 2$, $j = 1, \ldots, \bar{\nu} - 1$, and $\omega_1(t) = F(s)[u](t)$, $\omega_2(t) = F(s)[y](t)$, $F(s) = A(s)/\Lambda(s)$, with $A(s) = [I, DI, \ldots, D^{\bar{\nu}-2}I]^T$, and $\Lambda(s)$ being a monic stable polynomial of degree $\bar{\nu} - 1$.

Error Model. With the specification of $\Lambda(s)$, $\xi_m(s)$, $P(s)$, and $Z(s)$, there exist Θ_1^*, Θ_2^*, $\Theta_3^* = K_p^{-1}$ such that $I - \Theta_1^{*T} F(s) - \Theta_2^{*T} F(s)G(s) - \Theta_{20}^* G(s) = \Theta_3^* W_m^{-1}(s)G(s)$ [32], from which, for any $u(t)$, we have

$$K_p \left(u(t) - \Theta_1^{*T} \omega_1(t) - \Theta_2^{*T} \omega_2(t) - \Theta_{20}^* y(t) - \Theta_3^* r(t) \right)$$
$$= \xi_m(s)[y - y_m](t). \tag{A.75}$$

With the controller (A.74), we write (A.75) as

$$\xi_m(s)[y - y_m](t) = K_p \tilde{\Theta}^T(t)\omega(t) \tag{A.76}$$

where $\tilde{\Theta}(t) = \Theta(t) - \Theta^*$ with $\Theta(t)$ the estimate of $\Theta^* = [\Theta_1^{*T}, \Theta_2^{*T}, \Theta_{20}^*, \Theta_3^*]^T$, and $\omega(t) = [\omega_1^T(t), \omega_2^T(t), y^T(t), r^T(t)]^T$.

Let $e(t) = y(t) - y_m(t)$ and introduce $\bar{e}(t) = h(s)\xi_m(s)[e](t)$ for $h(s) = 1/f(s)$ with $f(s)$ being a monic and stable polynomial such that $h(s)\xi_m(s)$ is proper and stable. Define the estimation error

$$\epsilon(t) = \bar{e}(t) + \Psi(t)\xi(t) \tag{A.77}$$

where $\Psi(t)$ is the estimate of K_p, and

$$\xi(t) = \Theta^T(t)\zeta(t) - h(s)[\Theta^T \omega](t) \tag{A.78}$$
$$\zeta(t) = h(s)[\omega](t). \tag{A.79}$$

It then follows from (A.76) (filtered by $h(s)\xi_m(s)$) and (A.77)–(A.79) that

$$\epsilon(t) = K_p \tilde{\Theta}^T(t)\zeta(t) + \tilde{\Psi}(t)\xi(t) \tag{A.80}$$

where $\tilde{\Psi}(t) = \Psi(t) - \Psi^*$. The development containing (A.73)–(A.80) can be done for the discrete-time case, using discrete-time transfer functions.

Adaptive Laws. We choose the *continuous-time* adaptive laws as

$$\dot{\Theta}^T(t) = -\frac{S_p \epsilon(t)\zeta^T(t)}{m^2(t)} \tag{A.81}$$

$$\dot{\Psi}(t) = -\frac{\Gamma \epsilon(t)\xi^T(t)}{m^2(t)} \tag{A.82}$$

where $K_p S_p = (K_p S_p)^T > 0$ (see Assumption (A.4-3)), $\Gamma = \Gamma^T > 0$, and

$$m(t) = \sqrt{1 + \zeta^T(t)\zeta(t) + \xi^T(t)\xi(t)}. \tag{A.83}$$

Then the time derivative of the positive definite function

$$V = \text{tr}[\tilde{\Theta}(t)\Gamma_p \tilde{\Theta}^T(t)] + \text{tr}[\tilde{\Psi}^T(t)\Gamma^{-1}\tilde{\Psi}(t)] \tag{A.84}$$

with $\Gamma_p = \Gamma_p^T > 0$ (see Assumption (A.4-3)), along (A.81) and (A.82), is

$$\dot{V} = -\frac{2\epsilon^T(t)\epsilon(t)}{m^2(t)} \leq 0. \tag{A.85}$$

Similarly, we choose the *discrete-time* adaptive laws as

$$\Theta^T(t+1) = \Theta^T(t) - S_p \epsilon(t)\zeta^T(t) \tag{A.86}$$

$$\Psi(t+1) = \Psi(t) - \Gamma \epsilon(t)\xi^T(t) \tag{A.87}$$

where $2I > K_p S_p = (K_p S_p)^T > 0$ and $2I > \Gamma = \Gamma^T > 0$.

Then, the time increment of

$$V = \text{tr}[\tilde{\Theta}(k)\Gamma_p\tilde{\Theta}^T(k)] + \text{tr}[\tilde{\Psi}^T(k)\Gamma^{-1}\tilde{\Psi}(k)] \tag{A.88}$$

with $\Gamma_p = K_p^T S_p^{-1} = \Gamma_p^T > 0$, along (A.86) and (A.87), is

$$V(\tilde{\Theta}(t+1), \tilde{\Psi}(t+1)) - V(\tilde{\Theta}(t), \tilde{\Psi}(t))$$

$$= -\frac{\epsilon^T(t)}{m^2(t)}\left(2I - \frac{\zeta^T(t)\zeta(t)S_p^T K_p^T + \xi^T(t)\xi(t)\Gamma}{m^2(t)}\right)\epsilon(t) \leq 0. \tag{A.89}$$

From (A.85) or (A.89), the closed-loop system stability and asymptotic tracking properties can be proved [123].

A.3   Adaptive Pole Placement Control

Consider the linear time-invariant plant described by

$$y(t) = G(s)[u](t) \tag{A.90}$$

where $y(t) \in R$ and $u(t) \in R$ are the measured plant input and output, respectively, and $G(s) = Z(s)/P(s)$ with

$$P(s) = s^n + p_{n-1}s^{n-1} + \cdots + p_1 s + p_0 \tag{A.91}$$

$$Z(s) = z_{n-1}s^{n-1} + z_{n-2}s^{n-2} + \cdots + z_1 s + z_0 \tag{A.92}$$

for some unknown but constant parameters p_i, z_i, $i = 0, 1, \cdots, n-1$.

The control objective is to find an output feedback control signal $u(t)$ such that all closed-loop signals are bounded and $y(t)$ asymptotically tracks a reference output $y_m(t)$ satisfying

$$Q(s)[y_m](t) = 0 \tag{A.93}$$

where $Q(s)$ is a monic polynomial of degree n_q with no zeros in $Re[s] > 0$ and only nonrepeated zeros on the $j\omega$-axis so that $y_m(t)$ is bounded.

For examples, $Q(s) = 1$ for $y_m(t) = 0$; $Q(s) = s$ for $y_m(t) = a \neq 0$; $Q(s) = s^2 + \sigma^2$ for $y_m(t) = a\sin(\sigma t) + b\cos(\sigma t)$ with $\sigma \neq 0$ and not both a, b being zero; $Q(s) = (s^2 + \sigma_1^2)(s^2 + \sigma_2^2)$ for $y_m(t) = a_1\sin(\sigma_1 t) + a_2\sin(\sigma_2 t) + b_1\cos(\sigma_1 t) + b_2\cos(\sigma_2 t)$ with $\sigma_1 \neq 0$, $\sigma_2 \neq 0$, $\sigma_1 \neq \sigma_2$, not both a_1, b_1 being zero, and not both a_2, b_2 being zero; and $Q(s) = s(s^2 + \sigma^2)$ for $y_m(t) = a\sin(\sigma t) + b\cos(\sigma t) + c$ with $c \neq 0$, $\sigma \neq 0$ and not both a, b being zero.

The following assumptions are needed for this control problem:

(A.5-1) $Q(s)P(s)$ and $Z(s)$ are co-prime, and
(A.5-2) The order n of $P(s)$ is known.

Design Procedure. We use the following indirect approach to design an adaptive controller: First estimate the plant parameters in

$$\theta_p^* = [p_0, p_1, \cdots, p_{n-1}]^T \in R^n, \ \theta_z^* = [z_0, z_1, \cdots, z_{n-1}]^T \in R^n \qquad (A.94)$$

and then use a design equation to calculate the parameters of an output feedback controller from the adaptive estimates of θ_p^* and θ_z^*.

Parameter estimation. Let the estimates of θ_p^* and θ_z^* be

$$\theta_p = [\hat{p}_0, \hat{p}_1, \cdots, \hat{p}_{n-1}]^T \in R^n, \ \theta_z = [\hat{z}_0, \hat{z}_1, \cdots, \hat{z}_{n-1}]^T \in R^n \qquad (A.95)$$

and choose a monic stable polynomial $\Lambda(s)$ as

$$\Lambda(s) = s^n + \lambda_{n-1}s^{n-1} + \cdots + \lambda_1 s + \lambda_0. \qquad (A.96)$$

Then, operating both sides of (A.90) by the stable filter $1/\Lambda(s)$ and using

$$\theta^* = [\theta_z^{*T}, (\theta_\lambda - \theta_p^*)^T]^T, \ \theta_\lambda = [\lambda_0, \cdots, \lambda_{n-1}]^T \in R^n \qquad (A.97)$$

$$\phi(t) = \left[\frac{1}{\Lambda(s)}[u](t), \frac{s}{\Lambda(s)}[u](t), \cdots, \frac{s^{n-2}}{\Lambda(s)}[u](t), \frac{s^{n-1}}{\Lambda(s)}[u](t), \right.$$

$$\left. \frac{1}{\Lambda(s)}[y](t), \frac{s}{\Lambda(s)}[y](t), \cdots, \frac{s^{n-2}}{\Lambda(s)}[y](t), \frac{s^{n-1}}{\Lambda(s)}[y](t) \right]^T \in R^{2n} \qquad (A.98)$$

we express the plant (A.90) as

$$y(t) = \frac{Z(s)}{\Lambda(s)}[u](t) + \frac{\Lambda(s) - P(s)}{\Lambda(s)}[y](t) = \theta^{*T}\phi(t). \qquad (A.99)$$

Denote $\theta(t)$ as the estimate of θ^*:

$$\theta(t) = [\theta_z^T(t), (\theta_\lambda - \theta_p(t))^T]^T \qquad (A.100)$$

and define the estimation error

$$\epsilon(t) = \theta^T(t)\phi(t) - y(t). \qquad (A.101)$$

Using (A.99) and (A.101), we have

$$\epsilon(t) = \tilde{\theta}^T(t)\phi(t), \ \tilde{\theta}(t) = \theta(t) - \theta^*. \qquad (A.102)$$

Based on the error model (A.102), we can use some adaptive algorithms to update $\theta(t)$. For example, the normalized gradient algorithm is

$$\dot{\theta}(t) = -\frac{\Gamma\phi(t)\epsilon(t)}{m^2(t)}, \ \theta(0) = \theta_0, \ \Gamma = \Gamma^T > 0 \qquad (A.103)$$

where $m(t) = \sqrt{1 + \alpha\phi^T(t)\phi(t)}$ with $\alpha > 0$.

This algorithm guarantees that $\theta(t)$, $\dot\theta(t)$, and $\epsilon(t)/m(t)$ are bounded, and $\dot\theta(t)$ and $\epsilon(t)/m(t)$ belong to L^2, as verified using the positive definite function $V(\tilde\theta) = \tilde\theta^T \Gamma^{-1} \tilde\theta$ whose time derivative is $\dot V = -2\epsilon^2(t)/m^2(t)$.

Control design. Let the estimates of $P(\lambda)$ and $Z(\lambda)$ be

$$\hat P(\lambda,\theta_p) = \lambda^n + \hat p_{n-1}\lambda^{n-1} + \cdots + \hat p_1\lambda + \hat p_0 \qquad (A.104)$$

$$\hat Z(\lambda,\theta_z) = \hat z_{n-1}\lambda^{n-1} + \hat z_{n-2}\lambda^{n-2} + \cdots + \hat z_1\lambda + \hat z_0 \qquad (A.105)$$

such that $Q(\lambda)\hat P(\lambda,\theta_p)$ and $\hat Z(\lambda,\theta_z)$ are co-prime polynomials.[3]

Choose a monic desired closed-loop characteristic polynomial $A^*(s)$ of degree $2n + n_q - 1$ which has all its zeros in $Re[s] < 0$.

Find the polynomials $C(\lambda,\psi_c)$ and $D(\lambda,\psi_d)$:

$$C(\lambda,\psi_c) = \lambda^{n-1} + c_{n-2}\lambda^{n-2} + \cdots + c_1\lambda + c_0 \qquad (A.106)$$

$$D(\lambda,\psi_d) = d_{n_q+n-1}\lambda^{n_q+n-1} + \cdots + d_1\lambda + d_0 \qquad (A.107)$$

where $\psi_c = [c_0, c_1, \cdots, c_{n-2}]^T \in R^{n-1}$ and $\psi_d = [d_0, d_1, \cdots, d_{n+n_q-1}]^T \in R^{n+n_q}$, by solving the Diophantine equation [4]

$$C(\lambda,\psi_c)Q(\lambda)\hat P(\lambda,\theta_p) + D(\lambda,\psi_d)\hat Z(\lambda,\theta_z) = A^*(\lambda) \qquad (A.108)$$

and obtain $C(s,\psi_c)$ and $D(s,\psi_d)$ by the substitution

$$C(s,\psi_c) = C(\lambda,\psi_c)|_{\lambda=s} = s^{n-1} + \cdots + c_1 s + c_0 \qquad (A.109)$$

$$D(s,\psi_d) = D(\lambda,\psi_d)|_{\lambda=s} = d_{n_q+n-1}s^{n_q+n-1} + \cdots + d_1 s + d_0. \qquad (A.110)$$

Note that as an operator, $c_i s^i$, $i = 1,\ldots,n-1$, is different from $s^i c_i$ if $c_i = c_i(t)$ is a time-varying signal, and so is $d_j s^j$, $j = 1,\ldots,n_q + n - 1$, different from $s^j d_j$ if $d_j = d_j(t)$ is a time-varying signal.

Then the pole placement controller structure is

$$u(t) = (\Lambda_1(s) - C(s,\psi_c)Q(s))\frac{1}{\Lambda_1(s)}[u](t) + D(s,\psi_d)\frac{1}{\Lambda_1(s)}[y_m - y](t) \quad (A.111)$$

where $\Lambda_1(s) = \Lambda_0(s)\Lambda_c(s)$, with $\Lambda_0(s)$ and $\Lambda_c(s)$ being some monic and stable polynomials of degrees $n_q - 1$ and n, respectively.

Under the condition that $Q(\lambda)\hat P(\lambda,\theta_p)$ and $\hat Z(\lambda,\theta_z)$ are co-prime, this adaptive control scheme ensures that all signals in the closed-loop system are bounded and that $\lim_{t\to\infty}(y(t) - y_m(t)) = 0$.

---

[3] The adaptive estimate $\theta(t)$ from (A.103) does not automatically have this property. Special modifications of such $\theta(t)$ are needed for this property [55].

[4] If $Q(\lambda)\hat P(\lambda,\theta_p)$ and $\hat Z(\lambda,\theta_z)$ are co-prime, then for any $A^*(\lambda)$ of degree n_q+n-1, this equation has a unique solution $(C(\lambda,\phi_c), D(\lambda,\psi_d))$.

References

1. F. Ahmed-Zaid, P. Ioannou, K. Gousman, and R. Rooney, "Accommodation of failures in the F-16 aircraft using adaptive control," *IEEE Control Systems Magazine*, vol. 11, no. 1, pp. 73–78, 1991.

2. E. G. Alcorta and P. M. Frank, "Deterministic nonlinear observer-based approaches to fault diagnosis: A survey," *Control Engineering Practice*, vol. 5, no. 5, pp. 663–700, 1997.

3. A. M. Annaswamy, F. P. Skantze, and A.P. Loh, "Adaptive control of continuous-time systems with convex/concave parametrization," *Automatica*, vol. 14, no. 1, pp. 33–49, 1998.

4. P. J. Antsaklis and A. N. Michel, *Linear Systems*, McGraw-Hill, New York, 1997.

5. Z. Artstein, "Stabilization with relaxed controls," *Nonlinear Analysis*, pp. 1163–1173, 1983.

6. K. J. Åström and B. Wittenmark, *Adaptive Control*, 2nd ed., Addison-Wesley, Reading, MA, 1995.

7. E.-W. Bai and S. Sastry, "Discrete-time adaptive control utilizing prior information," *IEEE Trans. on Automatic Control*, vol. 31, no. 8, pp. 779–782, 1986.

8. D. S. Bayard, "A modified augmented error algorithm for adaptive noise cancellation in the presence of plant resonances," *Proc. of the 1998 American Control Conference*, pp. 137–141, Philadelphia, PA, 1998.

9. M. Blanke, "Fault-tolerant control systems," in *Advances in Control, Highlights of ECC99*, pp. 171–196, Springer-Verlag, New York, P. M. Frank, ed., 1999.

10. M. Blanke, R. Izadi-Zamanabadi, R. Bogh, and Z. P. Lunan, "Fault-tolerant control systems—a holistic view," *Control Engineering Practice*, vol. 5, no. 5, pp. 693–702, 1997.

11. M. Blanke, M. Staroswiecki, and N. E. Wu, "Concepts and methods in fault-tolerant control," *Proc.of the 2001 American Control Conference*, pp. 2606–2620, Arlington, VA, 2001.

12. M. Bodson, "Performance of an adaptive algorithm for sinusoidal disturbance rejection in high noise," *Automatica*, vol. 37, no. 7, pp. 1133–1140, 2001.

13. M. Bodson and J. E. Groszkiewicz, "Multivariable adaptive algorithms for reconfigurable flight control," *IEEE Trans. on Control Systems Technology*, vol. 5, no. 2, pp. 217–229, 1997.

14. M. Bodson, J. Jensen, and S. Douglas, "Active noise control for periodic disturbances," *IEEE Trans. on Control Systems Technology*, vol. 9, no. 1, pp. 200–205, 2001.

15. J. D. Boskovic, S. Li, and R. K. Mehra, "Intelligent control of spacecraft in the presence of actuator failures," *Proc. of the 38th IEEE Conference on Decision and Control*, pp. 4472–4477, Phoenix, AZ, December 1999.

16. J. D. Boskovic and R. K. Mehra, "A multiple model-based reconfigurable flight control system design," *Proc. of the 37th IEEE Conference on Decision and Control*, pp. 4503–4508, 1998.

17. J. D. Boskovic and R. K. Mehra, "Stable multiple model adaptive flight control for accommodation of a large class of control effector failures," *Proc. of the 1999 American Control Conference*, pp. 1920–1924.

18. J. D. Boskovic and R. K. Mehra, "An adaptive scheme for compensation of loss of effectiveness of flight control effectors," *Proc. of the 40th IEEE Conference on Decision and Control*, pp. 2448–2453, Orlando, FL, 2001.

19. J. D. Boskovic, S.-H. Yu, and R. K. Mehra, "A stable scheme for automatic control reconfiguration in the presence of actuator failures," *Proc. of the 1998 American Control Conference*, pp. 2455–2459.

20. J. D. Boskovic, S.-H. Yu, and R. K. Mehra, "Stable adaptive fault-tolerant control of overactuated aircraft using multiple models, switching and tuning," *Proc. of the 1998 AIAA Guidance, Navigation and Control Conference*, vol. 1, pp. 739–749, 1998.

21. A. J. Calise, S. Lee, and M. Sharma, "Development of a reconfigurable flight control law for tailless aircraft," *AIAA Journal of Guidance, Control, and Dynamics*, vol. 25, no. 5, pp. 896–902, 2001.

22. D. G. Chen and B. Paden, "Nonlinear adaptive torque-ripple cancellation for step motors," *Proc. of the 29th IEEE Conference on Decision and Control*, pp. 3319–3324, Honolulu, HI, 1990.

23. H. F. Chen and L. Guo, *Identification and Stochastic Adaptive Control*, Birkhauser, Boston, MA, 1991.

24. J. Chen and R. Patton, *Robust Model-based Fault Diagnosis for Dynamic Systems*, Kluwer, New York, 1998.

25. S. H. Chen, G. Tao and S. M. Joshi, "Adaptive actuator failure compensation for a transport aircraft model," *Proc. of the 2001 American Control Conference*, pp. 1827–1832, Arlington, VA, 2001.

26. S. H. Chen, G. Tao, and S. M. Joshi, "On matching conditions for adaptive state tracking control of systems with actuator failures," *Proc. of the 40th IEEE Conference on Decision and Control*, pp. 1479–1484.

27. A. Datta and P. A. Ioannou, "Performance improvement versus robust stability in model reference adaptive control," *IEEE Trans. on Automatic Control*, vol. 39, no. 12, pp. 2370–2388, 1994.

28. M. De Mathelin and M. Bodson, "Frequency domain conditions for parameter convergence in multivariable recursive identification," *Automatica*, vol. 26, no. 4, pp. 757–767, 1990.

29. M. A. Demetriou and M. M. Polycarpou, "Incipient fault diagnosis of dynamical systems using online approximators," *IEEE Trans. on Automatic Control*, vol. 43, no. 11, pp. 1612–1617, November 1998.

30. T. E. Duncan and B. Pasik-Duncan, "Adaptive control of continuous-time linear stochastic systems," *Mathematics of Control, Signals, and Systems*, vol. 3, pp. 45–60, 1990.

31. D. C. G. Eaton, "An overview of structural acoustics and related high-frequency-vibration activities," *ESA Bulletin*, no. 92, November 1997.

32. H. Elliott and W. A. Wolovich, "A parameter adaptive control structure for linear multivariable systems," *IEEE Trans. on Automatic Control*, vol. AC-27, no. 2, pp. 340–352, 1982.

33. A. F. Filippov, "Differential equations with discontinuous right hand sides," *American Mathematical Society Translations*, vol. 42, pp. 199–231, 1964.

34. T. E. Fortmann and K. L. Hitz, *An Introduction to Linear Control Systems*, Marcel Dekker, New York, 1977.

35. P. M. Frank, "Fault diagnosis in dynamic systems using analytical and knowledge-based redundancy—A survey and some new results," *Automatica*, vol. 26, no. 3, pp. 459–474, 1990.

36. P. M. Frank, "Enhancement of robustness in observer-based fault detection," *International Journal of Control*, vol. 59, no. 4, pp. 955–981, 1994.

37. G. F. Franklin, J. D. Powell, and A. Emami-Naeini, *Feedback Control of Dynamic Systems*, 3rd ed., Addison-Wesley, Reading, MA, 1994.

38. J. P. Gao, B. Huang, Z. D. Wang, and D. G. Fisher, "Robust reliable control for a class of uncertain nonlinear systems with time-varying multistate time delays," *International Journal of Systems Science*, vol. 32, no. 7, pp. 817–824, 2001.

39. S. S. Ge, T. H. Lee, and C. J. Harris, *Adaptive Neural Network Control of Robot Manipulators*, World Scientific, River Edge, NJ, 1998.

40. S. S. Ge, C. C. Hang, T. H. Lee, and T. Zhang, *Stable Adaptive Neural Network Control*, Kluwer Academic, Boston, 2001.

41. J. J. Gertler, *Fault Detection and Diagnosis in Engineering Systems*, Marcel Dekker, New York, 1998.

42. G. C. Goodwin and R. S. Long, "Generalization of results on multivariable adaptive control," *IEEE Trans. on Automatic Control*, vol. AC–25, no. 6, pp. 1241–1245, 1980.

43. G. C. Goodwin, P. J. Ramadge, and P. E. Caines, "Discrete-time multivariable adaptive control," *IEEE Trans. on Automatic Control*, vol. AC-25, no. 6, pp. 449–456, 1980.

44. G. C. Goodwin and D. Q. Mayne, "A parameter estimation perspective of continuous time model reference adaptive control," *Automatica*, vol. 23, no. 1, pp. 57–70, 1987.

45. G. C. Goodwin and K. S. Sin, *Adaptive Filtering Prediction and Control*, Prentice-Hall, Englewood Cliffs, NJ, 1984.

46. M. Gopinathan, J. D. Boskovic, R. K. Mehra, and C. Rago, "A multiple model predictive scheme for fault-tolerant flight control design," *Proc. of the 37th IEEE Conference on Decision and Control*, pp. 1376–1381, 1998.

47. H. Hammouri, M. Kinnaert, and E. H. El Yaagoubi, "Observer-based approach to fault detection and isolation for nonlinear systems," *IEEE Trans. on Automatic Control*, vol. 44, no. 10, pp. 1879–1884, 1999.

48. L. Hsu, R. Costa, A. Imai, and P. Kokotović, "Lyapunov based adaptive control of MIMO systems," *Proc. of the 2001 American Control Conference*, pp. 4808–4813, 2001.

49. M. Idan, M. Johnson, and A. J. Calise, "A hierarchical approach to adaptive control for improved flight safety," *Proc. of the 2001 AIAA Guidance, Navigation, and Control Conference*, Montreal, Canada.

50. M. Idan, M. Johnson, A. J. Calise, and J. Kaneshige, "Intelligent aerodynamic/propulsion flight control for flight safety: A nonlinear adaptive approach," *Proc. of the 2001 ACC*, pp. 2918–2923.

51. F. Ikhouane and M. Krstić, "Robustness of the tuning functions adaptive backstepping design for linear systems," *Proc. of the 34th IEEE CDC*, pp. 159-164, New Orleans, LA, 1995.

52. A. K. Imai, R. Costa, L. Hsu, G. Tao, and P. Kokotović, "Multivariable MRAC using high frequency gain matrix factorization," *Proc. of the 40th IEEE Conference on Decision and Control*, pp. 1193–1198, 2001.

53. P. A. Ioannou and P. V. Kokotović, *Adaptive Systems with Reduced Models*, Springer-Verlag, Berlin, 1983.

54. P. A. Ioannou and P. V. Kokotović, "Instability analysis and improvement of robustness of adaptive control," *Automatica*, vol. 20, no. 5, pp. 583–594, 1984.

55. P. A. Ioannou and J. Sun, *Robust Adaptive Control*, Prentice-Hall, Upper Saddle River, NJ, 1996.

56. P. A. Ioannou and K. Tsakalis, "A robust direct adaptive controller," *IEEE Trans. on Automatic Control*, vol. AC-31, no. 11, pp. 1033–1043, 1986.

57. P. A. Ioannou and K. Tsakalis, "Robust discrete time adaptive control," in *Adaptive and Learning Systems: Theory and Applications*, K. S. Narendra, ed., Plenum Press, New York, 1986.

58. R. Iserman, "On the applicability of model based fault detection for technical process," *Control Engineering Practice*, vol. 2, no. 3, pp. 439–450, 1997.

59. R. Iserman and P. Balle, "Trends in the application of model-based fault detection and diagnosis of technical processes," *Control Engineering Practice*, vol. 5, pp. 709–719, 1997.

60. A. Isidori, *Nonlinear Control Systems*, 3rd ed., Springer-Verlag, New York, 1995.

61. C. A. Jacobson and C. N. Nett, "An integrated approach to controls and diagnostic using the four parameter controller," *IEEE Control Systems Magazine*, vol. 11, pp. 22–29, 1991.

62. Z. P. Jiang, "A combined backstepping and small-gain approach to adaptive output feedback control," *Automatica*, vol. 35, no. 6, pp. 1131–1139, 1999.

63. Z. P. Jiang, "Decentralized and adaptive nonlinear tracking of large-scale systems via output feedback," *IEEE Trans. on Automatic Control*, vol. 45, no. 11, pp. 2122–2128, 2000.

64. P. Kabore and H. Wang, "On the design of fault diagnosis filters and fault tolerant control," *Proc. of the American Control Conference*, San Diego, CA, June 1999.

65. T. Kailath, *Linear Systems*, Prentice-Hall, Englewood Cliffs, NJ, 1980.

66. I. Kanellakopoulos, P. V. Kokotović, and A. S. Morse, "Systematic design of adaptive controllers for feedback linearizable systems," *IEEE Trans. on Automatic Control*, vol. 36, no. 11, pp. 1241–1253, 1991.

67. H. K. Khalil, *Nonlinear Systems*, 2nd ed., Prentice Hall, Upper Saddle River, NJ, 1996.

68. H. K. Khalil, "Adaptive output feedback control of nonlinear systems represented by input-output model," *IEEE Trans. on Automatic Control*, vol. 41, no. 2, pp. 177–188, 1996.

69. B. S. Kim and A. J. Calise, "Nonlinear flight control using neural networks," *AIAA Journal of Guidance, Control, and Dynamics*, vol. 20, no. 1, pp. 26–33, 1997.

70. M. Krstić, I. Kanellakopoulos, and P. V. Kokotović, "Nonlinear design of adaptive controllers for linear systems," *IEEE Trans. on Automatic Control*, vol. AC-39, no. 4, pp. 738–752, 1994.

71. M. Krstić, I. Kanellakopoulos, and P. V. Kokotović, *Nonlinear and Adaptive Control Design*, John Wiley & Sons, New York, 1995.

72. Y. D. Landau, *Adaptive Control: The Model Reference Approaches*, Marcel Dekker, New York, 1979.

73. Y. D. Landau, R. Lozano, and M. M'Saad, *Adaptive Control*, Springer-Verlag, London, 1998.

74. J. Leitner, A. J. Calise, and J. V. R. Prasad, "Analysis of adaptive neural networks for helicopter flight controls," *AIAA Journal of Guidance, Control, and Dynamics*, vol. 20, no. 5., pp. 972–979, September–October 1997.

75. F. L. Lewis, S. Jagannathan, and A. Yesildirek, *Neural Network Control of Robot Manipulators and Nonlinear Systems*, Taylor and Francis, Philadelphia, PA, 1999.

76. Y.-W. Liang, D.-C. Liaw, and T. C. Lee, "Reliable control of nonlinear systems," *IEEE Trans. on Automatic Control*, vol. 45, no. 4, pp. 706–710, 2000.

77. F. Liao, J. L. Wang, and G.-H Yang, "Reliable robust flight tracking control: An LMI approach," *IEEE Trans. on Control Systems Technology*, vol. 10, no. 1, pp. 76–89, 2002.

78. W. Lin and C. Qian, "Adaptive regulation of high-order lower-triangular systems: Adding a power integrator technique," *Systems and Control Letter*, vol. 39, pp. 353–364, 2000.

79. W. Lin and C. Qian, "Adaptive control of nonlinearly parametrized systems: Smooth feedback domination design," submitted to *IEEE Trans. on Automatic Control*.

80. X. C. Lou, A. S. Willsky, and G. C. Verghese, "Optimal robust redundancy relations for failure detection in uncertain systems," *Automatica*, vol. 22, no. 3, pp. 333–344.

81. B. Lui and J. Si, "Fault isolation filter design for linear time-invariant systems," *IEEE Trans. on Automatic Control*, vol. 42, no. 5, pp. 704–706, 1997.

82. J. M. Maciejowski, "Reconfigurable control using constrained optimization," *Proc. of ECC97*, pp. 107–130, Brussels, Belgium, 1997.

83. R. Mangoubi, *Robust Estimation and Failure Detection: A Concise Treatment*. Springer–Verlag, New York, 1998.

84. R. Marino and P. Tomei, *Nonlinear Control Design: Geometric, Adaptive and Robust*, Prentice-Hall, Englewood Cliffs, NJ, 1995.

85. M. A. Massoumnia, G. C, Verghese, and A. S. Willsky, "Failure detection and identification," *IEEE Trans. on Automatic Control*, vol. 34, no. 3, pp. 316–321, 1989.

86. R. H. Miller, "On-line detection of aircraft icing: An application of optimal fault detection and isolation," Ph.D. dissertation, University of Michigan, Ann Arbor, 2000.

87. R. H. Miller and B. R. William, "The effects of icing on the longitudinal dynamics of an icing research aircraft," *The 37th Aerospace Sciences Conference*, AIAA, 1999.

88. A. S. Morse, "Global stability of parameter adaptive control systems," *IEEE Trans. on Automatic Control*, vol. 25, no. 6, pp. 433–439, 1980.

89. A. S. Morse, "A gain matrix decomposition and some of its applications," *Systems and Control Letters*, vol. 21, no. 1, pp. 1–10, 1993.

90. A. S. Morse, D. Q. Mayne, and G. C. Goodwin, "Applications of hysteresis switching in parameter adaptive control," *IEEE Trans. on Automatic Control*, vol. AC-37, no. 9, pp. 1343–1354, 1992.

91. W. D. Morse and K. A. Ossman, "Model-following reconfigurable flight control system for the AFTI/F16," *Journal of Guidance, Control and Dynamics*, vol. 14, no. 6, pp. 969–976, 1990.

92. L. H. Mutuel and J. L. Speyer, "Fault-tolerant estimation," *Proc. of the 2000 ACC*, pp. 3718–3722, Chicago, IL, 2000.

93. "Linearized Dynamic Models of Boeing 737-100 Transport Aircraft," NASA Langley Research Center, 2000.

94. S. M. Naik, P. R. Kumar, and B. E. Ydstie, "Robust continuous-time adaptive control by parameter projection," in *Foundations of Adaptive Control*, P. V. Kokotović, ed., pp. 153–199, Springer-Verlag, Berlin, 1991.

95. K. S. Narendra and A. M. Annaswamy, *Stable Adaptive Systems*, Prentice-Hall, Englewood Cliffs, NJ, 1989.

96. K. S. Narendra and J. Balakrishnan, "Adaptive control using multiple models," *IEEE Trans. on Automatic Control*, vol. 42, no. 2, pp. 171–187, February 1997.

97. H. Noura, D. Sauter, F. Hamelin, and D. Theilliol, "Fault-tolerant control in dynamic systems: application to a winding machine," *IEEE Control Systems Maganize*, vol. 20, pp. 33–49, 2000.

98. R. D. Nussbaum, "Some remarks on a conjecture in parameter adaptive control," *Systems & Control Letters*, vol. 3, pp. 243–246, 1983.

99. R. Ortega, L. Hsu, and A. Astolfi, "Adaptive control of multivariable systems with reduced prior knowledge," *Proc. of the 40th IEEE Conference on Decision and Control*, pp. 4198–4203, Orlando, FL, 2001.

100. R. Ortega and Y. Tang, "Robustness of adaptive controllers: A survey," *Automatica*, vol. 25, no. 5, pp. 651–677, 1989.

101. Z. Pan and T. Basar, "Adaptive controller design for tracking and disturbance attenuation in parametric strict-feedback nonlinear systems," *IEEE Trans. on Automatic Control*, vol. 43, no. 8, pp. 1066–1083, 1998.

102. R. J. Patton, "Fault tolerant control: The 1997 situation," *Proc. of IFAC Symp. on SAFEPROCESS*, pp. 1033-1055, Hull, UK, 1997.

103. R. J. Patton and Kanguette, "Robust fault diagnosis using eigenstructure assignment of observers," in *Fault Diagnosis in Dynamic Systems: Theory and Application*, R. J. Patton, P. M. Frank, and R. N. Clark, eds., pp. 99–154, Prentice-Hall, Englewood Cliffs, NJ, 1989.

104. M. M. Polycarpou, "Fault accommodation for a class of multivariable nonlinear dynamical systems using a learning approach," *IEEE Trans. on Automatic Control*, vol. 46, no. 5, pp. 736–742, May 2001.

105. M. Polycarpou and P. A. Ioannou, "On the existence and uniqueness of solutions in adaptive control systems," *IEEE Trans. on Automatic Control*, vol. 38, no. 3, pp. 474–479, 1993.

106. M. M. Polycarpou and A. B. Trunov, "Learning approach to nonlinear fault diagnosis: detectability analysis," *IEEE Trans. on Automatic Control*, vol. 45, no. 4, pp. 806–812, April 2000.

107. M. M. Polycarpou and A. T. Vemuri, "Learning approaches to fault tolerant control: An overview," *Proc. of the IEEE Intl. Symp. on Intelligent Control*, pp. 157–162, Gaithersburg, MD, 1998.

108. R. T. Rysdyk and A. J. Calise, "Nonlinear adaptive flight control using neural networks," *IEEE Control Systems Magazine*, vol. 18, no. 6, December 1998.

109. S. Sastry and A. Isidori, "Adaptive control of linearizable systems," *IEEE Trans. on Automatic Control*, vol. 34, no. 11, pp. 1123–1131, 1989.

110. S. Sastry and M. Bodson, *Adaptive Control: Stability, Convergence, and Robustness*, Prentice-Hall, Englewood Cliffs, NJ, 1989.

111. L. V. Schmidt, *Introduction to Aircraft Flight Dynamics*, American Institute of Aeronautics and Astronautics, Reston, VA, 1998.

112. M. Staroswiecki and A.-L. Gehin, "From control to supervision," *Annual Review in Control*, vol. 25, pp. 1–11, 2001.

113. B. L. Stevens and F. L. Lewis, *Aircraft Control and Simulation*, John Wiley & Sons, New York, 1992.

114. X. D. Tang, G. Tao, and S. M. Joshi, "Adaptive control of parametric-strict-feedback nonlinear systems with actuator failures," *Proc. of the 40th IEEE Conference on Decision and Control*, pp. 1613–1614, Orlando, FL, 2001.

115. X. D. Tang, G. Tao, and S. M. Joshi, "Adaptive actuator failure compensation control of parametric strict-feedback systems with zero dynamics," *Proc. of the 40th IEEE Conference on Decision and Control*, pp. 2031–2036, Orlando, FL, 2001.

116. G. Tao, "Inherent robustness of MRAC schemes," *Systems and Control Letters*, vol. 29, no. 3-11, pp. 165–174, November 1996.

117. G. Tao, "Friction compensation in the presence of flexibility," *Proc. of the 1998 American Control Conference*, pp. 2128–2132, Philadelphia, PA.

118. G. Tao, *Adaptive Control Design and Analysis*, John Wiley and Sons, New York, 2003.

119. G. Tao, S. H. Chen, and S. M. Joshi, "An adaptive control scheme for systems with unknown actuator failures," *Proc. of the 2001 American Control Conference*, pp. 1115–1120.

120. G. Tao, S. H. Chen, and S. M. Joshi, "An adaptive actuator failure compensation controller using output feedback," *Proc. of the 2001 American Control Conference*, pp. 3085–3090.

121. G. Tao, S. H. Chen, and S. M. Joshi, "An adaptive control scheme for systems with unknown actuator failures," Technical report UVA-ECE-ASC-01-03-01, Department of Electrical and Computer Engineering, University of Virginia, Charlottesville, VA, March 2001.

122. G. Tao and P. A. Ioannou, "Robust model reference adaptive control for multivariable plants," *Intl. J. Adaptive Control and Signal Process*, vol. 2, no. 3, pp. 217–248.

123. G. Tao and P. A. Ioannou, "Stability and robustness of multivariable model reference adaptive control schemes," in *Advances in Robust Control Systems Techniques and Applications*, Academic Press, C. T. Leondes, ed., vol. 53, pp. 99–123, 1992.

124. G. Tao, S. M. Joshi, and X. L. Ma, "Adaptive state feedback and tracking control of systems with actuator failures," *IEEE Trans. on Automatic Control*, vol. 46, no. 1, pp. 78–95, January 2001.

125. G. Tao and P. V. Kokotović, *Adaptive Control of Systems with Actuator and Sensor Nonlinearities*, John Wiley & Sons, New York, 1996.

126. G. Tao, X. L. Ma, and S. M. Joshi, "Adaptive output tracking control of systems with actuator failures," *Proc. of the 2000 American Control Conference*, pp. 2654–2658, Chicago, IL, 2000.

127. G. Tao, X. L. Ma, and S. M. Joshi, "Adaptive state feedback control of systems with actuator failures," *Proc. of 2000 American Control Conference*, pp. 2669–2673, Chicago, IL, 2000.

128. G. Tao, X. D. Tang, and S. M. Joshi, "Output tracking actuator failure compensation control," *Proc. of the 2001 American Control Conference*, pp. 1821–1826, Arlington, VA, 2001.

129. M. Tian and G. Tao, "Adaptive dead-zone compensation for out-feedback canonical systems," *International Journal of Control*, vol. 67, pp. 791–812, 1997.

130. M. Tian and G. Tao, "Adaptive control of a class of nonlinear systems with unknown dead-zones," *Proc. of the 13th World Congress of IFAC*, vol. E, pp. 209–213, San Francisco, CA, July 1996.

131. K. Tsakalis, "Robustness of model reference adaptive controllers: An input-output approach," *IEEE Trans. on Automatic Control*, vol. 37, no. 5, pp. 556–565, 1992.

132. K. Tsakalis and P. A. Ioannou, *Linear Time Varying Systems: Control and Adaptation*, Prentice-Hall, Englewood Cliffs, NJ, 1993.

133. K. S. Tsakalis, "Model reference adaptive control of linear time-varying plants: The case of 'jump' parameter variations," *International Journal of Control*, vol. 56, no. 6, pp. 1299–1345.

134. V. I. Utkin, *Sliding Modes in Control Optimization*, Springer-Verlag, Berlin, 1991.

135. A. T. Vemuri and M. M. Polycarpou, "Robust nonlinear fault diagnosis in input-output systems," *International Journal of Control*, vol. 68, no. 2, pp. 343–360, 1997.

136. R. J. Veillette, "Reliable linear-quadratic state-feedback control," *Automatica*, vol. 31, no. 1, pp. 137–143, 1995.

137. R. J. Veillette, J. V. Medanic, and W. R. Perkins, "Designs of reliable control systems," *IEEE Trans. on Automatic Control*, vol. 37, no. 3, pp. 290–304.

138. M. Vidyasagar, *Nonlinear Systems Analysis*, 2nd ed., Prentice-Hall, Englewood Cliffs, NJ, 1993.

139. H. Wang and S. Daley, "Actuator fault diagnosis: An adaptive observer-based technique," *IEEE Trans. on Automatic Control*, vol. 41, no. 7, pp. 1073–1078, 1996.

140. H. Wang, Z. J. Huang, and S. Daley, "On the use of adaptive updating rules for actuator and sensor fault diagnosis," *Automatica*, vol. 33, pp. 217–225, 1997.

141. S. R. Weller and G. C. Goodwin, "Hysteresis switching adaptive control of linear multivariable systems," *IEEE Trans. on Automatic Control*, vol. 39, no. 7, pp. 1360–1375, 1994.

142. K. Wise, J. S. Brinker, A. J. Calise, D. F. Enns, and M. R. Elgersma, "Direct adaptive reconfigurable flight control for a tailless advanced fighter aircraft," *Intl. J. Robust and Nonlinear Control*, vol. 9, pp. 999–1009, 1999.

143. N. E. Wu and T. J. Chen, "Feedback design in control reconfigurable systems," *International Journal of Robust and Nonlinear Control*, vol. 6, pp. 560–570, 1996.

144. N. E. Wu, Y. Zhang, and K. Zhou, "Detection, estimation, and accommodation of loss of control effectiveness," *Intl. Journal of Adaptive Control and Signal Processing*, vol. 14, pp. 775-795, 2000.

145. N. E. Wu, K. Zhou, and G. Salomon, "Reconfigurability in linear time-invariant systems," *Automatica*, vol. 36, pp. 1767–1771, 2000.

146. H. Xu and M. Mirmirani, "Robust adaptive sliding control for a class of MIMO nonlinear systems," *Proc. of the 2001 AIAA Guidance, Navigation and Control Conference*, Montréal, Québec, Canada, 2001.

147. G-H. Yang, J. L. Wang, and Y. C. Soh, "Reliable H_∞ controller design for linear systems," *Automatica*, vol. 37, pp. 717–725, 2001.

148. Y. Yang, G.-H. Yang, and Y. C. Soh, "Reliable control of discrete-time systems with actuator failures," *IEE Proceedings: Control Theory and Applications*, vol. 147, no. 4, pp. 424–432, 2000.

149. B. Yao and M. Tomizuka, "Adaptive robust control of MIMO nonlinear systems in semi-strict-feedback forms," *Automatica*, vol. 37, no. 9, pp. 1305–1321, 2001.

150. G. Yen and L. Ho, "Fault tolerant control: An intelligent sliding model control strategy," *Proc. of the American Control Conference*, pp. 4204–4208, Chicago, IL, June 2000.

151. Y. M. Zhang and J. Jiang, "Design of proportional-integral reconfigurable control systems via eigenstructure assignment," *Proc. of the American Control Conference*, pp. 3732–3736, Chicago, IL, June 2000.

152. Y. M. Zhang and J. Jiang, "Integrated active fault-tolerant control using IMM approach," *IEEE Trans. on Aerospace and Electronic Systems*, vol. 37, no. 4, pp. 1221–1235, 2001.

153. Y. M. Zhang and J. Jiang, "Bibliographical review on reconfigurable fault-tolerant control systems," *Proc. of the 5th IFAC Symp. on SAFEPROCESS*, pp. 265–276, Washington, DC, 2003.

154. Q. Zhao and J. Jiang, "Reliable state-feedback control systems design against actuator failures," *Automatica*, vol. 30, no. 10, pp. 1267–1272, 1998.

Index